ICE Themes Geothermal Energy, Heat Exchange Systems and Energy Piles

The *ICE Themes* series showcases cutting-edge research and practical guidance in all branches of civil engineering. Each title focuses on a key issue or challenge in civil engineering, and includes research from the industry's finest thinkers and influencers published through the ICE Publishing programme. Themes in the series include climate change resilience, advances in construction management, developments in renewable energy, and innovations in construction materials plus many more.

Other titles in the *ICE Themes* series:

ICE Themes Flood Resilience (2018)
Edited by M. Escarameia and A. Tagg. ISBN 978-0-7277-6393-8
ICE Themes Wind Turbine Foundations (2018)
Edited by K. Gavin and W. Craig. ISBN 978-0-7277-6396-9

ICE Themes
Geothermal Energy, Heat Exchange Systems and Energy Piles

Edited by William Craig and Kenneth Gavin

Published by ICE Publishing, One Great George Street, Westminster, London SW1P 3AA.

Full details of ICE Publishing representatives and distributors can be found at: www.icebookshop.com/bookshop_contact.asp

Other titles by ICE Publishing:

Sustainable Buildings.
Elisabeth Green, Tristram Hope and Alan Yates. ISBN 978-0-7277-5806-4
Sustainable Water
Charles Ainger and Richard Fenner (eds.). ISBN 978-0-7277-5773-9
Sustainable Infrastructure: Principles into Practice
Charles Ainger and Richard Fenner. ISBN 978-0-7277-5754-8

www.icebookshop.com

A catalogue record for this book is available from the British Library

ISBN 978-0-7277-6398-3
ISSN (Print) 2516-4899
ISSN (Online) 2516-4902

Chapters from this title originally appeared in *Environmental Geotechnics*, Volume 3; *Proceedings of the Institution of Civil Engineers – Energy*, Volumes 168 and 169; *Proceedings of the Institution of Civil Engineers – Geotechnical Engineering*, Volume 166; and *Proceedings of the Institution of Civil Engineers – Engineering Sustainability*, Volumes 167 and 168.

© Thomas Telford Limited 2018

ICE Publishing is a division of Thomas Telford Ltd, a wholly-owned subsidiary of the Institution of Civil Engineers (ICE).

All rights, including translation, reserved. Except as permitted by the Copyright, Designs and Patents Act 1988, no part of this publication may be reproduced, stored in a retrieval system or transmitted in any form or by any means, electronic, mechanical, photocopying or otherwise, without the prior written permission of the Publisher, ICE Publishing, One Great George Street, Westminster, London SW1P 3AA.

This book is published on the understanding that the author is solely responsible for the statements made and opinions expressed in it and that its publication does not necessarily imply that such statements and/or opinions are or reflect the views or opinions of the publishers. While every effort has been made to ensure that the statements made and the opinions expressed in this publication provide a safe and accurate guide, no liability or responsibility can be accepted in this respect by the authors or publisher.

While every reasonable effort has been undertaken by the authors and the publisher to acknowledge copyright on material reproduced, if there has been an oversight please contact the publisher and we will endeavour to correct this upon a reprint.

Cover photo: Geothermal sustainable energy: pipes in a power plant

Commissioning Editor: Michael Fenton
Production Editor: Madhubanti Bhattacharyya
Market Development Executive: April-Asta Brodie

Typeset by The Manila Typesetting Company
Index created by Matthew Gale
Printed and bound in Great Britain by TJ International, Padstow

Contents

Foreword by William Craig and Kenneth Gavin   ix

Section 1 – Geothermal energy

01   **Geothermal energy in loess**   3
Introduction: ground-source heat pump systems
and loess deposits   4
Main characteristics of loess   6
Site description and thermal characterisation   7
Numerical modelling   9
Results and discussions   12
Conclusions   17
References   19

02   **Parsimonious numerical modelling of
deep geothermal reservoirs**   23
1. Introduction   24
2. Background   25
3. Conceptual modelling – a tripartite
 modelling approach   27
4. Well doublet production scenario   34
5. Discussion and conclusion   37
6. Recommendations for future work   38
References   39

03   **Geothermal subsidence study at
Wairakei–Tauhara, New Zealand**   41
1. Introduction   42
2. Geotechnical investigation   45
3. Subsidence mechanisms   48
4. Subsidence mitigation   55
5. Conclusion and discussion   56
References   57

Section 2 – Heat exchange

04   **Energy harvesting on road pavements:
state of the art**   61
1. Introduction   61
2. Road pavement energy harvesting technologies   62
3. Technical analysis   70
4. Conclusions   72
References   73

05   **Uncertainties in the design of ground
heat exchangers**   79
Introduction   80
Methodology   81

v

	Case study	85
	Conclusions	96
	References	97
06	**The role of ground conditions on energy tunnels' heat exchange**	**99**
	Introduction	100
	Energy tunnels	101
	Set-up of the numerical model	102
	Parametric numerical analyses for various ground conditions	106
	Preliminary design charts	111
	Conclusions	113
	References	113
07	**Simulations of a photovoltaic-thermal ground source heat pump system**	**115**
	1. Introduction	115
	2. Model	116
	3. Results	119
	4. Discussion	124
	5. Conclusion	128
	References	129
08	**The design of thermal tunnel energy segments for Crossrail, UK**	**131**
	1. Introduction	131
	2. Assessment of the heat inside the tunnel	132
	3. Design of the TES system	133
	4. Potential market for the tunnel heat	139
	5. Modelling of TES heat transfer	140
	6. Tunnel cooling study	144
	7. Durability and operational considerations	146
	8. Operational and commercial benefits	151
	9. Case study: Fisher Street to Tottenham Court Road	152
	10. Summary and conclusion	154
	References	155
09	**Thermal response testing through the Chalk aquifer in London, UK**	**157**
	1. Introduction	158
	2. Test details	160
	3. Test results	163
	4. Discussion	173
	5. Conclusion	174
	Appendix: Thermal resistance of ground loop pipes	175
	References	176

Section 3 – Energy piles

10   **Thermal performance of thermoactive continuous flight auger piles**   **181**
Introduction   182
Construction techniques   183
Thermal performance assessment   184
Results   190
Discussion   201
Conclusions   202
Appendix: Calculation of the concrete resistance   203
References   203

11   **City-scale perspective for thermoactive structures in Warsaw**   **205**
Introduction   206
Analysed data set   207
City-scale analysis for structures equipped with thermoactive foundation elements   213
Results   219
Conclusions   220
References   221

12   **Energy piles: site investigation and analysis**   **223**
1. Introduction   224
2. Energy pile installations   225
3. Thermal test analysis   227
4. Thermal test results   234
5. Discussion   242
6. Conclusion and recommendations   243
References   244

13   **Pile heat exchangers: thermal behaviour and interactions**   **249**
1. Introduction   250
2. Heat transfer concepts   252
3. Thermal performance of borehole heat exchangers   256
4. Thermal performance of pile heat exchangers   260
5. Thermomechanical interactions and pile behaviour   270
6. Practical constraints   274
7. Conclusion   275
References   276

Index   **281**

Foreword

In 2013 we acted as editors for a themed issue of the Institution of Civil Engineers Proceedings journal *Geotechnical Engineering* on the broad topic 'Geotechnical Challenges for Renewable Energy Developments'. In a single (April) issue there were eleven papers on a range of subjects from site investigation for foundations of a tidal power plant in the Bay of Fundy in North America, to the settlements associated with geothermal energy abstraction in the Wairakei–Tauhara field in New Zealand. The European focus was mainly related to foundation issues for wind turbine installations and to so-called energy piles. Subsequently other ICE journals have published a wide array of papers in this expanding and developing field.

Our aim in this and a companion volume is to collect together some of the most prominent and frequently accessed papers published by the Institution in *Geotechnical Engineering* and other journals in the years 2013-16. Such has been the interest in the particular problems of wind farm foundations, especially those offshore, that these have been brought together in a separate volume, to be published as *ICE Themes Renewable Energy: Wind Turbine Foundations*. In this present volume, we have gathered thirteen chapters in three main groups dealing with high temperature geothermal energy abstraction, energy extraction studies from pavements and low temperature gradient situations, and in particular, energy piles. The chapters are drawn from the original and later issues of ICE Proceedings *Geotechnical Engineering* and also from ICE Proceedings: *Energy*, ICE Proceedings: *Environmental Sustainability*, and from the journal *Environmental Geotechnics*. They encompass field observations on sites in several countries as well as computational and laboratory studies. We have sought to keep a broad international coverage and include chapters with authors from Australia, Ireland, Italy, New Zealand, Poland, Portugal as well as the United Kingdom. Ground conditions vary from hard rock to chalk, loess to London Clay.

In the field of geothermal energy, the chapter by **Bromley et al.** assesses the effect of more than 50 years of high temperature fluid extraction from a site at Wairakei-Tauhara in New Zealand, where power of the order of 250MWe is produced. In geotechnical terms the abstraction of fluids in excess of 2.5km^3 has led to maximum ground liquid pressure drawdown of the order of 25bar, though fluid reinjection has reduced this somewhat. The energy benefits have been offset by widespread settlement which has varied across the geothermal field but has reached 15m in places. The analysis indicates that geological conditions in the area are complex, resulting from hydrothermal alteration of porous and permeable formations, but that the settlement is related to consolidation of the ground as a result of effective stress increases. Such case histories are invaluable and provide a record of cause and effect which can be used in forward developments.

Fairs *et al.* look to the future and possible development of a deep geothermal system in the Weardale area of the UK with a potential output of tens of megawatts, either for direct use or for electrical power generation. Their approach is via a novel numerical modelling technique and should input to the design/development of this and other future systems as more data become available. With a view to broadening the range of geotechnical scenarios under study, inclusion of the chapter by **Bidarmaghz *et al.*** covers geothermal energy opportunities in loess deposits.

A linked set of chapters relates to the abstraction/dissipation of energy as a complementary by-product of the structural benefits of deep foundations in the form of so-called energy piles. Taken together these chapters cover many aspects of generic problems in assessing and designing such structures, usually in a specific context, though the lessons can be applied more widely. There is detailed consideration of site investigation and testing for two sites in Ireland by **Hemmingway and Long** and the key factors influencing heat transfer as well as thermal-mechanical interactions are covered by **Loveridge and Powrie**. While they indicate that there is unlikely to be significant mechanical disadvantage in the structural performance of such piles this aspect should be addressed explicitly on a case-by-case basis. **Loveridge and Cecinato** investigate the particular issues arising in determining the thermal behaviour of continuous flight auger piles. **Loveridge *et al.*** look specifically at thermal response testing within the broader context of site investigation and highlight the difficulties of characterising the thermal properties of the ground, with particular reference to the chalk aquifer beneath London. Similar difficulties are to be expected elsewhere and there is, as yet, no substantial database to guide system designers. **Ryżyński and Bogusz** consider the possibilities for thermoactive structures including energy piles at a city-wide scale in relation to Warsaw, Poland.

In the UK, energy piles remain quite rare and their use reflects much individual input to specific projects rather than a widespread consideration as the norm. In an era when demonstrable climate change will be a reality within the lifetime of most significant structural foundations currently being designed, it is likely that there will be significant uptake of the potential benefits of the new technologies provided that the longevity and robustness of the systems installed can be demonstrated and the inertia of established practice, which treats the geo-mechanical performance of foundations in isolation, can be overcome. The industry needs more case histories and long-term performance records in the public domain.

Energy piles are not the only low temperature gradient energy systems and there are other opportunities at ground surface and below to harvest energy. Transportation networks provide a number of renewable energy options. Road surfaces absorb energy from solar radiation as well as from traffic and some of this can be recovered (**Duarte and Ferreira**). There is scope for considerable development in this area. Underground metro systems around the world report slow rise in ambient temperatures as a result of absorption of energy from the prime vehicle power dissipated in traction and braking as well as from the millions of person hours per day spent within the sub-surface areas. Such difficulties can be reversed by heat exchange systems (**Di Donna and Barla, Nicholson *et al.***) In other contexts ground source heat pumps and ground heat exchangers can provide renewable opportunities at more local levels as discussed by **Mikhaylova *et al.* Varney and Vahdati** consider combining ground source heat pumps with photo-voltaic technologies at small scale in computer simulations.

This is not a comprehensive treatment of the subject matter but rather an introduction to many issues which will serve as an entry point to those who may venture further.

William Craig and Kenneth Gavin

Section 1
Geothermal energy

ICE Themes Geothermal Energy, Heat Exchange Systems and Energy Piles

Craig and Gavin
ISBN 978-0-7277-6398-3
https://doi.org/10.1680/gehesep.63983.003
ICE Publishing: All rights reserved

Chapter 1
Geothermal energy in loess

Asal Bidarmaghz PhD, MSc, BSc
Department of Engineering, The University of Cambridge, UK

Nikolas Makasis MSc, MEng
Department of Infrastructure Engineering, The University of Melbourne, Parkville, Australia

Guillermo A. Narsilio PhD, MSc, CEng
Australian Research Council Future Fellow and Senior Lecturer, Department of Infrastructure Engineering, The University of Melbourne, Parkville, Australia
(corresponding author: narsilio@unimelb.edu.au)

Franco M. Francisca PhD, CEng
Institute for Advanced Studies in Engineering and Technology (IDIT), Universidad Nacional de Córdoba and Consejo Nacional de Investigaciones Científicas y Técnicas (Conicet), Córdoba, Argentina

Magalí E. Carro Pérez PhD, MSc, CEng
IDIT, Universidad Nacional de Córdoba and Conicet, Córdoba, Argentina

Ground-source heat pump (GSHP) systems efficiently heat and cool buildings by using sustainable geothermal energy accessed by way of ground heat exchangers (GHEs). Loess covers vast parts of the world, about 10% of the landmass; therefore, the use of piles or 'micropiles' is extensive in these areas, particularly where the thickness of loessic soils is significant. These deep foundations have the potential to be used as 'energy piles' in GSHP systems, with a minimal additional cost. This chapter presents a case study of a representative real building in Córdoba, Argentina, where foundations are also used as GHEs. The thermal properties of local soils were experimentally measured and used in recently developed detailed state-of-the-art finite-element models. Results from the realistic simulations show that the partial substitution of electrical heating and cooling systems with geothermal systems could significantly reduce energy consumption and the size of associated infrastructure, despite the relatively low thermal conductivity of local loess. Moreover, the effects of surface air temperature fluctuations, which are routinely ignored in GHE design, are accounted for in these simulations. This case study shows the potential of GSHP technology in loessic environments and gives incentives to engineers to start considering the technology in their designs and practices.

Notation

A	inner cross-sectional area of a pipe: m^2
A_s	earth surface temperature annual swing above and below the average ground temperature: °C
$C_{p,m}$	specific heat capacity of solid material: J/(kg K)
$C_{p,w}$	specific heat capacity of carrier fluid: J/(kg K)
d_h	hydraulic diameter of pipe: m
f_D	Darcy friction factor
k_v	vegetation coefficient
p	pressure: Pa
Q_{wall}	external heat exchange rate through pipe walls: W/m
T	fluid temperature: °C
$T(z, t)$	depth and time-varying temperature: °C
$T_{\text{far-field}}$	undisturbed ground temperature: °C
T_g	annual average ground temperature: °C
T_{inlet}	average inlet pipe temperature: °C
T_m	temperature of solid material: °C
$T_{m,\text{pipe wall}}$	temperature of solid material at the pipe outer wall: °C
T_{outlet}	average outlet pipe temperature: °C
t	time: s (Equations 1–4), d (Equation 5)
t_0	number of days after 1 January to the minimum earth surface temperature: d
v	fluid velocity field: m/s
x, y, z	Cartesian coordinates, with z denoting depth: m
z	depth below the ground surface: m
α	soil thermal diffusivity: $\times 10^{-2}$ cm^2/s
λ_m	thermal conductivity of solid material: W/(m K)
λ_w	thermal conductivity of carrier fluid: W/(m K)
ρ_m	density of solid material: kg/m^3
ρ_w	density of carrier fluid: kg/m^3

Introduction: ground-source heat pump systems and loess deposits

Ground-source heat pump (GSHP) systems efficiently heat and cool buildings by using sustainable geothermal energy accessed by way of ground heat exchangers (GHEs). In closed-loop systems, GHEs comprise high-density polyethylene (HDPE) pipes embedded in specifically drilled boreholes or trenches or even built into foundations, all within a few tens of metres of the surface (Figure 1). GSHP systems operate at a coefficient of performance of about 4 throughout the year, basically delivering 4 kW of heating or cooling for every kilowatt input into running the heat pumps (Amatya *et al.*, 2012; Brandl, 2006; Johnston *et al.*, 2011; Preene and Powrie, 2009).

Loess is an unstable soil that develops collapse due to water content and/or load increases. Loessial soils are loosely cemented aeolian sediments composed mainly by fine sand, silt and clay particles that accumulate when deposited by wind (*primary loess*) and sometimes particles are retransported by either water or snow (*secondary loess*). In both cases, particles are frequently cemented by either precipitated calcium carbonate or silicates (Francisca, 2007;

Figure 1 Schematic diagram of a GSHP system in heating (winter) and cooling (summer) modes (Note: not to scale)

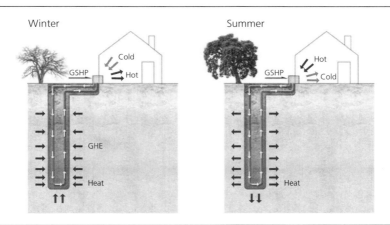

Moll and Rocca, 1991; Rocca et al., 2006). Loess covers vast parts of the world, about 10% of the landmass, and is encountered across New Zealand, from Western Europe to China (including Russia), across North America and in regions across South America (Quintana Crespo, 2005; Rocca et al., 2006; Zárate, 2003). In particular, loess deposits cover large areas of Córdoba, Argentina, and are tens of metres thick.

To support the structures constructed on this type of soil, the use of piles (>300 mm diameter) for multistorey buildings or 'micropiles' (<300 mm diameter) for lighter residential buildings is quite extensive. These foundations, which are already required for construction purposes, can serve a dual purpose both as foundations and as GHEs (i.e. energy piles). No extra costs (except the minor pipe costs) for drilling and installation are added as the already built structural piles are used as GHEs. However, there are virtually no studies conducted on the feasibility and design of geothermal systems in loessial soils and there is a missed opportunity in current practice to use deep foundations as structural elements and as GHEs when dealing with loessial sites. The GSHP alternative becomes even more attractive in areas where natural gas infrastructure is not available and only expensive liquefied petroleum gas or wood is used for heating. GSHP systems can significantly reduce the use of these fuels in these (predominantly rural) areas and at the same time can provide an alternative clean energy for conditioning of buildings.

To exemplify the benefits of GSHP systems in loess deposits and highlight potential issues, this chapter presents a numerical case study of a representative real building in Córdoba, where foundations are also used as GHEs. While this study focuses on the thermal performance of energy piles, the effect of cyclical heating and cooling on the structural and geotechnical strength of piles is a necessary aspect to also consider in geothermal projects; however, this latter aspect is beyond the scope of this work. Nevertheless, it is important to note that in the case of loess, structural concerns may arise from rupture or leakage of HDPE pipes and water filtration into this collapsible soil. It is therefore crucial to ensure a high standard in quality assurance and quality control of the thermo-fused pipe joints. This is routinely achieved by way of visual inspections and subsequent pressure tests of the U loops before and after casting and during commissioning of the GSHP systems, as outlined in various International Ground Source Heat Pump Association guidelines (Oklahoma State University, 2009).

Main characteristics of loess

Loess is a quaternary sediment composed mainly of fine particles that form an open microstructure. Therefore, this soil is characterised by an internal structure that controls the thermo-hydro-mechanical behaviour of loessial formations. Loess is an unstable soil that develops collapse due to water or internal pressure increases. This produces significant settlements that affect civil infrastructure and structures that suffer significant distortions (Rocca *et al.*, 2006). Loessial soils cover approximately 10% of the continents, including North America, Europe, Asia and South America (Rinaldi *et al.*, 2007). The most significant loess formations are found in Argentina (Iriondo, 1997; Zárate, 2003), the Czech Republic (Marschalko *et al.*, 2013), China (Kukla and An, 1989), Russia (Little *et al.*, 2002), Spain (Günster *et al.*, 2001) and the USA (Leighton and Willman, 1950). All these formations are characterised by a macroporous structure with very high void ratio (typically from 0·9 to 1·25), predominantly fine granulometry mainly composed of silt- and clay-size particles and poorly accommodated structure. As a consequence, loess collapses when water and pressure increases. This phenomenon produces significant damage to buildings constructed on shallow foundations (Rocca, 1985). Different loessial soils can be found depending on whether they suffered changes in the original structure. Primary loesses are those that preserve the original structure generated when they were deposited, while secondary loesses arise when particles of primary loess are transported by any other action such as gravity, water or snow.

Loess formations cover approximately 35% of Argentina (Terzariol, 2009) with a thickness that varies from 10 m to more than 60 m, thus making these formations a good candidate to investigate GSHP system applicability in loess. In general, the mineralogy of Argentinean loess is characterised by an abundance of plagioclase (20–60%), relatively little quartz (20–30%) and a significant percentage of volcanic glass (15–30%), and calcium carbonate (2–10%) (Teruggi, 1957). The Argentinean loess also has significant fraction of volcanic glass as a major component of the silt and sand fractions in comparison with other loess formations around the world (Kostić and Protić, 2000; Kröhling and Iriondo, 1999).

The open structure, large void ratio and mineralogy of loess control the heat flow in these formations. In addition, tillage systems and compaction influence soil thermal properties as well as volumetric water content (Guan *et al.*, 2009; Usowicz *et al.*, 1996). There is limited literature about the thermal characteristics of loess. Nevertheless, Table 1 summarises the most significant thermal properties of different loessial soils around the world. The most frequent thermal conductivity of Argentinean loess is close to 0·65 W/(m K) and tends to increase with the water content and bulk density of the soil (Dec *et al.*, 2009; Gogół *et al.*, 1973; Guan *et al.*, 2009; He *et al.*, 2000; Johnson and Lorenz, 2000; Kodešová *et al.*, 2013; Narsilio *et al.*, 2015; Usowicz *et al.*, 1996; Wang *et al.*, 2007; Zuo *et al.*, 2011).

Table 1 Loess thermal conductivity ranges in different places around the world

Site/location	Thermal conductivity: W/(m K)	Volumetric heat capacity: MJ/(m³ K)	Specific heat capacity[a]: J/(kg K)	Bulk density: kg/m³	Volumetric moisture content: %	Source
Poland	0·5–1·75			1230–1820	0·0–20·0	Gogól et al. (1973)
Felin, Poland	0·47–1·98	1·39–2·42		1320–1410	10·5–24·8	Usowicz et al. (1996)
Fangshan County, China	0·352, 0·344, 0·311	1·44, 1·39, 1·27		—	—	He et al. (2000)
Alaska, USA:						
CRREL Permafrost Tunnel	0·07–0·18			700–1000	5·7–8·0	Johnson and Lorenz (2000)
Birch Hill field sample	0·15	—		1160	1·9	
Chena Spur Road sample	0·73, 0·8			1350, 1360	3·8, 6·5	
China	0·29–1·65	1·497–2·95		1050–1650	0·5–43·5	Wang et al. (2007)
Loess Plateau, China	0·4–1·5	0·01–2·5		—	5·5–16·0	Guan et al. (2009)
Northern Germany	0·91–1·08	2·00–2·90		1470–1530	18·0–42·5	Dec et al. (2009)
Gansu, China	1·0	—		1200–1240	—	Zuo et al. (2011)
Suchdol, Czech Republic	0·25–1·5	1·1–3·1		1310	—	Kodešová et al. (2013)
Córdoba, Argentina	0·36–0·88	—		1200–1400	13·5–33·0	Narsilio et al. (2015)

[a] Estimated based on the reported data
CRREL: Cold Regions Research and Engineering Laboratory

Site description and thermal characterisation

The case study, a residential building, is located 2 km west of Córdoba City (Argentina) and consists of a typical residential two-storey building with a footprint of approximately 85 m² in a 475 m² block of land (17·6 m × 27 m) (Figure 2(a)). The geological formation corresponds to a wind plain aggradation mid-Pleistocene to early Holocene. This domain is characterised in the field of Córdoba City and its periphery, on both sides of the valley of the Suquía River, by a gently undulating plane of sedimentary cover – loessoial silt with a regional tilt to the east of the order of 0·5%.

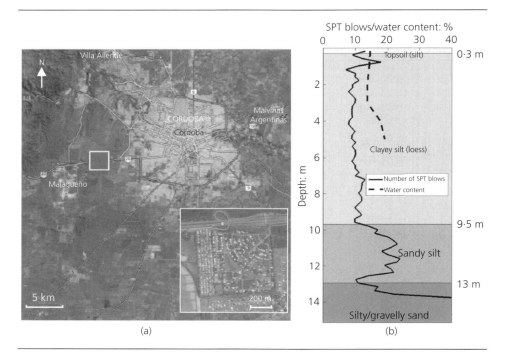

Figure 2 (a) Site location; (b) representative soil profile, water content and Standard Penetration Test (SPT) profile within the site. Data courtesy of GEoS and Profundar

The formation consists of four distinguishable layers (Figure 2(b)): non-plastic silt with some organic matter ($z = 0$–0.3 m below ground surface); collapsible non-plastic silt with some clay and sand ($z = 0.3$–9.5 m) (primary loess); low-plasticity silt with sand ($z = 9.5$–13 m) and a sand with silt and gravel layer ($z > 13$ m).

The water table was not found within 14 m below the ground surface at the time of the geotechnical site investigation, which was conducted during the dry season (winter). However, the groundwater table in this region is known to be located deeper than 20 m. In the loess layer, the natural water content was found to be about 17% and the dry density was about 1400 kg/m^3 (thus, a total density of the order of 1600 kg/m^3). The water content profile measured up to 5 m in depth is shown in Figure 2(b) and for modelling purposes ('Numerical modelling' section), it is assumed constant throughout the year, with the above value taken as representative.

The foundations of this case study involve a total of 13 bored piles, 9·5 m deep, with nine piles of 0·4 m diameter and four (central) piles of 0·6 m diameter. Due to the relatively short length of the piles, the groundwater table is expected to be deeper than the depth of the piles throughout the year.

Figure 3 Measured Córdoba loess thermal conductivity for a range of water contents and dry densities

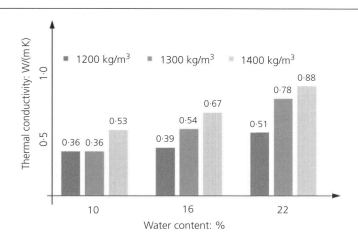

The thermal needle probe method was used for the determination of thermal conductivity of 'typical' loessial silts from Córdoba following ASTM recommendations (ASTM, 2008). Figure 3 summarises the measured thermal conductivities for a range of dry densities and water contents, resulting between 0·36 and 0·88 W/(m K).

Numerical modelling

The aforementioned information is used in detailed finite-element simulations of the GHE field corresponding to the building foundations using a recently developed state-of-the-art model, details of which can be found in Bidarmaghz (2014) and Bidarmaghz *et al.* (2016). The model can account for the local geology, surface thermal recharge and the local weather for a more realistic representation of GHEs' thermal behaviour. An overview of the model and the geometry and initial and boundary conditions of the case study follows.

Governing equations

A three-dimensional (3D) numerical model based on fundamental principles has been developed and implemented using finite-element methods. The governing equations for fluid flow and heat transfer are coupled numerically within the finite-element package Comsol Multiphysics to evaluate the thermal performance of GHEs. Although important, the mechanical behaviour of the energy piles is outside the scope of this work and this study focuses on only the thermal responses of energy piles under realistic heating and cooling cycles.

Heat transfer around and in the GHEs is modelled primarily by conduction and convection. Heat conduction occurs in the soil, GHE backfilling material (concrete) and HDPE pipe wall and partially in the carrier fluid (water). Heat convection dominates in the carrier fluid circulating in the pipes as there is no groundwater flow in the ground (the groundwater table was not found at the site). In this modelling, the fluid flow and heat transfer in the carrier fluid are coupled to the heat transfer in the pipes, the GHEs and the surrounding soil.

The fluid-flow regime in the long pipes is considered as fully developed (e.g. the entrance length of a typical 25 mm round pipe is only about 0·5 m); thus, the fluid flow and the heat transfer in the fluid are simulated using one-dimensional (1D) elements and are coupled to the 3D heat transfer in the GHEs and the surrounding soil.

To model the 1D fluid flow inside the pipes, the continuity and momentum equations for incompressible fluid are used as follows (Barnard *et al.*, 1966).

$$\nabla(A\rho_w \mathbf{v}) = 0 \tag{1}$$

$$\rho_w \left(\frac{\partial \mathbf{v}}{\partial t}\right) = -\nabla p - f_D \frac{\rho_w}{2d_h} |\mathbf{v}|\mathbf{v} \tag{2}$$

where A is the inner cross-section of the HDPE pipe, ρ_w is the carrier fluid density, $\mathbf{v}$ represents the fluid velocity field, t shows the time, p is the pressure, f_D represents the Darcy friction factor and d_h is the hydraulic diameter of the pipe. The energy equation for the fluid flow to describe the convective–conductive heat transfer for an incompressible fluid is (Lurie, 2008)

$$\rho_w A C_{p,w} \frac{\partial T}{\partial t} + \rho_w A C_{p,w} \mathbf{v} \nabla T = \nabla(A\lambda_w \nabla T) + f_D \frac{\rho_w A}{2d_h} |\mathbf{v}|\mathbf{v}^2 + Q_{wall} \tag{3}$$

where $C_{p,w}$ is the specific heat capacity of the fluid, λ_w represents the thermal conductivity of the fluid and Q_{wall} is the external heat exchange rate through the pipe wall. The above equations are solved for pressure p, velocity field $\mathbf{v}$ and temperature field T in the carrier fluid and are coupled to the temperature field T_m obtained from the conductive heat transfer equations solved for the soil, the GHEs and the pipes (Equation 4). The coupling arises from the Q_{wall} term in Equation 3, as Q_{wall} is directly proportional to the temperature inside the HDPE pipe T and the temperature on the outer wall of the HDPE pipe $T_{m,pipe\ wall}$. It should be noted that heat transfer in the soil, the GHEs and the pipe wall is purely conductive due to the absence of ground water flow.

$$\rho_m C_{p,m} \frac{\partial T_m}{\partial t} = \nabla(\lambda_m \nabla T_m) \tag{4}$$

where ρ_m represents the solid material density and $C_{p,m}$ and λ_m are the specific heat capacity and the thermal conductivity of the given solid material (soil, concrete, and HDPE) respectively. This model, completed with appropriate initial and boundary conditions, has recently been validated against available analytical solutions, and full-scale experimental data (Bidarmaghz, 2014; Bidarmaghz *et al.*, 2016; Narsilio *et al.*, 2016). The model is capable of accurately predicting heat transfer between the ground and the GHEs in both laminar and turbulent regimes and homogeneous and heterogeneous soil profiles and flexible in exploring a number of different pipe placement configurations in steady-state and transient conditions.

Case study: geometry, initial and boundary conditions

The finite-element model and the key initial and boundary conditions, together with the GHEs' (piles) configuration, are shown in Figure 4. The 13 GHEs are spaced between ~2 and 4·4 m. Only the four piles in the middle row are 0·6 m in diameter, allowing the insertion of three HDPE U loops, whose spacing between inlet and outlet pipes is 0·16 m (a detail is shown in the figure as well). The remaining nine piles can accommodate only two of these U loops, given their reduced diameter of 0·4 m. It should be noted that these GHEs are bored piles; therefore, full contact between the concrete and the soil is expected and considered in these simulations. However, isolated air pockets (incomplete concreting or imperfections in the pile) may reduce GHE (structural and thermal) efficiency. The size of the ground domain in the numerical model(s) is based on a sensitivity analysis conducted on the distance between the outer GHEs and the 'external' boundaries of the ground, such that the far-field boundaries are far enough from the GHEs in all models to avoid artificial edge effects (minimum distance of 8 m).

The parameters used in all numerical models are shown in Table 2. The initial and far-field ground temperature $T_\text{far-field}$ varies with depth z and time t of the year; these variations derive from the surface air fluctuations throughout the year, particularly in the upper few metres of the soil (i.e. about 4 m in Córdoba, Argentina). However, the average soil temperature deeper than about 4 m tends to be very close to the mean ambient temperature throughout the year at this location (i.e. 18°C).

Figure 4 (a) Finite-element model used in the simulations, with soil limits far enough to minimise artificial boundary effects and the building footprint; (b) a detail of a 0·6 m energy pile

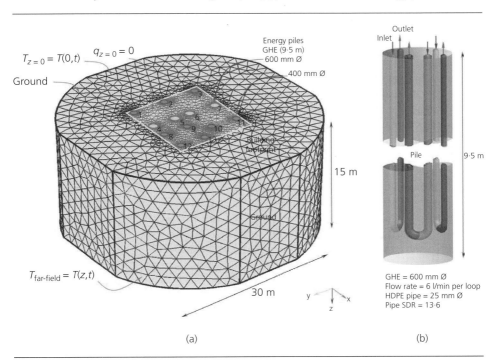

Table 2 Summary of the key input parameters used in the numerical models

Material	Thermal conductivity: W/(m K)	Specific heat capacity: J/(kg K)	Density: kg/m^3	$T_{\text{far-field}}$: °C	Diameter: m	Spacing: m
Loess (soil)	0·700	1200	1600	$T(z, t)$	—	—
Concrete (piles)	2·100	850	2350	—	0·4, 0·6	2–4·4
HDPE (pipe)	0·400			—	0·025	0·16
Water (fluid)	0·582	4180	1000	—	—	—

Thus, some heat is exchanged between the air and the ground (surface thermal recharge/discharge) and it is accounted for in additional simulations following Baggs's adjusted empirical formulations (Baggs et al., 1991).

$$T(z,t) = T_\text{g} + 1\cdot 07 k_\text{v} A_\text{s} \exp\left[-0\cdot 0031552 z \left(\frac{1}{\alpha}\right)^{0\cdot 5}\right]$$
$$\cos\left\{\frac{2\pi}{365}\left[(t-t_0 - 0\cdot 18335 z)\left(\frac{1}{\alpha}\right)^{0\cdot 5}\right]\right\}$$
(5)

where $T(z,t)$ is the ground temperature at depth of z metres after t days from 1 January taken as $T_{\text{far-field}}$ and as initial condition, T_g is the average annual ground temperature (18°C in Córdoba, Argentina), k_v is the vegetation coefficient taken equal to 0·50 in this study (according to Baggs et al., 1991, for ground in full sun $k_\text{v} = 1$ and for a long grass-covered area $k_\text{v} = 0\cdot 25$), A_s is the earth surface temperature annual swing above and below the average ground temperature (14°C), t_0 represents the number of days after 1 January to the minimum earth surface temperature (238 d in Cordoba) and α stands for the soil thermal diffusivity ($0\cdot 3645 \times 10^{-2}$ cm^2/s).

For simplicity, the annual peak building thermal load was estimated based on rules of thumb for typical gated community residencies in Córdoba (heating = 0·035 kW/m^3 or 30 kcal/(h m^3); cooling = 0·058 kW/m^3 or 50 kcal/(h m^3)) rather than detailed energy balance calculations. This case study comprises a total of 160 m^2 covered area, of which 20 m^2 is for a double garage that does not need conditioning. Thus, for 2·6 m high ceilings, the peak heating and cooling demands result in approximately 12·7 and 21·1 kW respectively.

Results and discussions

There are a number of design strategies that could be implemented when deciding on a residential GSHP system, including aiming to provide 100% of the heating and cooling required or to combine geothermal with auxiliary systems (i.e. designing a hybrid system) in all cases avoiding ground freezing that could lead to ground heave or damage to concrete in the pile

(Loveridge and Powrie, 2014). Perhaps the first question to be answered is what the geothermal installed capacity of this residential building would be if all 13 of its piles were used as GHEs. Figure 5(a) shows the total building heating and cooling demand distribution throughout a year, based on the peak demands (12·7 and 21·1 kW; 'Case study: geometry, initial and boundary conditions' section) and the air temperature recorded in 2014.

The newly developed numerical model revealed that the 13 piles are unable to satisfy 100% of both heating and cooling thermal demands of the building. Thus, one alternative is to consider hybrid systems: the GSHP system covers part of the demand and the auxiliary systems (e.g. smaller gas heater(s), smaller air conditioner unit(s) and thermal solar panels and busters) are

Figure 5 (a) Thermal load distributions throughout a year and (b) GHE response over a 20-year period (including typical GSHP operating EWT range)

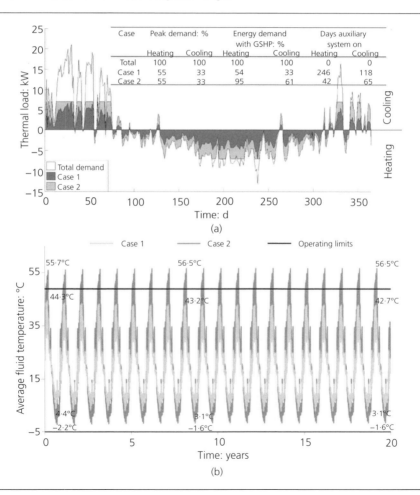

either used to top up the supply of heating or cooling (case 1) or just turned on when this base demand exceeds the geothermal capacity (case 2). Case 1 implies that the auxiliary system is running all the time concurrently with the GSHP, while case 2, only a number of days each year.

As presented in Figure 5(a), case 1 indicates that the GSHP system is designed to provide half and one third of total heating and cooling demand of the house respectively, with an auxiliary system being used 246 d in heating mode and 118 d in cooling mode. The geothermal system satisfies up to 7 kW of the peak heating and cooling demand of the house. In the other case (case 2), the GSHP system also has the same 7 kW peak capacity; however, due to the different thermal load distribution, the geothermal system is aimed at satisfying almost all heating and more than half of the cooling demand of the house with significantly fewer days of auxiliary systems being used.

To highlight how the GHE field will perform and react to the defined thermal loads, the modelled average fluid temperature variations (i.e. $(T_{inlet} + T_{outlet})/2$) in the most demanded GHE (pile 5; Figure 4) obtained from 20-year numerical simulations are presented in Figure 5(b). It is observed that satisfying case 1's thermal load distribution results in a reasonable temperature range (3·1–44·3°C) compatible with most commercially available heat pumps. On the other hand, case 2 results in a fluid temperature drop to a minimum of $-2·2$°C in cold seasons (which is acceptable with an antifreeze solution used as carrier fluid instead of plain water) and a rise to the maximum of 56·5°C in warm seasons. This maximum fluid temperature for case 2 is higher than typical GSHP recommended temperature for water entering the heat pump (EWT). It follows that this GHE field, using the structural piles, is not capable of delivering the more desirable thermal load defined for case 2. Thus, this article focuses on case 1 from here on.

Figure 6 shows the temperature field at middepth of the energy piles (about 5 m below ground surface) (year 10). The figure shows that freezing is avoided in the GHE field for case 1, with no point within the soil or the energy piles reaching less than 4·8°C, even during the coolest seasons when the GHEs extract more of the thermal heat from the ground. Therefore, potential structural problems due to concrete freezing and/or soil heaving are circumvented.

Figure 6 Temperature field in the 13 piles and surrounding soil at 5 m below surface (year 10)

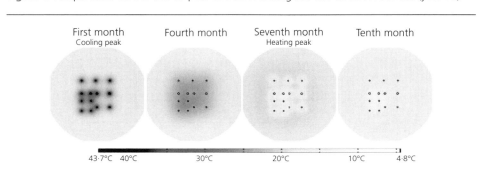

From these results, it follows that even though the low thermal conductivity of loess (0·7 W/(m K)) and relatively shallow GHEs (~10 m) affect the thermal efficiency of the GSHP systems, the incorporation of a GSHP system into the essential 13 structural piles (described in 'Site description and thermal characterisation' section) leads to about 30% savings in conventional energy usage (e.g. electricity or fossil fuels) and in energy bills, with the GSHP satisfying at least 54% of the heating needs and 33% of the cooling needs. Although this GHE field cannot provide 100% of the required heating and cooling demand, it can be used to provide part of the building thermal demand with little additional capital cost.

To increase the capacity of the GSHP system conditioning the case study building, another alternative would be drilling additional GHEs outside the footprint of the house in the available block of land. Increasing the number of GHEs may lead to some extra drilling costs; however, it may also have significant influence on further reducing running energy costs. Therefore, 12 9·5 m long GHEs are added to the original GHE-field, bringing the total number to 25. These additional GHEs do not serve any structural purpose. Surface thermal recharge is also accounted for. Figure 7 shows a schematic diagram of the GHEs and their locations within the available block of land with 12 GHEs being added to the original GHE field. The extra GHEs are of 0·6 m diameter, contain three U loops, and are of the same depth as the rest of the GHE field (9·5 m long GHEs).

Figure 7 Schematic diagram of the expanded GHE field and location of the GHEs within the block of land

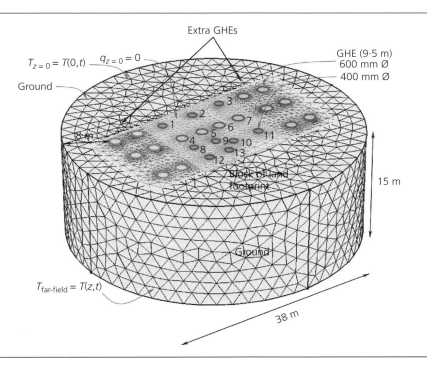

The building thermal load applied as one of the boundary conditions on the model is shown in Figure 8 with 12·7 kW of peak heating and cooling demand now aimed to be satisfied (instead of 7 kW).

The average fluid temperature obtained from one of the most thermally influenced GHEs (pile 5) is shown in Figure 9. Satisfying a peak load of 12·7 kW heating and cooling leads to the fluid temperature varying within a reasonable range (i.e. maximum of 37·2°C and minimum of 6·8°C). The extra 12 GHEs in the GHE field results in 100% of the heating demand and 60·2% of the cooling demand being provided by the GSHP system. A significant improvement in heating and cooling capability of the system is obtained in comparison to the cases investigated in cases 1 and 2. Moreover, the even smaller (and cheaper) auxiliary conditioning system is required for 118 d per year and only in cooling mode. The savings in the installation of auxiliary systems and on running costs would determine the economic feasibility of the additional drilling costs.

GSHP system designers and potential users of these systems should be aware that inadequate designs (or drastic changes in the thermal load patterns) may lead to more pronounced overheating or overcooling of the soil in the medium term until reaching a periodic thermal equilibrium. This change or drift in the overall ground temperature from year to year inevitably influences the efficiency of the system over time, thus affecting the running costs and saving levels of the GSHP systems during their life span.

Loessial soils in Córdoba, Argentina, are of relatively low thermal conductivities. However, loess covers extensive areas in Asia, Europe and America, for which higher thermal conductivities have been reported (see Table 1). To investigate the effect of higher soil thermal

Figure 8 Building thermal load applied on the model with extended number of GHEs (25 GHEs)

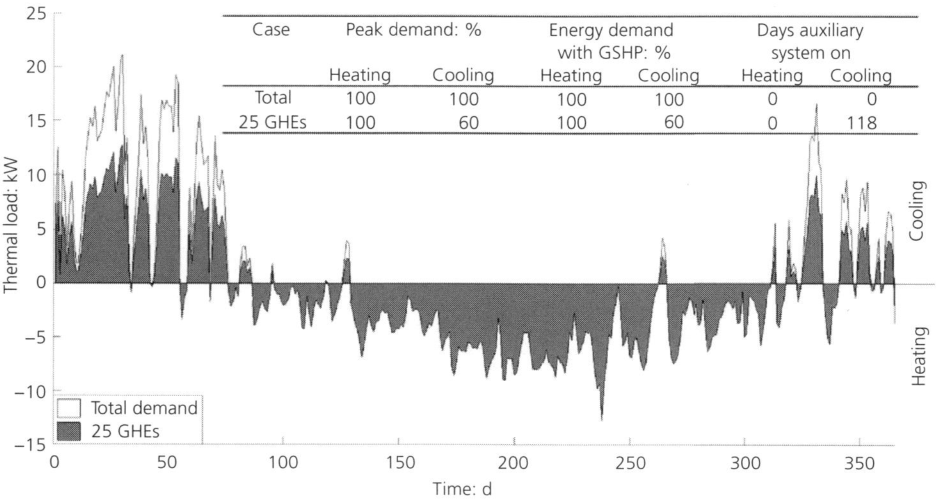

Case	Peak demand: %		Energy demand with GSHP: %		Days auxiliary system on	
	Heating	Cooling	Heating	Cooling	Heating	Cooling
Total	100	100	100	100	0	0
25 GHEs	100	60	100	60	0	118

Figure 9 Average GHE fluid temperature in the expanded GHE field during 20 years of operation

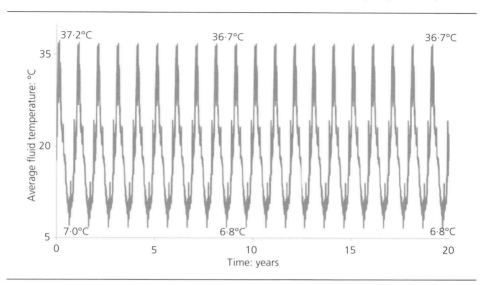

conductivities on thermal exchange capacity of GHEs in loess, numerical models with thermal conductivities of 1·1 and 1·6 W/(m K) have been built and solved and results of 20 years of system operation are shown in Figure 10. The building thermal load distribution applied on this model follows the pattern shown in Figure 5(a) indicated as case 1, which is applied on 13 9·5 m-long GHEs. It is observed that loessial soil with higher thermal conductivity will significantly improve the thermal performance of GHEs. The maximum fluid temperature in the most thermally affected GHE is about 5·4°C lower than in case 1 when the soil thermal conductivity is equal to 1·1 W/(m K) (38·9°C against 44·3°C). Increasing the thermal conductivity to 1·6 W/(m K), which can be considered as an average value of highly conductive loessial soils based on Table 1, further decreases the maximum fluid temperature to about 33·6°C, which is 10·7°C lower than the maximum fluid temperature observed in case 1. These results suggest that with higher thermal conductivity loess, the GHE field is not working at its full capacity and either the thermal load aimed to be satisfied by the GSHP system can be increased or alternatively the number of additional GHEs can be reduced.

Conclusions

Pile and micropiles in loessial areas can be easily converted into GHEs while being built. Results from realistic detailed modelling show that the partial substitution of electrical heating and cooling systems with geothermal systems consisting of 13 energy piles could significantly reduce energy consumption and the size of associated infrastructure (electricity grid). The temperate climate dominating in Córdoba presents ideal conditions for GSHP systems, allowing thermal recharge of the ground between seasons and thus maximizing the heating and cooling capacity that energy piles can achieve; similar conditions apply to other parts of the world. Results shows that the 13 energy piles embedded under the residential case study

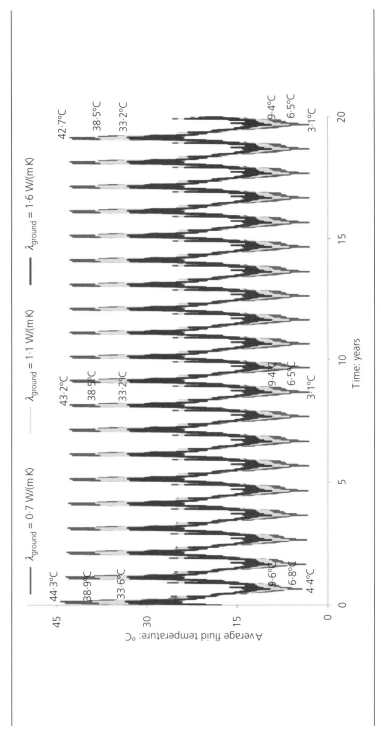

Figure 10 Average GHE fluid temperature during 20 years of operation for higher thermally conductive loessial soils (based on case 1)

building are capable of delivering approximately 54% and 33% of the heating and cooling demand of the building, respectively. The addition of 12 extra GHEs to the GHE field allows satisfying 100% of the heating demand and 60% of the cooling demand. Consequently, additional savings are achieved on the energy bills (running costs) and on the (smaller) auxiliary systems still needed to cover the remaining 40% cooling demand (capital costs). This case study shows the potential of GSHP technology in a local environment dominated by loess and gives some bases to geotechnical engineers to start considering the technology in their designs and practices.

A parametric analysis of thermal conductivity of loessial soil confirms that the higher thermal conductivity of the soil significantly helps to reduce the number of GHEs in the system. As the loessial soils around the world show thermal conductivities varying between 0·1 and about 2 W/(m K), numerical results suggest that GSHP systems in loessial soils of higher thermal conductivity (e.g. Europe and Asia) show a great potential to economically satisfy conditioning of buildings and encourages further application of this technology in loessial soils.

Acknowledgements

Funding from Australian Research Council (ARC) FT140100227 and the University of Melbourne is much appreciated. The authors would also like to acknowledge the contributions of H. Ferrero (Profundar), C. Serrano (GEoS and UCC) and E. Delacoste (UNC) regarding information about the residential case study and loess thermal properties.

REFERENCES

Amatya B, Soga K, Bourne-Webb P and Laloui L (2012) Thermo-mechanical performance of energy piles. *Geotechnique* **62(6)**: 503–519.

ASTM (2008) D 5334: Standard test method for determination of thermal conductivity of soil and soft rock by thermal needle probe procedure. *ASTM International, West Conshohocken, PA, USA*.

Baggs S, Baggs D and Baggs JC (1991) *Australian Earth-Covered Buildings*. New South Wales University Press, Kensington, Australia.

Barnard A, Hunt W, Timlake W and Varley E (1966) A theory of fluid flow in compliant tubes. *Biophysical Journal* **6(6)**: 717–724, http://dx.doi.org/10.1016/S0006-3495(66)86690-0.

Bidarmaghz A (2014) *3D Numerical Modelling of Vertical Ground Heat Exchangers*. PhD thesis, The University of Melbourne, Australia.

Bidarmaghz A, Narsilio G, Johnston I and Colls S (2016) The importance of surface air temperature fluctuations on long-term performance of vertical ground heat exchangers. *Geomechanics for Energy and the Environment* **6**: 35–44, http://dx.doi.org/10.1016/j.gete.2016.02.003.

Brandl H (2006) Energy foundations and other thermo-active ground structures. *Geotechnique* **56(2)**: 81–122, http://dx.doi.org/10.1680/geot.2006.56.2.81.

Dec D, Dörner J and Horn R (2009) Effect of soil management on their thermal properties. *Revista de la Ciencia del Suelo y Nutrición Vegetal* **9(1)**: 26–39.

Francisca FM (2007) Evaluating the constrained modulus and collapsibility of loess from standard penetration test. *International Journal of Geomechanics* **7(4)**: 307–310, http://dx.doi.org/10.1061/(ASCE)1532-3641(2007)7:4(307).

Gogół W, Gogół E and Artecka E (1973) Thermal conductivity investigations of moist soils. *Journal of Power Technologies* **40**: 49–69.

Guan X, Huang J, Guo N, Bi J and Wang G (2009) Variability of soil moisture and its relationship with surface albedo and soil thermal parameters over the Loess Plateau. *Advances in Atmospheric Sciences* **26(4)**: 692–700, http://dx.doi.org/10.1007/s00376-009-8198-0.

Günster N, Eck P, Skowronek A and Zöller L (2001) Late Pleistocene loess and their paleosols in the Granada Basin, Southern Spain. *Quaternary International* **76–77**: 241–245, http://dx.doi.org/10.1016/S1040-6182(00)00106-3.

He K, Wang B and Zhang G (2000) Study on soil thermal properties in forest land of catchment in loess. *Journal of Beijing Forestry University* **22(3)**: 27–32.

Iriondo MH (1997) Models of deposition of loess and loessoids in the Upper Quaternary of South America. *Journal of South American Earth Sciences* **10(1)**: 71–79, http://dx.doi.org/10.1016/S0895-9811(97)00006-0.

Johnson JB and Lorenz RD (2000) Thermophysical properties of Alaskan loess: an analog material for the Martian polar layered terrain. *Geophysical Research Letters* **27(17)**: 2769–2772, http://dx.doi.org/10.1029/1999GL011077.

Johnston I, Narsilio G and Colls S (2011) Emerging geothermal energy technologies. *KSCE Journal of Civil Engineering* **15(4)**: 643–653, http://dx.doi.org/10.1007/s12205-011-0005-7.

Kodešová R, Vlasakova M, Fer M et al. (2013) Thermal properties of representative soils of the Czech Republic. *Soil and Water Research* **8(4)**: 141–150.

Kostić N and Protić N (2000) Pedology and mineralogy of loess profiles at Kapela-Batajnica and Stalać, Serbia. *Catena* **41(1)**: 217–227, http://dx.doi.org/10.1016/S0341-8162(00)00102-8.

Kröhling DM and Iriondo MN (1999) Upper Quaternary palaeoclimates of the Mar Chiquita area, North Pampa, Argentina. *Quaternary International* **57–58**: 149–163, http://dx.doi.org/10.1016/S1040-6182(98)00056-1.

Kukla G and An Z (1989) Loess stratigraphy in central China. *Palaeogeography, Palaeoclimatology, Palaeoecology* **72(3–4)**: 203–225, http://dx.doi.org/10.1016/0031-0182(89)90143-0.

Leighton MM and Willman HB (1950) Loess formations of the Mississippi Valley. *The Journal of Geology* **58(6)**: 599–623.

Little EC, Lian OB, Velichko A et al. (2002) Quaternary stratigraphy and optical dating of loess from the east European plain (Russia). *Quaternary Science Reviews* **21(14)**: 1745–1762, http://dx.doi.org/10.1016/S0277-3791(01)00151-2.

Loveridge F and Powrie W (2014) 2D thermal resistance of pile heat exchangers. *Geothermics* **50**: 122–135, http://dx.doi.org/10.1016/j.geothermics.2013.09.015.

Lurie MV (2008) *Modeling and Calculation of Stationary Operating Regimes of Oil and Gas Pipelines*. Wiley, Weinheim, Germany.

Marschalko M, Yilmaz I, Fojtova L, Lamich D and Bednarik M (2013) Properties of the loess sediments in Ostrava region (Czech Republic) and comparison with some other loess sediments. *The Scientific World Journal* **2013**, Article ID 529431, http://dx.doi.org/10.1155/2013/529431.

Moll L and Rocca R (1991) Properties of loess in the center of Argentina. In *XI Pan American Conference on Soil Mechanics and Foundation Engineering*. Viña del Mar, Chile, pp. 1–14.

Narsilio G, Francisca F, Ferrero H et al. (2015) Geothermal energy in loess: a detailed numerical case study for Cordoba. In *XV Panamerican Conference on Soil Mechanics and Geotechnical Engineering*. Buenos Aires, Argentina.

Narsilio G, Bidarmgahz A, Colls S and Johnston I (2016) Geothermal energy: detailed modelling of ground heat exchangers. *Computers and Geotechnics*.

Oklahoma State University (2009) *Ground Source Heat Pump Residential and Light Commercial Design and Installation Guide*. International Ground Source Heat Pump Association, Stillwater, OK, USA.

Preene M and Powrie W (2009) Ground energy systems: from analysis to geotechnical design. *Geotechnique* **59(3)**: 261–271, http://dx.doi.org/10.1680/geot.2009.59.3.261.

Quintana Crespo E (2005) *Relación entre las Propiedades Geotécnicas y los Componentes Puzolánicos de los Sedimentos Pampeanos*, PhD thesis, Universidad Nacional de Córdoba, Córdoba, Argentina (in Spanish).

Rinaldi V, Rocca R and Zeballos M (2007) *Geotechnical Characterization and Behavior of Argentinean Collapsible Loess*. Taylor & Francis, Boca Raton, FL, USA.

Rocca R (1985) *Review of Engineering Properties of Loess*, MSc thesis, University of California, Berkeley, CA, USA.

Rocca RJ, Redolfi ER and Terzariol RE (2006) Características geotécnicas de los loess de Argentina. *Revista Internacional de Desastres Naturales, Accidentes e Infraestructura Civil* **6(2)**: 149–166 (in Spanish).

Teruggi ME (1957) The nature and origin of Argentine loess. *Journal of Sedimentary Research* **27(3)**: 322–332.

Terzariol RE (2009) Años de estudio de los suelos loessicos en Córdoba, Argentina. *Desafíos y Avances de la Geotecnia Joven en Sudamérica* **1**: 323–337 (in Spanish).

Usowicz B, Kossowski J and Baranowski P (1996) Spatial variability of soil thermal properties in cultivated fields. *Soil and Tillage Research* **39(1)**: 85–100, http://dx.doi.org/10.1016/S0167-1987(96)01038-0.

Wang TH, Liu ZC and Lu J (2007) Experimental study on coefficient of thermal conductivity and specific volume heat of loess. *Yantu Lixue (Rock and Soil Mechanics)* **28(4)**: 655–658.

Zárate MA (2003) Loess of southern South America. *Quaternary Science Reviews* **22(18)**: 1987–2006, http://dx.doi.org/10.1016/S0277-3791(03)00165-3.

Zuo J, Wang J, Huang J *et al.* (2011) Estimation of ground heat flux and its impact on the surface energy budget for a semi-arid grassland. *Sciences in Cold and Arid Regions* **3(1)**: 41–50, http://dx.doi.org/10.3724/SP.J.1226.2011.00041.

Craig and Gavin
ISBN 978-0-7277-6398-3
https://doi.org/10.1680/gehesep.63983.023
ICE Publishing: All rights reserved

Chapter 2
Parsimonious numerical modelling of deep geothermal reservoirs

Tim H. Fairs MSc, MSc, CGeol FGS
Consultant Geologist, Subterra Ltd, Manchester, UK

Paul L. Younger PhD, CGeol FGS, CEng FICE, FREng
Rankine Chair of Engineering and Professor of Energy Engineering, School of Engineering, University of Glasgow, Glasgow, UK

Geoff Parkin PhD
Senior Lecturer, School of Civil Engineering and Geosciences, Newcastle University, Newcastle upon Tyne, UK

Numerical modelling has been undertaken to help improve understanding of a deep geothermal system being considered for development in the vicinity of Eastgate (Weardale, County Durham, UK). A parsimonious numerical modelling approach is used, which allows the possibility to develop a workable formal framework, rigorously testing evolving concepts against data as they become available. The approach used and results presented in this study are valuable as a contribution to a wider understanding of deep geothermal systems. This modelling approach is novel in that it utilises the mass transport code MT3DMS as a surrogate representation for heat transport in mid-enthalpy geothermal systems. A three-dimensional heat transport model was built, based on a relatively simple conceptual model. Results of simulation runs of a geothermal production scenario have positive implications for a working geothermal system at Eastgate. The Eastgate Geothermal Field has significant exploitation potential for combined heat and power purposes; it is anticipated that this site could support several tens of megawatts of heat production for direct use and many megawatts of electrical power using a binary power plant.

Notation

c_s	specific heat capacity of host medium (J/(kg K))
C_{ss}	source or sink concentration (kg/m^3)
C^k	dissolved mass concentration of species k (kg/m^3)
D_h	thermal diffusivity (m^2/s)
D_m	molecular diffusion coefficient (m^2/s)
K_d	distribution coefficient (m^3/kg)
n	porosity
q_h	heat source/sink (W/m^3)
q_{ss}	fluid source or sink (s^{-1})
R	retardation factor

T	temperature (K)
t	time (s)
v_a	specific discharge (m/s)
α	dispersivity tensor (m)
λ_m	effective thermal conductivity of the porous media (W/(m K))
ρ_b	bulk density (mass of sold divided by total volume) (kg/m^3)
$\rho_m c_m$	volumetric heat capacity of the porous medium (J/(m^3 K))
$\rho_w c_w$	volumetric heat capacity of the water (J/(m^3 K))

1. Introduction

The development of engineering models for geological systems is beginning to reach considerable levels of sophistication (e.g. Parry *et al.*, 2014). It is now widely accepted that the supreme model of any system is the conceptual model, with observational and computational models essentially serving to probe the consistency of the conceptual model with available data (cf Brassington and Younger, 2010; Konikow and Bredehoeft, 1992; Parry *et al.*, 2014). Viewed in this light, it is never too early to establish a computational model, so that concepts can be rigorously tested for consistency with data as soon as these become available. Such an approach can be particularly useful in systems of complex geometry and potentially non-linear behaviour that usually defy intuitive identification of likely system responses to engineered changes. This chapter reports just such an exercise: an attempt to develop a preliminary mathematical model of a deep geothermal system which has been discovered in the vicinity of Eastgate in Weardale, County Durham, UK. Additionally, this study attempts to utilise these modelling simulations to assess the capability of the well-known mass transport modelling software MT3DMS for simulating heat transport in the deep subsurface.

Two exploratory wells drilled in the Eastgate area in recent years provided the experimental database for this study. These wells targeted a high natural permeability vein/fault structure known as the Slitt Vein, that penetrates the radiothermal Weardale granite and acts as a conduit for thermal hypersaline fluids.

Limited well and regional geological data were used to construct a conceptual model of the hydrogeological system of the Eastgate area and this guided fluid flow modelling using MODFLOW and heat transport modelling using MT3DMS. MODFLOW is a modular finite difference flow model written in Fortran code that solves the general equation governing groundwater flow and is based on both Darcy's law and the law of conservation of mass (McDonald and Harbaugh, 2003). MT3DMS is a modular finite-difference three-dimensional (3D) mass transport model, widely used for simulation of contaminant transport in porous media and remediation assessment studies (Zheng and Wang, 1999). The application of MT3DMS to simulation of thermal transport phenomena in saturated aquifers is possible because the governing equations for solute transport are mathematically equivalent to those for heat transport (Hecht-Mendez *et al.*, 2010).

The specific objectives of the study were firstly to calibrate a flow model for the Eastgate geothermal system in MODFLOW using data obtained from Eastgate boreholes 1 and 2, and then secondly, to simulate a hypothetical well doublet with a shallow injection well (Eastgate No. 1) and a deepened production well (Eastgate No. 2) in MT3DMS. This was undertaken so as to explore a sustainable production scenario over long time periods (100 years), in order to observe any temperature decline and predict any potential thermal breakthrough from reinjected water into the production well.

While there have been previous models built of other geothermal systems, this modelling exercise is unique for the following reasons. Firstly, as it attempts to model a fault-controlled geothermal system in low-to-mid-enthalpy granite, where there is sparse data control. Secondly, it attempts to use mass transport modelling code to simulate heat transport in a deep geothermal system.

2. Background
2.1 Geological setting
The North Pennines of England have many geological and hydrogeological attributes that are favourable for the development of geothermal energy. Firstly, there is a known natural heat source at depth, namely the Weardale granite which is distinctly radiothermal. Secondly, there is an apparent plentiful supply of deep hydrothermal brines and a 'plumbing system' in the form of the Slitt Vein. Thirdly, there is a natural 'lagging' with the overlying Carboniferous Limestone and the Whin Sill dolerite, providing insulation for the system. The study area lies in the axis of Weardale, one of the principal valleys in the North Pennines, ~40 km west southwest of the city of Durham. The geological features of interest at Eastgate form part of the North Pennine Orefield, a regionally uplifted and domed structure comprising of a series of horst blocks (Kimbell *et al.*, 2010) (Figure 1).

2.2 Geothermal developments
In addition to the mineralisation that is associated with the North Pennine Batholith, recent interest has concentrated on its geothermal potential, with investigations of the hydrogeologically active fracture systems, specifically the Slitt Vein in the Eastgate area (Manning *et al.*, 2007).

2.3 Eastgate drilling/testing history
Exploration for this geothermal resource commenced in December 2004 with the drilling of Eastgate No. 1 well, reaching a total depth of 995 m. The well trajectory was designed to track the Slitt Vein and associated splays vertically downwards for up to 1 km (Figure 2). The borehole successfully penetrated 723 m of Weardale granite, overlain by 272 m of Quaternary and Lower Carboniferous cover rocks. At 410 m depth, a major open fissure was encountered, which was assumed to be a splay fault associated with the Slitt Vein. The borehole received a large influx of warm formation water, with a high electrical conductivity. Alkali geothermometry was conducted on water samples, suggesting that the water achieved equilibrium with respect to Na, K and Ca at depths of between 3 and 4 km, based on a geothermal gradient of 40°C/km. This is a significant point, as it implies that the formation water encountered possibly forms part of a deep circulation system that appears to be still active (Manning *et al.*, 2007).

Following the drilling of the Eastgate No. 1 borehole, extensive hydraulic testing was carried out during March 2006. In the first phase, the entire open section of the borehole from 403 to

Figure 1 Geological map of the North Pennines of England (simplified after Kimbell et al. (2010))

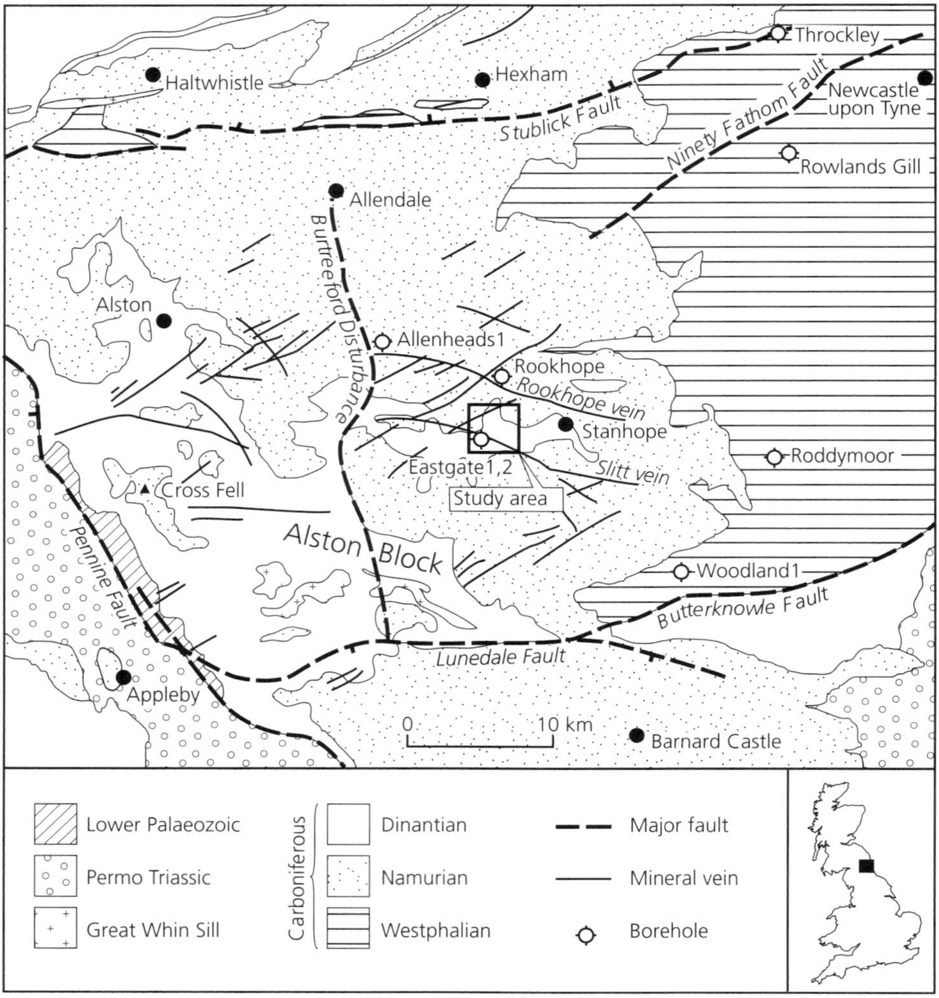

995 m depth was test pumped at a rate of 888 m^3/d. The pumping rate during the second phase of testing in the lower fracture zone of the granite below 431 m was 518 m^3/d. The pumping tests and their respective drawdown were revealing as the first test suggested that the flow was by way of the Slitt Vein, whereas the hydraulic response from the second test suggested that the lower fracture system was not in hydraulic connection with the Slitt Vein (Younger and Manning, 2010). These test results also confirmed the high permeability of the Slitt Vein, with an intrinsic transmissivity in excess of 4000 darcy m.

The Eastgate No. 2 borehole was drilled in March 2010 to a total depth of 420 m and successfully penetrated 134 m of Weardale granite, in addition to 286 m of recent and Lower

Figure 2 Schematic diagram of cross-section of Eastgate Borehole No. 1 (after Manning *et al.*, 2007)

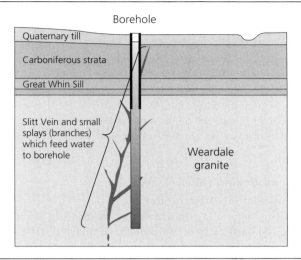

Carboniferous cover rocks. The Weardale granite at this location was found to be hydraulically tight, with an average intrinsic transmissivity of around 6 darcy m, obtained from a rising head test. This was in line with expectations, as Eastgate No. 2 was located 300 m from the Slitt Vein and thus confirmed the high permeabilities in Eastgate No. 1 as being due to the borehole intercepting the Slitt Vein.

3. Conceptual modelling – a tripartite modelling approach

The first critical step in the process of constructing a robust numerical model is to establish a conceptual model or a theory-based description that satisfies the various boundary conditions and assumptions made with regard to the hydrogeological processes (Brassington and Younger, 2010).

This conceptual model attempts to define the Eastgate geothermal system that essentially consists of a fracture-hosted hydrothermal brine circulation system, represented in this study area as the Slitt Vein, cross-cutting the Weardale granite in a west northwest to east southeast orientation. Fluid and heat flow are assumed to be predominantly along the main Slitt Vein, with associated splays and fractures deeper within the Weardale granite providing a secondary conduit as observed from the Eastgate No. 1 pump test results. Apart from convective heat transport, there is also thought to be a contribution from conductive heat transport through the host granite, which will be discussed further in Section 3.3. The Weardale granite is overlain unconformably and blanketed by over 270 m of Lower Carboniferous strata, including the Whin Sill, which are also intersected by the Slitt Vein fracture system to the surface. No recharge was included in the model, but sensitivity to recharge is considered later in section 4.2. The model contains the two Eastgate wells – Eastgate No. 1 is considered as a shallow injection well and Eastgate No. 2, deepened in the model to intersect the Slitt Vein at 2500 m, acts as a production well.

In terms of model domain size, the horizontal dimensions were chosen as 3000 m × 3000 m, in consideration of the radii of influence of the two boreholes, along with a vertical depth of 3000 m that covers the deepened production well at Eastgate No. 2.

A modelling strategy was adopted that began with a very simple model containing only the Slitt Vein, with geological and hydrogeological complexity progressively added to the model, with respect to the surrounding fracture zone and the host granitic body. This approach led to the creation of a tripartite modelling process, whereby the first model is a two-dimensional (2D) representation of only the Slitt Vein, the second is a 3D 'limited-extent' model that includes a fracture zone surrounding the Slitt Vein and the third is a 3D 'full-extent' model that includes the main granitic host (Figures 3–5, respectively); these are discussed in more detail below.

3.1 Fluid flow modelling (MODFLOW)

3.1.1 Two-dimensional Slitt Vein model

This model is a relatively simple 2D representation of only the Slitt Vein that is assumed to be vertical and with the two Eastgate wells intersecting the vein at discrete intervals (Figure 3). The major assumption of this model is that active flow only occurs within the Slitt Vein. The model grid consists of a single row of 30 cells of total length 3000 m and an individual cell size of 100 m. It has a total vertical depth of 3000 m and a layering scheme that conforms generally to the major stratigraphic horizons, as well as the zones of intersection of the two Eastgate boreholes with the Slitt Vein. The model was run first in steady state with the Eastgate No. 1 well set as a producer at a rate of 888 m³/d, so that an attempt could be made to calibrate the

Figure 3 Two-dimensional Slitt Vein model showing intersection of Eastgate well No. 1 (injector) and deepened Eastgate well No. 2 (producer) as shaded cells

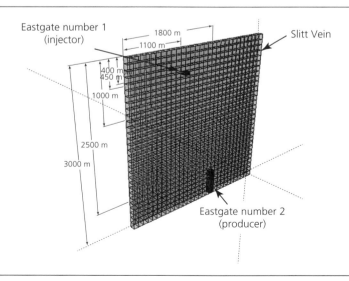

model against the drawdown results reported from the testing. The model was then simulated as a well doublet system, with Eastgate No. 1 acting as an injector and a deepened Eastgate No. 2 as a producer, both pumping at 888 m³/d.

3.1.2 Three-dimensional limited-extent model

This model comprises the 2D Slitt Vein model surrounded by a limited 200 m zone of enhanced permeability within the fractured host granite and overlying Carboniferous Limestone (Figure 4). The major assumption of this model is that active flow occurs mainly within the Slitt Vein, but with a contribution from the surrounding fracture zone. As with the 2D Slitt Vein model, the 3D limited-extent model was calibrated against the drawdown results reported from the first test phase. Calibration of the model was also carried out using the drawdown results from the second test phase for the lower 500 m interval within the surrounding fracture zone. However, the well intersections were limited to the Slitt Vein for the well doublet scenario, with Eastgate No. 1 acting as an injector and Eastgate No. 2 as a producer, as described previously.

Figure 4 Three-dimensional limited-extent model showing Slitt Vein with surrounding fracture zone of enhanced permeability

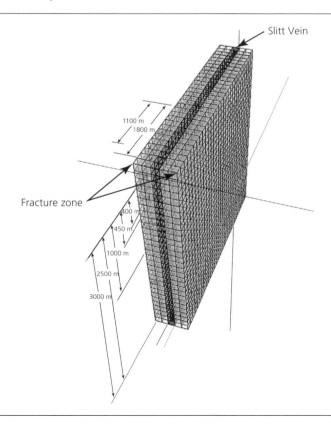

Figure 5 Three-dimensional full-extent model showing Slitt Vein with surrounding fracture zone of enhanced permeability, and host Weardale granite and Carboniferous Limestone. Well intersection intervals for Eastgate well Nos. 1 and 2 into the Slitt Vein are indicated as shaded cells

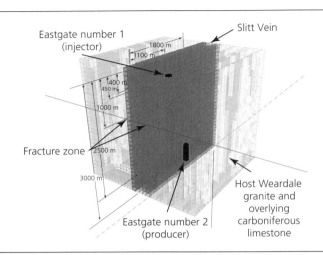

3.1.3 Three-dimensional full-extent model

This model comprises the Slitt Vein and surrounding limited-extent enhanced permeability fracture zone surrounded by a much larger volume of the host granite and overlying Carboniferous Limestone that have negligible permeability (Figure 5). The major assumption of this model is that the host granite beneath the blanket of Carboniferous Limestone is effectively impermeable, but that this large volume of rock may be significant in the modelling if it is a major contributory factor as a heat source. The model grid consists of 1250 m of the host country rock laterally on either side of the Slitt Vein/fracture zone. In terms of model calibration and simulation, the same procedures were applied as for the previously described 3D limited-extent model.

3.2 Fluid flow calibration results (MODFLOW)

Calibration was carried out by attempting to match the model drawdown values with the observed drawdown values from the two Eastgate No. 1 well tests. Drawdown is defined as the difference between the initial water level in a given well and the observed water level at any specific time during a period of pumping. Model drawdown was varied by changing the value of the hydraulic conductivity parameter. Horizontal and vertical hydraulic conductivities were both initially set as equal for all cells in the model and based on values reported from the hydraulic responses from Eastgate No. 1 testing results (Younger and Manning, 2010). This approach to calibration can be considered as a form of inverse modelling or parameter identification. This is analogous to pumping test analysis, where radial flow represented by an analytical equation is used to determine transmissivity and hence, hydraulic conductivity. For this study, a different geometrical configuration is represented, but the response is similarly controlled by the ability of the active flow zones to convey flow rates that are affected primarily

Table 1 Fluid flow model calibration (hydraulic conductivity values represent the zones Slitt Vein/fracture zone/host rock)

Model	Well test drawdown: m	Initial model drawdown: m	Initial hydraulic conductivities: m/d	Revised hydraulic conductivities: m/d	Final model drawdown: m
Two-dimensional Slitt Vein	0·5	0·0013	3200	320	0·5
Three-dimensional limited extent	27·0	0·0016	3200/170	3·2/0·17	1·9
Three-dimensional full extent	27·0	0·0004	3200/170/21	320/17/0·21	12·0

by the hydraulic conductivity and the zone widths. Hence, the drawdown response is sensitive to both hydraulic conductivity and the assumptions of the geometry and extent of the active flow zones as represented in the models.

Table 1 summarises calibration results for the three models. It can be observed that the initial model drawdowns were very small in comparison with the well test drawdowns, and that significant changes were required in the hydraulic conductivity values in order to get a reasonable match. The numbers in the columns for the initial and revised hydraulic conductivities relate to the values that were used for the specific zones of the model in question. The difficulty in achieving equivalent drawdown values reflects the challenges of both optimising model parameters whilst at the same time reconciling the complexity of the fracture network in this system.

Despite this, it should be appreciated that a match within the same order of magnitude is considered reasonable when dealing with hydraulic conductivity; this can naturally vary over several orders of magnitude for a given rock type in water-supply aquifers (Younger, 2007). In addition, it was difficult to obtain a 'typical' value for hydraulic conductivity for comparison purposes due to the unique nature of this environment.

3.3 Heat transport modelling (MT3DMS)

MT3DMS uses the following partial differential equation to solve for solute transport in transient groundwater flow systems (Zheng and Wang, 1999)

$$(1 + \rho_b K_d/n)\partial C^k/\partial t = \text{div}[(D_m + \alpha_{V_a})\,\text{grad}\,C^k] - \text{div}(\alpha_{V_a} C^k) + q_{ss} C_{ss}/n \qquad (1)$$

Equation 1 is the general governing equation for solute transport.

The left-hand side of the equation is the product of the transient term and the retardation factor (R), where $R = (1 + \rho_b K_d/n)$. For solute transport, retardation is caused by adsorption of solutes by the aquifer matrix material. On the right-hand side of the equation, the first term defines hydrodynamic dispersion, which includes pure molecular diffusion (D_m) and mechanical dispersion (α_{V_a}). The second term defines advection and the third term defines the source and sink processes.

The heat transport equation invokes the second law of thermodynamics (i.e. conservation of thermal energy in this case), considering both conduction and convection (De Marsily, 1986), and can be simplified to the following form (Hecht-Mendez et al., 2010)

$$(\rho_m c_m / n\rho_w c_w) \cdot \partial T / \partial t = \mathrm{div}[(\lambda_m / n\rho_w c_w + \alpha_{V_a}) \mathrm{grad}\ T] - \mathrm{div}(\alpha_{V_a} T) + q_h / n\rho_w c_w \qquad (2)$$

Equation 2 is the general governing equation for heat transport.

In comparing the two above equations, coefficients needed for heat transport can be readily substituted for their solute transport counterparts, so that MT3DMS can be used without modification to model heat transport. The following coefficients are described with their implementation in MT3DMS, as originally detailed in Hecht-Mendez et al. (2010).

Retardation factor (R) and the distribution coefficient (K_d) in the solute transport equation represent solute adsorption by the aquifer matrix; so in the heat transport equation, retardation is a result of heat transfer between the fluid and solid aquifer matrix. MT3DMS represents thermal retardation by calculating the distribution coefficient (K_d) for the temperature species as a function of thermal properties as follows: $K_d = c_s / \rho_w c_w$, where c_s is the specific heat capacity of the solids and $\rho_w c_w$ is the volumetric heat capacity of the water.

The value for the distribution coefficient is input in MT3DMS in the 'Chemical Reaction Package', as the slope of the isotherm, with the type of sorption set to 'linear isotherm' (ISOTHM = 1), so that the temperature exchange rate between the solid and water is independent of any temperature changes.

As described above, there are two parts in the solute transport equation to describe hydrodynamic dispersion; molecular diffusion (D_m) and mechanical dispersion (α_{V_a}). Heat conduction is mathematically equivalent to molecular solute diffusion. Whereas in solute transport, molecular diffusion is a function of the concentration gradient, in heat transport it is a function of the temperature gradient and is equivalent to the following: $D_h = \lambda_m / n\rho_w c_w$.

This value is input in MT3DMS in the 'Dispersion Package' as the molecular diffusion coefficient (DMCOEF). The terms for hydrodynamic dispersion (α_{V_a}) describe the differences in flow velocity at a pore scale; the specific dispersivity coefficients in MT3DMS are longitudinal dispersivity and the ratios of transverse horizontal and vertical dispersivity to longitudinal dispersivity. These coefficients are directly applied and input as heat dispersivity coefficients for heat transport in MT3DMS.

In the heat transport equations, the source and sink term represents heat input or extraction, and the temperature value (K) is equivalent to concentration (kg/m^3). Therefore, temperature is

Table 2 Heat transport variables, values, units and reference sources

Symbol	Variable	Value	Units	Reference source
K_d	Distribution coefficient	2.10×10^{-4}	m³/kg	
λ_m	Effective thermal conductivity of the host rock (granite/Carboniferous Limestone)	3.4	W/(m K)	Banks (2008: p. 35) Downing and Gray (1986)
n	Porosity	0.05	—	Younger (personal communication)
$\rho_m c_m$	Volumetric heat capacity of granite/limestone	$2.4 \times 10^{+6}$	J/(m³ K)	Banks (2008: p. 35)
$\rho_w c_w$	Volumetric heat capacity of water	$4.18 \times 10^{+6}$	J/(m³ K)	Hecht-Mendez *et al.* (2010)
c_s	Specific heat capacity of host (Weardale granite)	845	J/(kg K)	Downing and Gray (1986)
a_l	Longitudinal dispersivity	0.5	m	Hecht-Mendez *et al.* (2010)
a_{th}	Transverse horizontal dispersity	0.5	m	Hecht-Mendez *et al.* (2010)
a_{tv}	Transverse vertical dispersity	0.5	m	Hecht-Mendez *et al.* (2010)
D_h	Thermal diffusivity	2.97×10^{-6}	m²/s	
R	Retardation factor	2.29	–	
ρ_b	Dry bulk density of host rock	1690	kg/m³	Downing and Gray (1986)
	Injection temperature	293	K	Younger (personal communication)
	Geothermal gradient	38	°C/km	Manning *et al.* (2007)

substituted directly for concentration as 'Source/Sink' with the type of source set to 'Well' (ITYPE0 = 1).

Table 2 lists heat transport variables with the values used as initial input in the numerical codes in MT3DMS for the modelling.

3.4 Assumptions and limitations of modelling approach

Table 3 is a summary of the main assumptions made in the modelling process, coupled with comments in terms of the resultant limitations to the modelling effort. It is envisaged that as the implications of these limitations are recognised, these can then guide further study, which will be discussed later in Section 6 of this chapter.

Table 3 Assumptions and limitations of modelling process

	Assumptions	Limitations
1	Subsurface geology is understood	Subsurface geology could be much more complex, in terms of faulting/fracturing and orientation of Slitt Vein, along with connectivity. Only one geological model is considered here
2	Conceptual model is comprehensive	Model may not have all potential boundary conditions and flow directions considered
3	Model dimensions are appropriate	Model may not be sufficiently large to capture all flow and heat transport volumes
4	Boundary conditions are appropriate	Model may not have correct boundary conditions in terms of flux
5	Subsurface fluid flow is almost entirely from the Slitt Vein	Significant fluid and heat transport may be coming from host granite by way of a larger fracture system
6	Recharge is uniform temporally and areally	Recharge may vary areally and temporally which will affect the flow budget in the model
7	Recharge is solely from precipitation	Recharge could be from surface run-off and indirectly by way of flooded mine workings
8	Porosity distribution is uniform	Fracture porosity may not be evenly distributed
9	Hydraulic conductivity and permeability is isotropic	Hydraulic conductivity may be anisotropic and heterogeneous
10	Heat transport modelling using MT3DMS is appropriate for use in deep geothermal systems	MT3DMS is run decoupled from MODFLOW, which may cause significant errors due to temperature variations affecting water viscosity and density, which in turn affect hydraulic conductivity

4. Well doublet production scenario

4.1 Heat transport simulation results (MT3DMS)

A well doublet production scenario was simulated with a transient state model in MT3DMS, using the flow simulation from the 3D full-extent model in MODFLOW. The heat transport simulation was run over a 100-year period, with temperature values monitored at two discrete points in the model; these were 200 m west of the Eastgate No. 1 (injector) and Eastgate No. 2 (producer) wells, respectively. It was observed that the temperature monitored in close proximity to the injector well (Eastgate No. 1) showed an overall 5–6°C decrease in temperature for the 100-year time period, with temperature monitored nearby to the producer well (Eastgate No. 2) showing a larger decrease in temperature of 14°C during this time period (Figures 6 and 7). The lower part of Figure 6 is a cross-sectional view along the Slitt Vein axis for this simulation at the end of 10 years and shows both Eastgate wells. It can be seen that the isotherms are relatively undisturbed across the model, except in the area of the Eastgate No. 1 well, where the model appears to be showing some perturbation by the reinjection of cooler

Figure 6 (a) Time-series plot for the model covering 100 years, with temperature measured at the cell 200 m to the west of the Eastgate No. 1 injector well. (b) Cross-sectional view along Slitt Vein axis at the end of 10 years (3650 d). Eastgate wells and temperature measurement point indicated. Note that the numbers on the y-axis in (b) are temperature contour labels, and each square spatially represents 100m

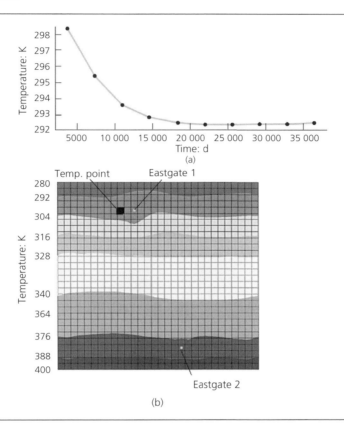

water at 293 K. The upper part of Figure 6 shows a time-series plot for the model covering the entire time period of 100 years, with temperature measured at the cell 200 m to the west of the Eastgate No. 1 injector well. It can be seen from the decline curve of the plot that the temperature decrease is more rapid at first and then levels-off around 20 000 d (54·8 years), with the temperature stabilising around 293 K (20°C). The lower part of Figure 7 is a cross-sectional view along the Slitt Vein axis for this simulation at the end of 100 years, showing both Eastgate wells. It can be seen that by this time, the isotherms have been extensively disrupted throughout, particularly in the area around the abstraction well (Eastgate No. 2). The upper part of Figure 7 shows a time-series plot for the entire time period of 100 years for the cell 200 m to the west of the Eastgate No. 2 abstractor well. It can be seen that the temperature decrease appears to be linear over this time period.

Figure 7 (a) Time-series plot for the model covering 100 years, with temperature measured at the cell 200 m to the west of the Eastgate No. 2 producer well. (b) Cross-sectional view along Slitt Vein axis at the end of 100 years (36 500 d). Eastgate wells and temperature measurement point indicated. Note that the numbers on the y-axis in (b) are temperature contour labels, and each square spatially represents 100m

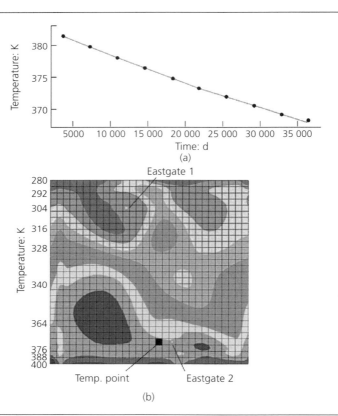

4.2 Sensitivity analysis

Sensitivity analysis gives an increased understanding of the relationship between input and output variables in a model and helps to identify the inputs that cause the most uncertainty in the outputs; these should then be key areas for study if confidence in a model is to be achieved.

A summary of the results of the sensitivity analysis is shown in Figure 8 in the form of a tornado diagram (this is a useful and graphical way of showing the relative importance or sensitivity of input variables to the output variable, which in this case is the temperature change (ΔK)). The chosen temperature value is at a nominal model cell location, 300 m to the west of Eastgate No. 1 well at the end of 7330 d (20·1 years). The input variables were all systematically varied from a range of −50% to +2000% and the output variable, in this case the change in temperature (ΔK), at the designated point in the model was recorded.

Figure 8 Tornado diagram showing sensitivity of model input parameters to output parameter of temperature change (ΔK). The chosen temperature value is at a model cell location, 300 m to the west of Eastgate No. 1 well at the end of 7330 d (20·1 years). Note that the low and high legend entries relate to parameter value extremes. Units are as in Table 2

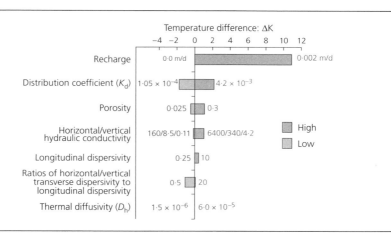

It can be interpreted from the results of the sensitivity analysis shown in Figure 8 that the inputs likely to cause the most uncertainty in the outputs in the model are, in order of decreasing significance, recharge, distribution coefficient (K_d), porosity and hydraulic conductivity.

5. Discussion and conclusion

A steady-state 3D flow model was constructed in MODFLOW and calibrated with observed drawdowns from the Eastgate boreholes. A transient state 3D heat transport model was constructed in MT3DMS and simulation runs indicate significant temperature decreases around the area of the producer well, with a slow linear decline in temperature over time. In contrast, the temperature around the injector well shows an initial rapid decline, which then stabilises to the injection temperature. These results from the heat transport simulations have positive implications and significance for a potential working geothermal production system operating under these conditions, as thermal decline is predicted to be sufficiently slow at the producer well, with no apparent thermal breakthrough from the injector well over the 100-year time period in which the model was run. The majority of heat transport in this model appears to occur within the Slitt Vein due to a combination of convective and conductive processes. Sensitivity analysis indicates that the heat transport model is most sensitive to parameters representing convective transport – those governing pore water velocities such as recharge, hydraulic conductivity, porosity and local heat exchange between water and rock. It is less sensitive to parameters related to large-scale conductive heat transport such as distribution coefficient, thermal diffusivity and dispersivity. This is a good example of a generic finding from this study which has wider applicability.

Through parsimonious numerical modelling, a self-consistent 'model' (in its general sense) of the behaviour of this type of environment has been developed, both in terms of process understanding and for quantifying appropriate parameters. In doing so, this modelling work contributes to the building of a portfolio of evidence, though at the present time this is very limited due to a scarcity of available datasets for equivalent geothermal systems. This makes this work all the more important not only in terms of its contribution through the methodology of the exploratory modelling approach, but also in the interpreted hydraulic conductivity values. It is also timely in terms of helping to inform future engineering strategies for developing deep geothermal energy.

This type of model cannot be proven or validated, it can only be tested and invalidated (Konikow and Bredehoeft, 1992). The most one can ever ask is that one achieves credible consistency between concepts, that is, those used to establish the model and the available data (cf Brassington and Younger, 2010; Parry *et al.*, 2014). It is believed that this has been generally achieved in this model. First, the concept of heat and fluid flow concentrated along a high permeability fault system was corroborated with the results of the drawdown calibrations. Second, the results of the sensitivity analysis indicated that the heat transport model is most sensitive to parameters related to convective heat transport and this supports the model of fault-related heat and fluid flow.

The modelling approach is valid for other circumstances, precisely because it is physically based – that is, it is based on the physics of the system. There is no reason why a similar fault-associated geothermal system could not be modelled using the same approach. Site-specific parameters are, of course, site specific, but the approach is generically valid.

6. Recommendations for future work

In addition to gaining more raw data to help the modelling, there are a number of issues surrounding assumptions made in this initial model that need to be addressed in future modelling. These are described below.

The assumption that heat transport modelling using MT3DMS is appropriate for use in deep geothermal systems has limitations and is clearly an area for further study, as any large temperature variation will affect water viscosity and density, which in turn affects hydraulic conductivity. These changes are not taken account of as MT3DMS is decoupled from MODFLOW. Fracture heterogeneity is clearly an issue that requires further study and future work in this area may involve constructing various geostatistical models to attempt to capture the variability inherent in this property, to support simulations in which permeability is allowed to be heterogeneous. In terms of the sensitivity analysis, work has been carried out to date and this could be taken further in terms of carrying out predictive sensitivity analysis. This would involve selecting variations of a particular scenario based on the sensitivity analysis and then running these as a range of simulations. Another important piece of future work could be the calculation of the Peclet number for a range of models, which would reflect convection and conduction-dominated scenarios; this would be very useful in terms of understanding the relative importance and implications of these heat transport processes. Finally, the current model was run over what was considered a sufficiently long time period of 100 years, but more simulation runs could be made for longer periods, to observe what further thermal decline occurs.

As and when further resources become available to develop the Eastgate Geothermal Field, the model presented here supports the notion that it could be exploited in combined heat and power mode to produce several tens of megawatts of direct-use heat, as well as many megawatts of electrical power by deployment of a binary power plant operating according to the principles of the organic Rankine cycle or similar approaches (cf Younger, 2013; Younger et al., 2012).

This chapter has practical relevance and potential application for any civil engineer interested in the quantification and feasibility of an area for prospective geothermal production assessment; the modelling techniques described in this chapter could be readily adopted and employed for other systems.

Acknowledgements

This chapter is based on work undertaken by the first author for a dissertation submitted as part of an MSc in Applied Hydrogeology at Newcastle University. The authors acknowledge funding of the Eastgate fieldwork from One NorthEast, the Geothermal Challenge Fund of the UK Government's Department of Energy and Climate Change, and helpful suggestions from Cluff Geothermal Ltd on field data interpretation, and funding from the Natural Environment Research Council (NERC grant number NE/I018905/2) for part of the geological work used to underpin the modelling. The assistance of Christine Jeans at Newcastle University in preparing the figures is gratefully acknowledged.

REFERENCES

Banks D (2008) *An Introduction to Thermogeology: Ground Source Heating and Cooling*. Wiley-Blackwell, London, UK.

Brassington FC and Younger PL (2010) A proposed framework for hydrogeological conceptual modelling. *Water and Environment Journal* **24(4)**: 261–273.

De Marsily G (1986) *Quantitative Hydrogeology*. Academic Press, Orlando, FL, USA.

Downing RA and Gray DA (1986) *Geothermal Energy. The Potential in the United Kingdom*. British Geological Survey, H.M. Stationary Office, London, UK.

Hecht-Mendez J, Molina-Giraldo N, Blum P and Bayer P (2010) Evaluating MT3DMS for heat transport simulation of closed geothermal systems. *Ground Water* **48(5)**: 741–756.

Kimbell GS, Young B, Millward D and Crowley QG (2010) The North Pennine batholith (Weardale granite) of northern England: new data on its age and form. *Proceedings of the Yorkshire Geological Society* **58(2)**: 107–128.

Konikow LF and Bredehoeft JD (1992) Ground–water models cannot be validated. *Advances in Water Resources* **15(1)**: 75–83.

Manning DAC, Younger P, Smith FW *et al.* (2007) A deep geothermal exploration well at Eastgate, Weardale, UK: a novel exploration concept for low-enthalpy resources. *Journal of the Geological Society, London* **164(2)**: 371–382.

McDonald MG and Harbaugh AW (2003) The history of MODFLOW. *Ground Water* **41(2)**: 280–283.

Parry S, Baynes FJ, Culshaw MG *et al.* (2014) Engineering geological models: an introduction: IAEG commission 25. *Bulletin of Engineering Geology & the Environment* **73(3)**: 689–706.

Younger PL (2007) *Groundwater in the Environment: An Introduction*. Blackwell, Oxford, UK.

Younger PL (2010) *Geothermal Borehole Works at Eastgate, County Durham Undertaken in 2010 with funding from the Deep Geothermal Challenge Fund*. Department of Energy and Climate Change, London, UK, Final Report.

Younger PL (2013) Renewable heat and binary power: the contribution of deep geothermal energy. Keynote Lecture. *Proceedings of the 13th UK Heat Transfer Conference (UKHTC2013), 2nd–3rd September 2013, Imperial College, London, UK.*

Younger PL and Manning DAC (2010) Hyper-permeable granite: lessons from test-pumping in the Eastgate Geothermal Borehole, Weardale, UK. *Quarterly Journal of Engineering Geology and Hydrogeology* **43(1)**: 5–10.

Younger PL, Gluyas JG and Stephens WE (2012) Development of deep geothermal resources in the UK. *Proceedings of the Institution of Civil Engineers – Energy* **165(EN1)**: 19–32, http://dx.doi.org/10.1680//ener.11.00009.

Zheng C and Wang PP (1999) *MT3DMS: A Modular Three Dimensional Multi-species Transport Model for Simulation of Advection, Dispersion and Chemical Reactions of Contaminants in Groundwater Systems; Documentation and User's Guide*. US Army Corps of Engineers Research and Development Center, Vicksburg, MS, USA, p. 202, Contract Report SERDP-99-1. See http://hydro.geo.ua.edu/mt3d/ (accessed 02/07/2015).

ICE Themes Geothermal Energy, Heat Exchange Systems and Energy Piles

Craig and Gavin
ISBN 978-0-7277-6398-3
https://doi.org/10.1680/gehesep.63983.041
ICE Publishing: All rights reserved

Chapter 3
Geothermal subsidence study at Wairakei–Tauhara, New Zealand

Chris Bromley MSc(Hons)
Geothermal geophysicist, Chairman IEA-GIA, GNS Science, Wairakei, New Zealand

Kerin Brockbank BE
Reservoir engineer, Contact Energy, Wellington, New Zealand

Trystan Glynn-Morris BSc(Hons)
Geotechnical/drilling/reservoir engineer, Contact Energy, Wellington, New Zealand

Michael Rosenberg MSc(Hons)
Geologist, GNS Science, Wairakei, New Zealand

Michael Pender PhD, FIPENZ, MASCE
Professor of Geotechnical Engineering, University of Auckland, Auckland, New Zealand

Michael O'Sullivan PhD, FIPENZ
Professor of Engineering Science, reservoir modeller, University of Auckland, Auckland, New Zealand

Steve Currie BSurv
Managing Director, Energy Surveys, Taupo, USA

Geothermal, as a source of renewable energy (power and heating), has the potential to meet 3–5% of global demand by 2050. For some high-temperature reservoirs, a technical challenge that may constrain deployment is ground subsidence caused by reservoir pressure decline. At the Wairakei–Tauhara geothermal system in New Zealand, an integrated geotechnical-geoscientific investigation of the causes of local subsidence anomalies (up to 15 m, accumulated over 50 years) has successfully identified and modelled the factors, mechanisms and processes involved. Zones of hydrothermally altered porous sediments and clays, at up to 400 m depth, display inelastic deformation behaviour, yielding to a highly compressible state once subjected to a fluid pressure decline. Monitoring shows that pressure dissipates slowly through these low-permeability capping formations. Adaptive management involves continued monitoring of subsidence rates, and sampling, analysis and predictive modelling where necessary. Simulation modelling indicates that long-term mitigation can be achieved by sustaining pressures through targeted shallow injection.

1. Introduction

Geothermal is a global source of renewable energy that is considerably underutilised (Goldstein et al., 2011). Indeed, it is projected that, by the year 2050, geothermal could be providing 3–5% of global power and heating requirements (Bromley and Beerepoot, 2011). This is a significant increase from the current 0·3%. With economic incentives, by 2100, up to 10% is possible. Already, the proportion of geothermal electricity generation ranges from 12% to 30% in six countries, including New Zealand and Iceland. These countries are favourably located for the presence of economically accessible, high-temperature geothermal systems (i.e. near plate tectonic boundaries and active volcanoes). In the coming decades the geothermal industry will be tasked with accelerating deployment, in order to help mitigate long-term global climate change effects. This may lead to some secondary environmental challenges. In the past, one such challenge, in several high-temperature geothermal systems, has been ground subsidence resulting from reservoir pressure decline, which is a consequence of hot fluid extraction. Understanding of the subsidence processes, in order to produce reliable forecasts, has been hampered by inadequate geotechnical and hydrological information regarding the compressing formations and their relationship to subsurface pressure transients within the anomalous subsidence zones. A comprehensive investigation of the observed subsidence anomalies at Wairakei–Tauhara, New Zealand, has recently been completed. Boreholes were specifically designed and drilled to recover and analyse cores taken from formations with anomalous geotechnical properties. This chapter provides an overview of the total investigation, and a companion paper describes the core analysis results (Pender et al., 2013).

1.1 Wairakei–Tauhara history

The Wairakei geothermal power station was commissioned in 1958. Over the past nearly 60 years, net fluid production has been in excess of 2·5 km^3. Three power plants across Wairakei–Tauhara currently use the steam produced to generate approximately 250 MWe of baseload power. A modern, 166 MWe geothermal plant is also under construction in the Te Mihi (north-west Wairakei) sector to replace and augment 45 MWe of the original Wairakei turbines. Although deep liquid pressure drawdown initially amounted to about 25 bar, after 1980 these pressures levelled off, indicating balancing inflows of induced natural fluid recharge (O'Sullivan et al., 2009). Since 2000, liquid pressures have been rising slightly, in response to increased amounts of fluid reinjection (~30% of mass withdrawn). Shallow steam-zone pressures, however, have declined steadily in response to cooling and direct steam extraction in the main Wairakei borefield areas.

Since 1955, subsidence has been a regular feature of the monitored environment of the 75 km^2 Wairakei–Tauhara geothermal system (Allis et al., 2009). In a delayed response to the initial pressure drawdown, four local subsidence areas (or 'bowls') formed at specific locations across the system, superimposed on background deformation. The bowls are of the order of 0·5–1 km^2 in area, and are centred at Wairakei Valley, and in the adjacent Tauhara field, at Spa Valley, Rakaunui Road and Crown Road (Figure 1). The environmental effects that resulted from these subsidence bowls included ground deformation and drainage effects from ponding, tilting, and ground strain.

Figure 1 Map of Wairakei–Tauhara geothermal system, showing geology cross-section location (Figure 3), subsidence investigation and monitor boreholes, subsidence rates at 5 mm/year contour intervals (2004–2009), and key benchmarks in main subsidence bowls: Wairakei, Spa, Rakaunui, and Crown. Aratiatia Dam is reference site

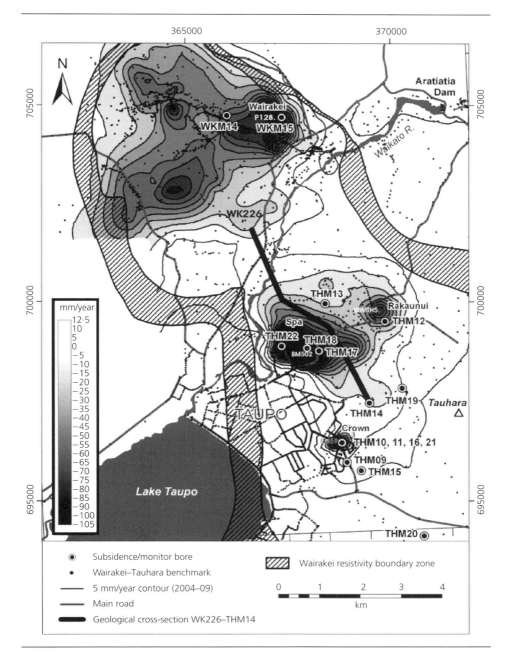

By 2009, a maximum 15 m of accumulated subsidence had occurred at the centre of the Wairakei Valley bowl, of which 12·6 m was measured and 2·4 m extrapolated (1955–1966). Rates peaked at 0·5 m/year in the 1970s. The well-documented effects included strain on pipelines and drains within the Eastern Wairakei borefield, some local ponding of the Wairakei Stream, and tilting of power pylons and the swimming pool and buildings at the Wairakei Resort Hotel (Allis, 2000). At Tauhara, within the spa bowl, the observed effects have included accumulated 1% tilt across a building, tension cracks across roads, and compression deformation of a wastewater pipeline (Bromley et al., 2010). At the Crown bowl, a tension crack has appeared in the tiled garage floor of a house located at the edge of the bowl (Bromley et al., 2009). These effects have been gradual, and have not required more than minor remedial work to date.

Subsidence at Wairakei–Tauhara has been an issue in a number of formal hearings related to environmental consents for expanded development of the system as a source of renewable energy. Approvals, granted in 2007, were qualified by the need for more research and analysis of local subsidence mechanisms. In 2008, a subsidence investigation programme was initiated by the operators, Contact Energy Ltd, to address the issues raised. A subsequent application to expand development of the Tauhara sector by a further 250 MWe was granted in 2011; project commencement currently awaits financial closure.

1.2   Deformation surveys

Commencing in 1955, an extensive benchmark network was progressively expanded across the entire Wairakei–Tauhara system, and re-levelled approximately every 4 years. Figure 1 shows a contour map of subsidence rates over the 2004–2009 period. Subsidence-level changes at key benchmarks within the subsidence bowls are plotted in Figure 2. Backward extrapolations in time used pro-rata rates from nearby benchmarks. There has been a trend of rising and then declining rates, but with peak times that varied significantly from bowl to bowl: Wairakei bowl was the first and Crown bowl the last. Repeat horizontal surveys were also undertaken in selected areas, particularly bowl edges, to identify horizontal rates of movement, which reached a maximum of about 30% of the vertical rates.

At the centre of the Wairakei bowl, the vertical rates rose steadily to about 480 mm/year in 1979, and have since declined steadily to 45 mm/year (2004–2009). Rates near the centre of the Spa bowl, in Tauhara, increased to about 100 mm/year (relative to an external benchmark) by 2009. Over the past 10 years, rates on the western side of Spa bowl (closest to the Waikato River) have been declining, while rates on the eastern side have levelled off. At the nearby Rakaunui bowl centre, relative rates have declined from a peak of about 60 mm/year in 1999 to 23 mm/year in 2011. Further south, at the Crown bowl, relative rates rose from 10 mm/year in 1999 to a peak of 62 mm/year in 2005. Annual repeat surveys showed a decline to about 23 mm/year in 2010, and near zero in 2011.

1.3   Background deformation

Over the past decade, permanent cGPS (continuous global positioning system) stations have been installed at sites surrounding the Wairakei–Tauhara geothermal system. The resulting regional ground deformation data show fluctuations in level of about 5–10 mm/year, lasting several months to years. These are mostly of tectonic origin (local strain adjustments to deep

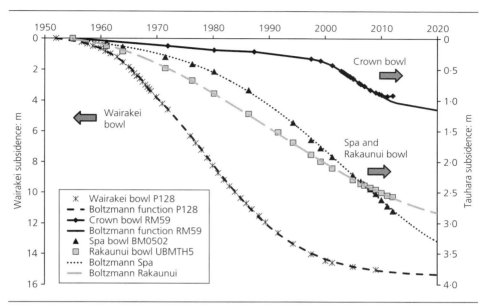

Figure 2 History (with forward and backward extrapolations) of subsidence at representative benchmarks near centres of Wairakei bowl and three Tauhara bowls: Crown, Spa and Rakaunui. Boltzmann functions are fitted to these data and projected out to 2020. Parameters given in Table 1

stress accumulation). They may account for some of the time variations seen in the local levelling data. Satellite-based ground deformation images (DInSAR – differential interferometric synthetic aperture radar) provide additional supporting information, confirming the locations and approximate subsidence rates of the known bowls. They also demonstrate that the existing benchmark network covers all significant anomalies (Hole et al., 2007; Samsonov et al., 2009).

Outside the main bowls, level changes, relative to a reference datum at Aratiatia Dam (Figure 1), show different amounts of background subsidence accumulating steadily since 1955: about 1–2 m in the Wairakei sector, 1 m in the northern Tauhara sector, about 0·4 m in central Tauhara, and about 0·1 m to 0·2 m of uplift in southern and eastern Tauhara. Over the initial 25 years, deep pressure declined throughout the Wairakei–Tauhara geothermal system, leading to boiling and steam zone creation. This was followed by gradual steam pressure decline (O'Sullivan et al., 2009). The implication of these observations is that the principal cause of the observed background deformation is a combination of both pressure decline in shallow steam zones and regional tectonic tilting.

2. Geotechnical investigation

In 2008, Contact Energy embarked upon a subsidence investigation programme that involved the drilling of 13 investigation wells, with a depth range of 200–800 m, in order to capture cores for geotechnical analysis, including compressibility testing. Most wells were also completed to allow for monitoring of changes in shallow aquifer conditions. Novel methods were used for drilling, continuous coring and 'intact' sampling in a geothermal setting. These are described by Ramsay *et al.* (2011). Triaxial (K_0) compressibility tests (pre-yield and post-yield constrained moduli [M], yield stress and Poisson's ratio) were undertaken on selected 'intact' core samples. These are discussed further by Pender *et al.* (2013). In addition to core inspection, description, rock-quality designation (RQD; relative measure of degree of fracturing observed in core) and photography of the continuous core, other geotechnical tests conducted on selected samples, as appropriate, included point load tests (to infer unconfined compressive strength), shear vane, pocket penetrometer, density, water content (porosity), Atterberg limits and particle size analysis. Geological samples were taken for methylene blue titration tests (smectite content), scanning electron microprobe and X-ray diffraction (for clay type and abundance).

Data were also collected on the physical geothermal conditions (pressure, temperature and permeability) at various locations and depths in and around the subsidence anomalies. Brockbank *et al.* (2011) describe these physical measurements, and the shallow aquifer characteristics. Correlations were noted between inferred feed zones (permeable horizons) based on drilling data, core measurements and well measurements. The depths of low RQD (high fracture density), low core recovery and drilling fluid loss zones were often correlated with permeable feed zones. Where fractures control the permeability and silica deposition is present, these zones were generally of low to average porosity. Low rock strength (from point load tests) was typically found near steam or two-phase feeds (weakened by clay alteration), and higher strengths near liquid feeds (strengthened by silica deposition). Permeability of the mudstone cap could not be measured in situ (Ramsay *et al.*, 2011), but was inferred by modelling (Allis and Zhan, 2000) to be 0·05–0·3 mD (10^{-15} m^2).

2.1 Geology and relationship to subsidence

The geological units, described in detail by Rosenberg *et al.* (2009), comprise a sequence of young volcanic products and derived sediments, capped by layers of high-porosity, but mostly low-permeability, lacustrine sediments (upper and lower Hula Falls formation, HFF). The middle unit of the HFF consists of a relatively permeable breccia. A cross-section (Figure 3) between Wairakei and Tauhara illustrates representative geology. Karapiti rhyolite lavas (KAR) and volcanic breccia formations (RBX and WAF) provide permeable aquifers beneath the HFF. These link deep upflows of geothermal fluid at Wairakei and Tauhara.

All geological units have geotechnical properties that vary widely – a consequence of their heterogeneous deposition and inter-layered hydrothermal alteration, characterised by pervasive hydrothermal clays (mostly kaolinite, smectite, illite or chlorite) with deposition of silica or calcite. The hydrothermal alteration causes either stiffening or weakening of the original lithology through mineral dissolution and deposition (Lynne *et al.*, 2011). All four subsidence bowls contain low-pressure steam zones that contribute to the weakening process through rock alteration by acidic, gas-rich, steam condensates. Permeability in parts of these weak layers is relatively high compared with that outside the bowls. Highly variable drilling conditions were also observed (Ramsay *et al.*, 2011), including hard pans, apparent cavities inferred from sudden drops of the drill-string, and intervals where core was not recoverable. Pockets of

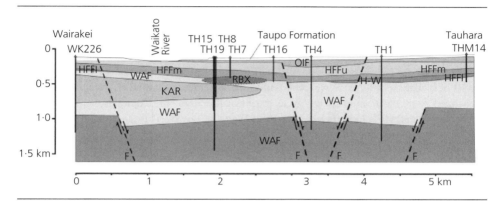

Figure 3 North-west–south-east interpretive geological cross-section between Wairakei and Tauhara. Formation abbreviations: OIF = Oruanui ignimbrite; HFF = Huka Falls (u, upper, m, middle, l, lower units); WAF = Waiora; H-W = transitional HFF to WAF; KAR = Karapiti 2A rhyolite lava; RBX = rhyolitic breccia. See Rosenberg et al. (2009) for more detail

extremely soft, hydrothermal clay-rich material are evident in the core, particularly within the Wairakei and Crown bowls.

At the Wairakei bowl, the anomalously compressible formations include tuff breccia in the WAF (230–330 m depth), subunits in the HFF at 75–230 m depth, and possibly decaying peat/vegetation layers at shallow depth (30–45 m) beneath the Taupo ignimbrite (TPO). At the Spa and Rakaunui bowls, the principal layers are at 130–400 m within subunits of the HFF, where it overlies a firm but permeable andesite lava flow (SPA). The Crown bowl anomalous material is a 35–200 m deep hydrothermal eruption breccia (CXB) and overlying Post Oruanui sequence (POS). There, the weakest material is hosted in boiling groundwater at 50–70 m depth.

2.2 Well measurements

Pressure and temperature measurements in the newly drilled monitor bores around Wairakei–Tauhara reveal a complex zone of interaction between deep and shallow aquifers (Brockbank et al., 2011). An uppermost groundwater aquifer (recharged by rainfall and steam condensate) is perched approximately 50 m above an intermediate liquid aquifer of mixed origin, which is hydraulically connected to the nearby Lake Taupo and Waikato River (Bromley, 2009). This, in turn, is perched about 100 m above the level of the deep geothermal brine. Pressure drawdown has increased boiling, which has formed expanding two-phase zones (with varying saturations of liquid and steam, of varying relative mobility). These zones occur between the liquid aquifers, across a range of pressures (4–22 bar (400–2200 kPa)). They are often trapped beneath capping layers such as the upper and lower HFF. Near the Spa bowl the lower HFF is absent, allowing the deep liquid to penetrate to relatively shallow depths (~200 m) and locally control pressures in the mid-HFF. The upper and intermediate liquid aquifers also reveal a lateral pressure gradient from east to west, reflecting the influence of rainfall recharge from the slopes of Mount Tauhara, and drainage towards the Waikato River.

47

2.3 Core property correlations

Comparison of the various measured material properties has contributed to an improved understanding of the mechanisms involved in the formation of these subsidence anomalies. Figure 4 shows an example for borehole THM16 (Crown bowl). The point-load test (PLT) rock failure strengths are highly variable across depth scales of just a few metres. However, they also reveal trends in average strength that show some positive correlation with denser lithology, silica deposition and constrained modulus values, and some negative correlation with porosity and hydrothermal clay content. Porosities generally decline with depth from about 60% in the near-surface pumice (TPO) to about 20% in competent reservoir rocks at depths below about 400 m (WAF), but within the subsidence anomalies exceptionally high porosities (>50%) were found at depths as great as 220 m. Notable examples were the CXB at Crown bowl, and HFF formations in Spa and Wairakei bowls. This supports an association between anomalous porosity and locally compressible material. A correlation was also observed between relatively high Poisson's ratios (0·3 ± 0·1), the presence of smectite clay, and low constrained modulus values and low yield stresses, particularly in the centre of the upper HFF mudstone, and in the upper portion of the CXB. Low pocket-penetrometer and shear-vane values (<30 kPa) on soil-like samples from 150 m to 300 m depth in the Wairakei and Spa bowls, and 40–70 m depth in the Crown bowl, also correlate well with sequences of low constrained modulus value and PLT strength. A sample from the top of the POS formation at 37 m depth was found to be 97% clay (particle-size analysis test), and to have anomalous porosity (72%) and Atterberg limits (92% liquid limit, 72% plastic limit), and a high liquidity index, indicating the presence of highly compressible or sensitive clay. Between 48 m and 160 m, samples of CXB had liquid limits of 51–61%, plastic limits of 24–39% and sand contents of 22–44%. These results all point to a clay-rich deposit of highly compressible material with a moderate plasticity index.

3. Subsidence mechanisms

Satisfactory subsidence mechanisms require an explanation for observed timing, amplitudes, and locations of subsidence anomalies. With respect to timing, some minor changes in deformation are of tectonic origin. At Wairakei–Tauhara, the majority, however, is interpreted to be of geothermal origin. Changes in the observed subsidence rates with time are explainable in terms of a combination of: (*a*) the timing of pressure changes (mostly within two-phase zones); (*b*) the delays in timing caused by vertical pressure diffusion through the inter-layered low-permeability formations; and (*c*) the effect of non-linear yield behaviour. Yielding of the more compressible materials involves transition to greater rates of compression, through the effect of declining pore pressure, after a local yield stress is exceeded by the effective vertical stress (EVS).

The effect on material properties of elevated formation temperatures (100–200°C, at up to 400 m depth) was not able to be assessed using the available apparatus. However, oedometer compressibility testing of local samples of HFF mudstone, in a heated water bath, and at temperatures of 15–70°C, had previously revealed no significant difference in behaviour with increasing temperature (Read *et al.*, 2001). The effect of formation cooling with time was also

Figure 4 Example of comparison of material properties, against depth, from Crown bowl cores. Yield stress plotted relative to in situ effective vertical stress (EVS) curve (for 1955 and 2009); note that pre-yield moduli apply if above the curve, post-yield if below. Average moduli for 25 m layers (broad bars) used to generate a one-dimensional model (Figure 6)

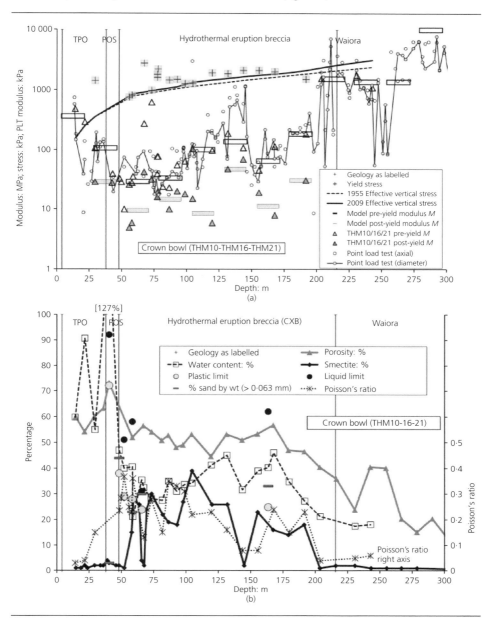

considered. It was simulated using an elastic thermal expansion coefficient of 8×10^{-6} per degree C, but was found to contribute an insignificant amount (<10%) to total observed subsidence rates, and could not have accounted for the presence of the subsidence bowls.

To help explain the amplitude, shape and location of the anomalies, core compressibility test results are summarised graphically in Figure 5, in the form of constrained modulus (K_0) as a function of reduced level (depth). Ellipses are drawn around groups of test results to illustrate the depth intervals of the softer material encountered within the bowls. The testing has also shown that the deformation of the softer material is irrecoverable, in most cases, because the process is not elastic. The reload compressibility is approximately 10–20% of the virgin (post-yield) compressibility. Consequently, an increase in fluid pressure due to injection or natural recharge does not lead to a significant reversal of the subsidence process.

The dominant subsidence mechanism is pressure decline, which occurs in low-pressure, shallow, two-phase zones or heated groundwater aquifers. This causes compression of local 'pockets' of intensely altered, highly porous and highly compressible sediments and volcanic tuffs. These 'pockets' occur at different depth intervals in each bowl. At the Wairakei bowl, Allis and Bromley (2009) postulated that a small part of the observed subsidence in the 1970s and 1980s may have originated from thermal degradation, by an increased flux of rising steam, of local peat deposits near the base of the TPO (30–45 m depth).

Figure 5 Constrained (K_0) moduli (M) at 2009 in situ effective vertical stress (EVS) conditions (either pre- or post-yield) against elevation (masl). Ellipses show intervals of relatively soft material found within bowls

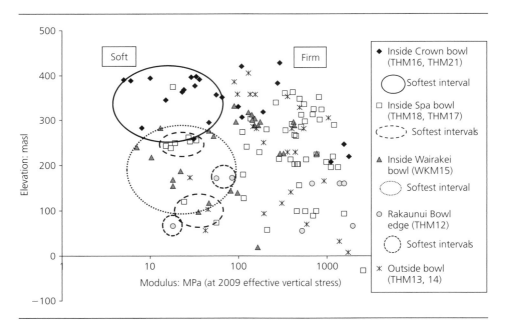

At the Crown bowl, the association of clay-rich acid alteration products and a hydrothermal eruption breccia with the measured bowl geometry suggests a mechanism whereby the subsidence could be confined to the buried eruption crater. X-ray diffraction revealed that the breccia source rocks were hydrothermally altered before the eruption, but preferential and strong alteration also occurred after the eruption had ceased. The explanation is that acidic steam from shallow boiling was able to permeate and condense in the eruption matrix more readily than it could in the surrounding country rock. Adding to the subsurface complexity, clay-rich material that formed in acidic mud pools or small caverns near the base and sides of the original crater may have become entombed by fresh eruptions or capped and preserved by subsurface silica deposits.

In several places across Wairakei–Tauhara, subsidence rates have reduced in recent years, and a possible mechanism to explain this is shallow pressure recovery, caused by lateral or downward inflows of cool groundwater. Specifically, these could have occurred at the Crown bowl (newly constructed stormwater soakage pond), on the western side of the Spa bowl (Waikato River water recharge), and on the eastern side of the Wairakei bowl (lateral groundwater recharge).

3.1 Subsidence modelling

Simulations of the patterns in time and space of anomalous subsidence at Wairakei–Tauhara have previously been undertaken using modelling software. These included application of a one-dimensional finite-element code for fully coupled consolidation and single-phase fluid flow (Allis and Zhan, 2000), and a two-dimensional simulation using the soil mechanics software Plaxis (Lawless *et al.*, 2003; White *et al.*, 2005). These methods were useful, but not entirely satisfactory when it came to matching observed subsidence and making future projections, because of large uncertainties, assumptions and limitations with some aspects of the codes. Arguably, the model outputs were based on too many assumptions regarding the non-linear physical properties of the anomalously compressible formations, and regarding changes in the local shallow pressure gradients, which in turn are controlled by unknown permeability structure and undersaturated (boiling) conditions.

In order to minimise unknown variables, a subsequent approach used a novel method to match historic subsidence trends in the main bowls. The method fitted Boltzmann functions (mathematical functions of sigmoidal type simulating diffusion through a layer) to simulate the effects of one-dimensional pressure diffusion through a compacting layer (Bromley, 2006). The Boltzmann equation and the best-fitting parameters for the four bowls (as used to generate the curves shown in Figure 2) are listed in Table 1.

This method accurately matches observed subsidence changes with time, and successfully predicts short-term changes (5–10 years), by assuming a status quo trend in aquifer pressure, along with continued diffusion of past pressure transients into the low-permeability, compressible layer. However, for longer-term predictions, more sophisticated codes are required. Ideally, a fully calibrated simulation of reservoir pressure changes, and their effects on shallow aquifers, using a multiphase code such as TOUGH2 (a finite-difference computer code (v. 2) simulating multiphase reservoir fluid flow) is needed, along with a fully coupled, non-linear, geomechanical modelling code.

Table 1 Parameters used to fit Boltzmann function curves, as plotted in Figure 2

Boltzmann:	$S = P(T)HC \times (1 - 1/\{1 + \exp[(T - T_m)/T_d]\})$			
Bowl (BM)	$P_{min}HC$: m	$T(P_{min})$	T_m: years	T_d: years
Wairakei (P128)	15·45	1968·6	1979·1	8·2
Spa (BM0502)	4·00	2018·0	1989·0	18·0
Rakaunui (TH5)	3·50	1994·0	1994·0	18·0
Crown (RM59)	0·85	1980·0	2005·7	3·5

S is subsidence (m); $P(T)$ is local pressure decline as a function of time (bar), reaching a minimum at $T(P_{min})$; H is formation thickness (m) and C its average compressibility (bar^{-1}); T_m is the mid-point time and T_d the diffusion time (years) for the Boltzmann function

In 2006, three-dimensional aspects of the shape of the subsidence anomalies were also investigated, using an analytical modelling approach attributed to Geertsma (1973). The depth and radius of a compacting cylinder are varied to fit the observed curvature and horizontal deformation at the edges of the bowls. Calculated depths to the top surface of a compacting disc were 34 m at Crown Road, 88 m at Spa and 100 m at Wairakei (Bromley et al., 2010). Depth and thickness information was supported by other independent data, such as reports of casing deformation in boreholes (Allis et al., 2009).

Following the 2008–2009 geotechnical drilling programme, simple one-dimensional layered models were again constructed for the bowl centres. These models used the measured compressibility values, averaged onto 10 m or 25 m thick layers to calculate subsidence histories. Non-linear yielding effects were applied at the appropriate times. These were based on the measured yield stress from the core samples, and the EVS at the layer depth based on overburden stress minus pore fluid pressure, from historical pressure records. Pressures with depth were interpolated from measured borehole data, by assuming hydrostatic or two-phase vertical pressure gradients, and by assuming smooth pressure changes with time in the various aquifers. Examples of this modelling process for the Crown bowl are illustrated in Figure 6. For the Wairakei bowl the inferred history of pressure–depth profiles is shown in Figure 7, the compressibility properties against depth in Figure 8, and the modelling results in Figure 9. Despite the uncertainties resulting from the assumptions, and the averaging processes, the calculated subsidence values provide a reasonable match to the observed trends (Bromley et al., 2010). Projections, based on these models, were developed further by using future pressure trends from a TOUGH2 reservoir model that simulates the effects of continued Wairakei production and a proposed 250 MWe development expansion at Tauhara (O'Sullivan et al., 2009). The results indicate that, with the possible exception of the Crown bowl, subsidence rates in the bowls are likely to reduce compared with background rates over the next 10–15 years.

Figure 6 Crown bowl subsidence and fitted one-dimensional model curve, using pressure trends interpolated from observations and labelled according to monitor well. Projected subsidence assumes further shallow pressure decline, without targeted injection. Post yielding occurs in 25 m layer (+250 m RL) from 1967, plus 50 m after 2009. Future pressure trends based on reservoir model for central Tauhara, including proposed 250 MWe expansion. Uplift from pressure rise calculated using 20% unload compressibility. 'Extrapolated' curve is from Boltzmann function fit, and assumes status quo pressure

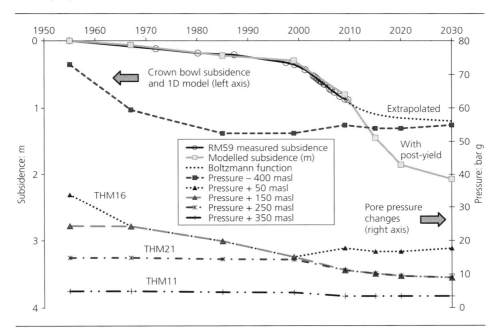

Advanced one-dimensional subsidence modelling of the anomalies was undertaken by Ajitha Wanninayake (pers. comm., R. Davidson, 2010) using FLAC, a non-linear code for soil and soft rock deformation. This model automatically accounts for the yielding, softening and hardening behaviour of the deforming strata as a function of pressure change and vertical stress state. For the compressible formations, a soft soil model was chosen. Selected K_0 triaxial test data were replicated using computer simulations to establish an appropriate range of non-linear soil properties for use in the one-dimensional subsidence simulations. Historic pressure trends with depth and time were input into the model using the same interpolated values referred to above. The soil parameter values were fine-tuned until a good match to the observed ground subsidence was achieved. Future deformation to 2030 was then calculated by running the model using expected future pressure trends. The projections are consistent with the projections in Figures 6 and 9, and indicate additional subsidence of about 15% in the Wairakei bowl, up to 8% at Spa and Rakaunui bowls, and ~100% at Crown bowl. The latter model result is uncertain, because it relies on a small amount of shallow pressure decline occurring in the

Figure 7 Inferred history of pressure changes against depth for WKM15 (Wairakei bowl). Pressures interpolated from measured data and then predicted to 2030 using reservoir model for Tauhara expansion. Interpolations use hydrostatic or two-phase pressure gradients to link aquifers of known pressure history

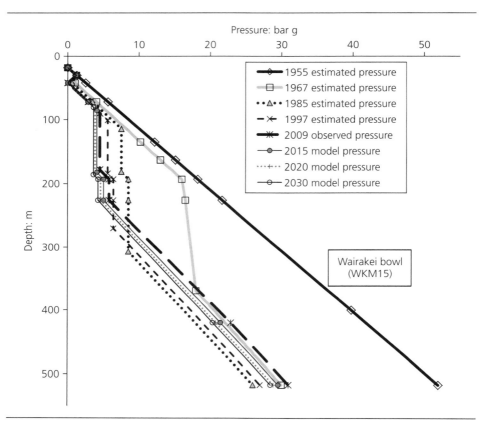

future (assuming expanded production), and triggering transition of a thick sequence of clay alteration to post-yield behaviour. In reality, the properties of the layers have been averaged, and significant variability in these parameters with depth will probably smooth out any response over time to a small pressure decline.

Three-dimensional subsidence modelling of each bowl has also been undertaken by coupling the TOUGH2 finite-difference geothermal reservoir fluid-flow model to a conventional, geotechnical, finite-element, stress–strain model (ABAQUS, finite-element computer code calculating ground deformation) (O'Sullivan et al., 2010). Compressibility moduli and other material properties were adjusted within a range of values, as assessed from core tests, to improve the match. Given the heterogeneity in properties, the uncertainties in timing of yielding behaviour, and difficulties with simulating local two-phase conditions inside the subsidence anomalies, using relatively coarse layer thicknesses (50 m), this modelling was not expected to provide more than an approximate history match, and indicative future trends.

Figure 8 Layered model compressibility parameters (pre- and post-yield constrained moduli (*M*) and yield stress values), and geology, against depth to 500 m (−135 masl), interpolated from core data from WKM15 (Wairakei bowl)

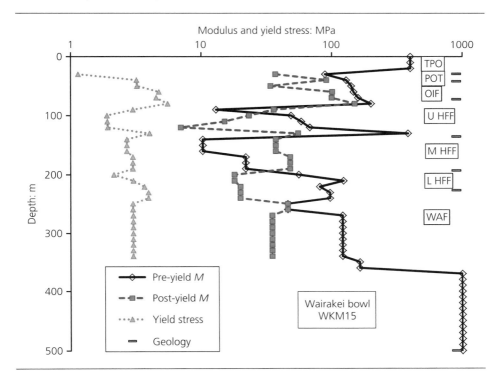

However, by adjusting effective compressibility of model blocks, the three-dimensional models have produced reasonable fits to observed subsidence anomaly amplitudes, shapes and locations. The detailed history matching of local rate changes with time is not yet optimised, and this aspect is expected to improve, along with the reliability of future projections, as the pressure trends in the shallow parts of the reservoir model are improved. This will be achieved through progressively more detailed reservoir simulations, and improved simulations of yielding behaviour.

4. Subsidence mitigation

Significant adverse effects from subsidence are not expected to occur in the future at Wairakei or Tauhara. However, ongoing monitoring will continue, and if subsidence rates do increase significantly with time, at critical locations, mitigation will be put in place. The best method for achieving this is targeted shallow injection. Optimum injection depths are bowl dependent, and range from 50 m to 400 m in the intermediate and shallow aquifers. The intention would be to sustain or raise pressures in the shallow zones adjacent to the most compressible material. Targeted injection is more direct, and should be more effective, than pressure management in the deeper liquid zone, which is the traditional location for injecting separated geothermal water or steam condensate. However, some indirect beneficial effect can still be anticipated where

Figure 9 Wairakei bowl (WKM15) one-dimensional model results (calculated and observed levels) with pore fluid pressure changes at representative levels (see Figure 7). Compressibility parameters shown in Figure 8. Post-yield behaviour applies after 1967 at 200–340 m depth, and 20% reload compressibility for rising pressures after 1985. This simulation requires 2·6 m additional compression of a steam-heated peat-layer between 1967 and 1999. Projections assume Tauhara expansion

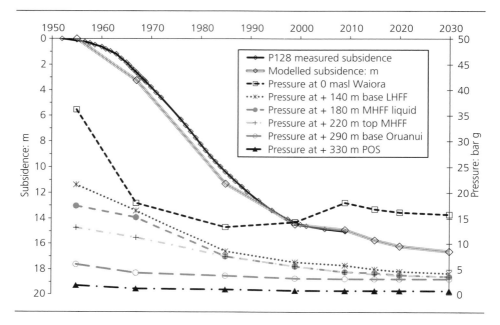

deep, in-field, liquid injection causes local pressure increases, and pushes liquid upflows into overlying steam-dominated aquifers. An adaptive management scheme provides for uncertainty in future projections, by ensuring continued monitoring of subsidence rates, sampling, analysis and predictive modelling where necessary, and mitigation where appropriate.

5. Conclusion and discussion

In summary, as a result of this investigation programme, there has been a substantial increase in subsurface data and knowledge about the causes and mechanisms of subsidence at Wairakei and Tauhara. The programme recovered and tested about 4 km of core material from within and outside the recognised subsidence anomalies ('bowls'). Within each of the bowls, material from different depth intervals, ranging up to 400 m, was found to be anomalously compressible, and of sufficient thickness to explain the bowl amplitudes. The bowl locations result from anomalous geological structures and properties related to intense hydrothermal clay alteration of porous and permeable formations. Typically, this is associated with boiling in subsurface feeder channels that are located adjacent to historic surface hot spring discharges.

At present, subsidence rates are generally declining. Some rate variations can be expected with future development expansion, if there is an effect on shallow pressures, but an overall reduction in rates is the long-term prognosis.

The principal outcome of the programme is that a subsidence mechanism has been identified using the new rock property data and borehole information. Anomalous subsidence is linked to the occurrence of anomalously compressible material, and to reducing pressures at relatively shallow depths. The timing of accelerated subsidence depends on the vertical effective stress exceeding the yield stress, and this varies with depth and location. Models are available that permit a reasonable assessment of the effect on subsidence of future pressure changes. The most important uncertainties arise from the interpolation of pressures, the variability in rock properties, the determination of yield stresses, and the timing of in situ yielding. Accordingly, subsidence projections, although significantly improved, are still subject to these uncertainties. Mitigation is an option by applying targeted fluid injection, and adaptive management provides the best way of addressing uncertainty by responding effectively and rapidly to new information collected during ongoing monitoring.

Acknowledgements

Contact Energy management is thanked for permission to publish the results of this work. Preparation of the chapter was made possible through research funding provided by the New Zealand Government (geothermal programme). Dr Rick Allis is thanked for his numerous contributions to the subsidence programme, and for his overall review role with respect to the interpretation of results. The integrated nature of the investigation would not have been possible without the collaborative efforts of the co-authors, together with project coordinators (Noel Kortright, Graham Ramsay), drilling advisor (Melvyn Griffiths), drilling engineer (Ralph Winmill), rig operators (Boat Longyear staff), geotechnical specialists (Richard Davidson, John Garvey), geologists (Bridget Lynne, Ernesto Ramirez, Geoff Kilgour), subsidence modellers (Emily Clearwater, Angus Yeh, Ajitha Wanninayake) and reservoir engineers (Warren Mannington, Sang-Goo Lee). These people are all gratefully acknowledged for their valuable contributions.

REFERENCES

Allis RG (2000) Review of subsidence at Wairakei field, New Zealand. *Geothermics* **29(4–5)**: 455–478.

Allis RG and Bromley CJ (2009) Unravelling the subsidence at Wairakei, New Zealand. *Transactions/Geothermal Resources Council* **33**: 299–306.

Allis RG and Zhan X (2000) Predicting subsidence at Wairakei and Ohaaki geothermal fields, New Zealand. *Geothermics* **29(4–5)**: 479–497.

Allis RG, Bromley CJ and Currie S (2009) Update on subsidence at the Wairakei–Tauhara Geothermal System. *Geothermics* **38(1)**: 169–180.

Brockbank K, Bromley CJ and Glynn-Morris T (2011) Overview of the Wairakei–Tauhara subsidence investigation program. *Proceedings of the 36th Workshop on Geothermal Reservoir Engineering, Stanford University, Stanford, CA, USA*, SGP-TR-191.

Bromley CJ (2006) Predicting subsidence in New Zealand geothermal fields: a novel approach. *Transactions/Geothermal Resources Council* **30**: 611–616.

Bromley CJ (2009) Groundwater changes in the Wairakei–Tauhara geothermal system. *Geothermics* **38(1)**: 134–144.

Bromley CJ and Beerepoot M (2011) Global geothermal deployment: the IEA Roadmap for the future. *Proceedings of the Australian Geothermal Energy Conference, Melbourne, Australia*, Geoscience Australia Record 2011/43 Geocat#73026.

Bromley CJ, Currie S, Manville V and Rosenberg MD (2009) Recent ground subsidence at Crown Road, Tauhara and its probable causes. *Geothermics* **38(1)**: 181–191.

Bromley CJ, Currie S, Ramsay G, Rosenberg M, Pender M, O'Sullivan M *et al.* (2010) *Tauhara Stage II Geothermal Project: Subsidence Report, GNS Science Consultancy Report 2010/151 (and appendices 1–10)*. Contact Energy Ltd, Wellington, New Zealand.

Geertsma J (1973) Land subsidence above compacting oil and gas reservoirs. *Journal of Petroleum Technology* **25(6)**: 734–744.

Goldstein B, Hiriart G, Bertani R and Bromley CJ (2011) Geothermal energy. In *IPCC Special Report on Renewable Energy Sources and Climate Change Mitigation* (Edenhofer O *et al.* (eds)). Cambridge University Press, New York, NY, USA, pp. 401–436.

Hole JK, Bromley CJ, Stevens NF and Wadge G (2007) Subsidence in the geothermal fields of the Taupo Volcanic Zone, New Zealand from 1996 to 2005 measured by InSAR. *Journal of Volcanology and Geothermal Research* **166(3/4)**: 125–146.

Lawless J, Okada W, Terzaghi S, Ussher G, White P and Gilbert C (2003) Two dimensional subsidence modelling at Wairakei–Tauhara, New Zealand. *Transactions/Geothermal Resources Council* **27**: 761–764.

Lynne B, Pender MJ and Glynn-Morris T (2011) Scanning electron microscopy and compressibility measurements: a dual approach providing insights into hydrothermal alteration and rock strength at Tauhara geothermal field. New Zealand. *Proceedings of the 33rd New Zealand Geothermal Workshop, University of Auckland,* paper No. 17.

O'Sullivan MJ, Yeh A and Mannington WI (2009) A history of numerical modelling of the Wairakei geothermal field. *Geothermics* **38(1)**: 155–168.

O'Sullivan M, Yeh A and Clearwater E (2010) *Three-dimensional model of subsidence at Wairakei–Tauhara*. Uniservices and Department of Engineering Science, University of Auckland, February 2010 report for Contact Energy Ltd (Appendix 10 of Bromley *et al.*, 2010). See http://www.contactenergy.co.nz/web/ourprojects/tauhara_phase_two_appendix10 (accessed 30/10/2012).

Pender M, Ramsay G, Glynn-Morris T, Lynne B and Bromley C (2013) Rock compressibility at the Wairakei–Tauhara geothermal field, New Zealand. *Proceedings of the Institution of Civil Engineers – Geotechnical Engineering* **166(2)**: 224–234.

Ramsay G, Glynn-Morris T, Pender MJ and Griffiths M (2011) Geotechnical investigations of subsidence in the Wairakei–Tauhara Geothermal Field. *Proceedings of the 33rd New Zealand Geothermal Workshop, University of Auckland, Auckland, New Zealand*, paper No. 49.

Read SAL, Barker PR and Reyes AG (2001) Consolidation properties of Huka Falls Formation – linkages to subsidence at Ohaaki and Wairakei. *Proceedings of the 23rd New Zealand Geothermal Workshop, University of Auckland, Auckland, New Zealand*, pp. 57–62.

Rosenberg MD, Bignall G and Rae AJ (2009) The geological framework of the Wairakei–Tauhara geothermal system, New Zealand. *Geothermics* **38(1)**: 72–84.

Samsonov S, Tiampo K, Beavan J, Bromley CJ, Scott B and Jolly G (2009) ALOS PALSAR interferometry of Taupo Volcanic Zone, New Zealand. *Proceedings of the 30th Canadian Symposium on Remote Sensing, Lethbridge, Alberta, Canada*, paper No. 478.

White PJ, Lawless JV, Terzaghi S and Okada W (2005) Advances in subsidence modelling of exploited geothermal fields. *Proceedings of the World Geothermal Congress 2005, Antalya, Turkey*, paper No. 0222.

Section 2
Heat exchange

ICE Themes Geothermal Energy, Heat Exchange Systems and Energy Piles

Craig and Gavin
ISBN 978-0-7277-6398-3
https://doi.org/10.1680/gehesep.63983.061
ICE Publishing: All rights reserved

Chapter 4
Energy harvesting on road pavements: state of the art

Francisco Duarte MSc
PhD Student, Pavement Mechanics Laboratory, CITTA, Department of Civil Engineering, University of Coimbra, Coimbra, Portugal

Adelino Ferreira PhD
Director, Pavement Mechanics Laboratory, CITTA, Department of Civil Engineering, University of Coimbra, Coimbra, Portugal

With the growing need for alternative energy sources, research into energy harvesting technologies has increased considerably in recent years. The particular case of energy harvesting on road pavements is a very recent area of research. This chapter deals with the development of energy harvesting technologies for road pavements, identifies the technologies that are being studied and developed, examines how such technologies can be divided into different classes and gives a technical analysis and comparison of those technologies, using the results achieved with prototypes.

Notation
ΔT thermal gradient of the road pavement
η energy conversion efficiency

1. Introduction
The increasing movement of human beings from rural areas to the city has led to an exponential increase in the consumption of the planet's resources in recent years. Energy, and in particular electrical energy, is one of the resources with greatly increased demand. Currently, and for the first time in history, more than 50% of the world's population lives in cities (Buhaug and Urdal, 2013) and, by 2050, this number will increase by more than 3 billion people, leading to global urbanisation, which will lead to further increases in energy consumption, especially in cities.

With the present energetic paradigm, most electrical energy production uses fossil fuel combustion, which makes economies dependent on fuel costs. This is also leading to irreversible environmental damage. According to the International Energy Agency (IEA, 2013), in 2011, globally, more than 80% of energy production came from fossil fuels. Urgent action is required to change the paradigm of electrical energy generation as, presently, energy is mostly produced outside cities, consuming non-renewable resources and inducing energy losses between the point of production and the point of consumption. Energy production must be based on renewable resources, decentralised, happen near the point of consumption and, preferably, when it is needed.

In the area of renewable energies, besides the major energy sources (hydro, solar, wind, waves), energy harvesting has recently been adopted on a micro-scale, where it is possible to generate electrical energy from small energy variations, such as thermal gradients, pressure, vibrations, radiofrequency or electromagnetic radiation, among others (Khaligh and Onar, 2010). Road surfaces are continuously exposed to two phenomena: solar radiation and vehicle loads. From both of these it is possible to extract energy, which, using specific technologies, can be transformed into electrical energy (Andriopoulou, 2012). Within cities, there are roads that carry vehicles, the main option for mobility. Vehicles consume energy to work their engines and release energy in different ways, by way of different components. Part of the energy released by vehicles goes into the road pavement. Around 15–21% of the energy is transferred to the vehicle's wheels (Hendrowati *et al.*, 2012; IEA, 2012). As vehicles abound in all cities in developed countries, this means that a considerable amount of energy is transferred to road pavements without being used. Roads are also exposed to solar radiation, which induces thermal gradients between their layers. This solar radiation and the resulting thermal gradients can also be transformed into useful energy. So road pavements represent a considerable source of energy ready to be harvested and converted into useful forms of energy, such as electrical energy, at the same time reducing the need to 'import' energy from distant places.

This chapter aims to review the energy harvesting technologies with possible implementation on road pavements, using both solar energy and vehicle-released energy as an energy source.

2. Road pavement energy harvesting technologies

2.1 Introduction

Energy harvesting is described as a concept by which energy is captured, converted, stored and utilised using various sources, by employing interfaces, storage devices and other units (Khaligh and Onar, 2010; Priya and Inman, 2009). Simplified, energy harvesting is the conversion of ambient energy present in the environment into other useful forms of energy – for example, electrical energy (Kazmierski and Beeby, 2009).

Energy harvesting is divided into two main groups: macro-energy harvesting sources, associated with solar, wind, hydro and ocean energy; and micro-energy harvesting, associated with electromagnetic, electrostatic, heat, thermal variations, mechanical vibrations, acoustic and human body motion as energy sources (Harb, 2010; Khaligh and Onar, 2010; Yildiz, 2009). Macro-energy harvesting is related to large-scale energy harvesting, usually in the order of kilojoules or more. Micro-energy harvesting is related to small-scale energy harvesting, usually in the order of a joule or less.

From the energy harvesting technologies identified by Harb (2010), two groups of technologies have a great potential for implementation on pavements: one uses solar radiation as an energy source and the other uses the mechanical energy from vehicle loads. Considering these energy sources, different technologies and systems have been developed and tested in recent

Figure 1 Road pavement energy harvesting technologies

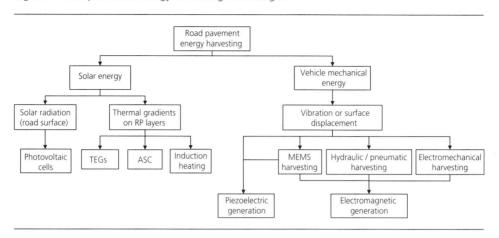

years. The main energy harvesting technologies applicable on road pavements can be divided into two main groups, as presented in Figure 1. The first group is related to technologies that make use of the solar exposure on the road pavement. Solar radiation can be directly harvested by photovoltaic cells and transformed into electrical energy; it can induce thermal gradients between the road pavement layers, which can be used to power thermoelectric generators (TEGs), which produce electrical energy, or be harvested by asphalt solar collectors (ASC), which extract the temperature accumulated on the road pavement. Induction heating is a concept in which introducing conductive particles in the asphalt mixture provides self-healing capacities autonomously at high temperatures by harvesting solar radiation. The second group is related to technologies that make use of the mechanical energy transferred from vehicles to the road surface. This can be harvested directly by piezoelectric harvesters, which generate electrical energy; or it can be harvested by hydraulic, pneumatic, electromechanical or micro-electromechanical systems (MEMS) that transfer the harvested energy to electromagnetic generators, which produce electrical energy. In the case of MEMS, they can also transfer the harvested energy to piezoelectric generators.

2.2 Solar energy harvesting on road pavements

2.2.1 Photovoltaic technology

Researchers from the Korea Institute (Kang-Won and Correia, 2010) have investigated the possibility of harvesting solar energy from road pavements, using solar cells embedded into the pavement infrastructure. They have concluded that the current thin-film solar cells are difficult to use on surfaces that receive mechanical loads and environmental conditions can cause premature corrosion and wear. For these reasons, the researchers are developing new thin-film solar cells that meet the requirements for use on road surfaces.

Julie and Scott Brusaw proposed a solar collector system to replace the upper layer of the road pavement, called Solar Roadway (SR, 2015). The Solar Roadway is a series of structurally engineered solar panels that are able to support traffic loads and are applied on the road surface transforming solar radiation into electrical energy. Each panel has an installed power of 36 W and measures $0·37$ m^2. They have patented the upper layer of the product with a design patent (Brusaw and Brusaw, 2014). The Brusaws indicate a conversion efficiency of the system of $11·2\%$ (SR, 2015), a value that cannot be confirmed in any scientific publication or product certification. One of the major challenges of this project is to offer safety and the appropriate conditions for the mobility of the vehicles passing over the panels. At the same time, the upper layer needs to guarantee the transmission of the solar radiation to the photovoltaic cells beneath it in order to present good conversion efficiency.

In the Netherlands, Toegepast Natuurwetenschappelijk Onderzoek (TNO), along with other technological partners, have developed a pilot project consisting of a modular cycle path system, named SolaRoad (SolaRoad, 2014; TNO, 2014). The cycle path is constructed of concrete elements that are covered by a $1·0$ cm-thick glass top layer. Underneath this glass layer, crystal silicon solar cells are laid, although the values of the installed power and energy conversion efficiency are not presented. The modules are embedded in concrete slabs and applied on the base layer of the pavement. The authors of the project indicate that the next step will be to adapt the system for application on roads (SolaRoad, 2014).

2.2.2 Thermoelectric technology

TEGs produce electrical energy due to the Seebeck effect, described by Schreier *et al.* (2013) as a voltage difference on a material, resulting from a temperature gradient imposed between its surfaces. The greater the temperature gradient, the more the energy generated.

Hasebe *et al.* (2006) developed a pavement-cooling system using a TEG incorporated into the pavement. Solar heat is collected by a water piping system embedded in the road pavement, and this water is cooled by river water. The water first passes on the hot side of the TEG, and then on the cold side of the TEG, installed under the road. The maximum power output of the prototype tested was 5 W, using 19 Bi-Te cells ($1·23$ cm^3 each), for a ΔT of $40·5°C$. For a ΔT of $25·9°C$, the maximum output power was $2·9$ W, and for a ΔT of $11·5°C$, the maximum output power was $0·9$ W. The conversion efficiency of the system was not presented.

Wu and Yu (2012) studied the implementation of TEGs on the surface of pavements. They propose the connection of the lower part of the module with the subgrade soil by way of highly thermal-conductive materials in order to facilitate heat conduction and thus increase electrical energy production. They have concluded that its maximum efficiency reached 4% using Quantum Well structured (QW) materials, instead of Bi-Te cells, but the problem was the storage of that energy. Tests were performed with a pavement ΔT of $50°C$, which generated 300 mV with one TEG cell. They have concluded that the efficiency of the electronic system was $41·3\%$, which results in a global efficiency of the TEG system of $1·6\%$. The prototype, which used a $7·7$ cm^3 TEG unit, was able to produce a maximum output power of $0·02$ W, for a thermal gradient of $6·44$ K on the TEG (Wu and Yu, 2013).

Due to their low conversion efficiency, none of these studies have led to a product available on the market. Kelk (2015), a Japanese company, has different products based on TEGs, and sells the TEG cells individually, but none of their available products have been developed for implementation on road pavements so far.

2.2.3 ASC technology

An alternative way of using the thermal gradients between the road pavement layers is to transform the pavement into a solar collector, using pipes and pumps, arranged specifically in order to capture solar energy and convert it into thermal energy. This is a method that has been studied in recent years by different authors, called ASC (Bobes-Jesus *et al.*, 2013).

Sullivan *et al.* (2007) proposed an ASC system for heating and cooling road pavements and other infrastructures, using the heat-absorbing property of asphalt concrete applied on the construction of flexible pavements. The ASC system consists of an asphalt pavement layer with water pipes in it. The ASC system is linked to two underground water storage reservoirs, one for cold water and another for warm water (RES, 2007), involving seasonal storage. Solar radiation induces an increase in the temperature of the road surface, which is absorbed by the road pavement materials and transferred to a water piping system applied under the road surface and is then stored in the soil or other storage tanks (Sullivan *et al.*, 2007).

The real-world performance of ASCs has been demonstrated by systems installed in different places around the world, working in different climatic conditions. In Switzerland, the Solar Energy Recuperation from the Road Pavement (SERSO) system, presented by Lund (2000), was successfully installed with the main purpose of melting ice on roads. In the Netherlands, Road Energy Systems, developed by Ooms (Sullivan *et al.*, 2007) and TNO (Loomans *et al.*, 2003), has been commercialised in recent years, with the main advantages presented by the company being a focus on increased road safety (RES, 2007).

In the UK, ICAX (ICAX, 2014) develops and implements ASC solutions. In Japan, the Gaia system has been developed during the last decade, with different studies being performed focused more on the snow-melting heat-storage capacity (Gao *et al.*, 2010; Morita and Tago, 2000). The authors say that the heat collecting capacity is between 150 and 250 W/m^2 under normal summer weather conditions (Gao *et al.*, 2010). Other applications have been developed in China (Tu *et al.*, 2010) and the USA (Mallick *et al.*, 2009).

2.3 Vehicle mechanical energy harvesting on road pavements

The mechanical energy transmitted by vehicle wheels to the road surface can induce two types of action on the road pavement: vibrations or surface displacement. This energy can be harvested using different methods and different technologies.

2.3.1 Piezoelectric technology

Piezoelectric materials fall within a class of multiple solid-state materials that can generate electrical energy with the application of pressure or vibrations (Beeby *et al.*, 2006). Both vehicle pressure and vibrations induced on the road pavement can be used to actuate piezoelectric transducers, in order to convert mechanical energy into electrical energy (Xiang *et al.*, 2013).

Zhao *et al.* (2010) proposed and studied the application of cymbal piezoelectric transducers on road pavements. The amount of generated energy was of 1·2 mW at 20 Hz for one vehicle passage. This was equivalent to 0·06 J per vehicle passage and a conversion efficiency of lower than 15%.

Zhao et al. (2013) studied different piezoelectric materials in order to determine how to adapt them for use on road pavements. They have identified that none of the typical piezoelectric transducers made of lead zirconate titanate were suitable for the asphalt pavement environment. They suggest a specific design and optimisation process to adapt the piezoelectric materials for road pavement application.

Xiong et al. (2012) have defined two coupling modes of piezoelectric components: 31-mode and 33-mode. With 31-mode, the piezoelectric material generates electrical energy from transverse displacement. On the 33-mode, the power output of the system increases linearly with the deflection of the pavement or the stress along the poling direction of the material. They have stated that the usual power generation capacity of piezoelectric transducers is about 300 $\mu W/cm^3$.

Wischke et al. (2011) studied the application of piezoelectric modules in road pavements in tunnels. They concluded that the vibrations caused by vehicles were small due to vehicle suspension.

The patent applications US20050127677 (Luttrull, 2005) and US20100045111 A1 (Abramovich et al., 2010a) relate to systems that make use of piezoelectric transducers on road pavements that produce electrical energy while being deformed due to the passage of vehicles. The system presented by Abramovich et al. (2010a) was tested in a real environment by Innowattech on a product called Innowattech Piezo Electric Generator (IPEG) (Innowattech, 2014). There are no published results for the generated energy, or for the conversion efficiency.

Abramovich et al. (2010b, 2012) have developed a system with new methodologies to increase energy generation and simplified the installation process, as well as using a new methodology to multiply the forces of the vehicles delivered to the piezoelectric transducers (Klein et al., 2012). Nevertheless, there were no scientific results presented for any of these systems.

Bowen and Near (2000) have patented a piezoelectric actuator for road pavements, which was developed recently (Near, 2013), presenting an energy harvesting product based on piezoelectric components for use on road pavements. With this product, Near (2013) created the Genziko company in the USA (Genziko, 2014).

Hill et al. (2014) have compared the products developed by Innowattech and Genziko using data provided by both companies. From Innowattech, they present an energy generation per module, per vehicle, of 5·76 J, while Genziko have presented an energy generation per module, per vehicle, of 40 J, almost seven times more than that of Innowattech. However, the authors of the study have concluded that none of these companies have enough real-environment validations to support the presented energy generation values.

2.3.2 Electromagnetic technology

Electromagnetic generators operate based on electromagnetic induction, known as Faraday's law, where, if an electric conductor is moved in relation to a magnetic field, electric current will be induced in the conductor (Beeby et al., 2006). These generators are mostly used in big power plants, based on both non-renewable and renewable sources. In energy harvesting, smaller electromagnetic generators have been developed over the last decade in order to convert environmental energy sources (mechanical vibrations, mostly) into electrical energy (Arroyo and Badel, 2011; Beeby *et al.*, 2007; Elliott and Zilletti, 2014; Munaz *et al.*, 2013; Peralta *et al.*, 2014; Saha, 2011).

Electromagnetic generators are different from piezoelectric systems in that they are not actuated directly by the mechanical energy of vehicles. Interfaces are applied where the harvester units are based on hydraulic or pneumatic systems, electromechanical systems or MEMS, which will be presented in the following sections.

2.3.2.1 Hydraulic and pneumatic harvesting systems. A hydraulic system consists of a drive or transmission system that uses a pressurised hydraulic fluid to transmit forces and actuate mechanical components, which are usually actuated by electric machines. In the case of pneumatic systems, the difference is in the working fluid; air is used instead of liquid (Parr, 2011). In road pavements, hydraulic systems can be used in the opposite way, transmitting the mechanical energy of the vehicles to actuate electric machines.

Some companies and individual inventors have registered patents where they use hydraulic or pneumatic mechanisms to harvest energy released from vehicles and convert it into electrical energy. These systems are designed to be implemented on roads, as are related patent applications US5634774 (Angel and Gomez, 1997), GB2290115A (Nakatsu, 1994), US6376925B1 (Galich, 2002), US20040130158A1 (Kenney, 2004), US20070246940A1 (Valon, 2007), WO2007045087 (Horianopoulos and Horianopoulos, 2007), US20100192561A1 (Hendrickson, 2010a), WO2010081113A1 (Hendrickson, 2010b), US20110215593A1 (Chang and Lee, 2011) and GB2476826A (Houghton, 2011) for hydraulic systems, and US4173431 (Smith, 1979) and GB2408074A (Morley *et al.*, 2005) for pneumatic systems.

Horianopoulos and Horianopoulos (2007) developed a hydraulic device that harvests energy on road pavements, claiming an energy generation capacity of 51 kWh with 10 000 vehicle passages along 50 m (Kinergy Power, 2014). This is proportional to 367 (J/m)/vehicle, an average of 91·8 J/wheel, which is a very high value. However, this value is not supported by any scientific evidence, and the average vehicle weight used in the study is not presented. In their patent (Horianopoulos and Horianopoulos, 2007), the working principles of the technology are described without reference to the conversion efficiency. On the basis of this system, they have created the product KinerBump, and the company KinergyPower International Corporation (Kinergy Power, 2014) in USA.

Moreover, Hendrickson (2010a, 2010b) has created the company New Energy Technologies with a product line called MotionPower (NewEnergyTechnologies, 2015), based on his patents. Real-environment tests were performed with this technology but the company has not published the results.

2.3.2.2 Electromechanical harvesting systems. In electromechanical systems, electrical devices are operated by mechanical components or vice versa. In the case of electromechanical energy harvesting systems, mechanical energy is used to actuate an electrical machine, which produces electrical energy.

In the case of road pavement energy harvesting, the electromechanical systems can be divided into four classes.

- Conversion of rotational motion of a surface into a rotational motion of an electric generator (Rot–Rot).
- Conversion of linear motion of a surface into a rotational motion of an electric generator (Lin–Rot).
- Conversion of linear motion of a surface into a linear motion of an electric generator (Lin–Lin).
- Conversion of rotational motion of a surface into a linear motion of an electric generator (Rot–Lin).

Table 1 summarises the electromechanical systems that are being developed according to the motion conversion principle. From this table, it can be seen that most of the systems have a rotational output motion.

On the basis of the systems presented in the patents WO2004067850A1 (Hughes *et al.*, 2004) and WO2009101448A1 (Hughes *et al.*, 2009), the company Highway Energy Services was created (HES, 2015) in the UK. Besides the information presented in their patents, no technical data are available on these systems.

Pirisi (2012), following his patent, has developed a prototype of the technology which, together with researchers from Politecnico di Milano, he has tested and presented the results (Pirisi *et al.*, 2012a, 2012b, 2013). The generator is described as a tubular permanent magnet linear generator, an electromechanical device able to convert linear motion into electrical energy; using a 1:10 scale prototype in the laboratory, they claim a conversion efficiency of 85% (Pirisi *et al.*, 2013) between the mechanical energy applied to the slider of the generator

Table 1 Patents related to electromechanical systems

		Input motion	
		Rotational	Linear
Output motion			
	Rotational	US4238687 (Martinez, 1980), US6767161B1 (Calvo and Calvo, 2004), US7102244B2 (Hunter, 2006), US20090315334A1 (Chen, 2009), US7714456B1 (Daya, 2010), US20110187125A1 (Jang, 2011) and WO2012099706A2 (Mansfield et al., 2012)	US4434374 (Lundgren, 1984), US20070181372A1 (Davis, 2007), WO2009101448A1 (Hughes et al., 2009), WO2011145057A2 (Duarte and Casimiro, 2011) and WO2013114253A1 (Duarte and Casimiro, 2013)
	Linear	US20120248788A1 (Pirisi, 2012)	WO2004067850A1 (Hughes et al., 2004)

and the electrical output efficiency. This value does not include power electronics, but presents only the conversion efficiency of the generator. In laboratory tests, the system was applied on the top of the road surface. The authors of this system have created the company Underground Power, which is developing a product called 'LYBRA' (Underground Power, 2014).

In the USA, the authors of the patent WO2012099706A2 (Mansfield et al., 2012), a class-1 system, have created the company Energy Intelligence. So far, there have been no scientific results presented for the project. The authors maintain that the system will be embedded in the road surface, replacing the upper layers of the road pavement (EnergyIntelligence, 2014).

Following the development of a suitable system to convert people-released energy into electrical energy (Duarte and Casimiro, 2011), with a 60% conversion efficiency (Duarte et al., 2013a, 2013b), Duarte and co-authors have developed the system presented in the patent WO2013114253A1 (Duarte and Casimiro, 2013), with a class-2 system, suited for application on roads. The authors, together with the company Waydip, have tested a real-scale prototype, naming the project Waynergy Vehicles (Waydip, 2015). Laboratory tests with the prototype obtained experimental energy generation data, achieving a conversion efficiency of about 50% for the mechanical energy delivered to the system and the electrical energy output delivered to an electric load (Duarte et al., 2014, 2016). The system was applied on the upper layer of the road pavement.

2.3.2.3 MEMS harvesting systems.
MEMS is a technology that is usually defined as miniaturised mechanical and electromechanical elements, made using the techniques of microfabrication, which can vary from relatively simple structures having no moving parts to extremely complex electromechanical systems with multiple moving parts, usually under the control of integrated microelectronics. There has been much recent interest in using MEMS to harvest energy from ambient vibration and transform it into electrical energy (Stephen, 2006).

To find optimal architectures for maximal power generation under the different operating constraints, analysis and verification by simulation of three classes of MEMS-based vibration-driven microgenerator architectures were presented by Mitcheson et al. (2004).

Harb (2010) studied and tested different MEMS systems in laboratory, actuating electromagnetic microgenerators. These generators presented a maximum energy conversion efficiency of 18%, with ten cells and a buck converter.

Zorlu and Külah (2013) developed a MEMS-based energy harvesting device to generate electrical energy from vibrations, with implementation on road pavements being one of the possible applications. In laboratory tests with a prototype, a maximum output of $3·2$ mW/cm^3 was achieved, which is a high power density for this type of application. However, when a prototype was developed to be tested in a real environment, the energy generation was $6·0$ μW/cm^3, 500 times lower than in laboratory tests. So, the technology presents some potential but, when applied in a real environment, its efficiency decreased considerably.

The patent US20130193930A1 (Baugher, 2013) presents a system consisting of a microstructure for implementation under the road surface, which uses vibrations to actuate piezoelectric materials to generate electrical energy. No commercial application of this system has been developed so far, nor have technical results of experiments been published, as the system is under development.

3. Technical analysis

3.1 Introduction

To perform a technical analysis and evaluate an energy generation technology, the most commonly used parameters are the installed power (per area or volume), conversion efficiency, power density and the energy generation of the technology in normal operating conditions (Table 2). In the specific case of road pavement energy harvesting, it is also important to classify the technologies according to the installation method (IM), as this is an important issue regarding the final cost of the solution, the driving and safety conditions and the maintenance operations of the equipment. Finally, as these are mostly new technologies, it is important to classify them regarding their development status – in this case, using technology readiness levels (TRLs).

Table 2 Parameters for performing a technical analysis

Parameter	Description
Installed power	The installed power of an electrical energy generation device is its energy generation capacity in nominal conditions – that is, the maximum theoretical power it can generate. It is related to the output power and is expressed in watts (W). In many cases, it is expressed by comparing the installed power with the occupied area of the device (W/m^2), or with the occupied volume of the device (W/m^3). In micro-energy harvesting devices, the analysis is usually done in regard to volume
Conversion efficiency	Energy conversion efficiency (η) is the ratio between the useful output of an energy conversion device and the energy input. In the case of electrical machines, the output is electrical energy measured in joules (J), or electrical power measured in watts (W). The energy conversion efficiency is a dimensionless parameter, usually expressed as a percentage
Energy generation	Energy generation is used to quantify the amount of electrical energy generated under the operating conditions. It gives the energy input of the system, its efficiency and the installed power. Usually, it is expressed in joules, but in some micro-energy harvesting devices it can also be related to the volume (J/m^3). In the analysis of energy harvesting devices, sometimes power generation is also presented, related to the volume of the device (W/m^3)
IM	The different energy harvesting devices can be installed in the road pavement using different techniques, and in different layers of the road pavement. Four main IMs were identified
TRL	TRLs are measures used to evaluate the maturity of a technology during its developmental stages. These levels were initially defined by NASA (Mankins, 1995), but are now commonly used in project evaluations

3.2 Comparison of technologies

Following the analysis of the different technologies presented in this study, the main characteristics of each one are presented in Table 3. For this analysis, the technologies that convert both solar and vehicle mechanical energy into useful electrical energy and which have been tested on road pavements were considered. So, ASC, induction healing and MEMS technologies were excluded from the analysis.

Table 3 Technical analysis of different road pavement energy harvesting technologies

Technology	Company/R&D institute	Installed power: W/m^2	Conversion efficiency: %	Energy generation	IM[a]	TRL[b]
Photovoltaic	Solar Roadways	97·3	11·2	NA	1	4
	TNO	NA	NA	NA	2	7[c]
TEG	Hasabe et al.	NA	NA	38·0 mW/cm^3	3	3
	Wu and Yu	NA	1·6	2·6 mW/cm^3	3	3
Piezoelectric	Innowattech	NA	NA	5·8 J/veh m	3	4
	Genziko	1942·0	NA	40·0 J/veh m	3	4
Hydraulic	Kinergy	NA	NA	188·0 J/veh m	1/2	4
	New Energy Technologies	NA	NA	NA	1/2	4
Electromechanical	Waydip	833·0	50·0	680·0 µW/cm^3 180·0 J/veh m	2	4
	Underground Power	NA	85·0[d]	NA	1/2	4
	HES	NA	NA	NA	2	3
	Energy Intelligence	NA	NA	NA	2	3

[a] IM 1 – on the road pavement surface, fixed to the upper layer (the device surface becomes the new road surface); IM 2 – embedded in the road pavement, upper layer, surface exposed (the device surface becomes the new road surface); IM 3 – embedded in the road pavement, upper layer, surface covered by road pavement material; IM 4 – embedded in the road pavement, lower layer, surface covered by road pavement material
[b] TRL 1 – basic principles observed and reported; TRL 2 – technology concept and/or application formulated; TRL 3 – analytical and experimental critical function and/or characteristic proof of concept; TRL 4 – component validation in laboratory environment; TRL 5 – component validation in relevant environment; TRL 6 – system/subsystem model or prototype demonstration in a relevant environment; TRL 7 – system prototype demonstration in an operational environment; TRL 8 – actual system completed and qualified through tests and demonstration; TRL 9 – actual system proven in operational environment
[c] For cycle lanes. For road pavements, it has only been conceptualised, not prototyped (TRL 1/2)
[d] Efficiency on a 1:10 scale, and not considering the losses of control, storage and deliver energy to an electrical load

From Table 3, it may be seen that most of the studies do not meet all the parameters required to perform a complete technical analysis, hindering a more detailed and direct comparison of all the technologies. Most researchers or companies present only the energy generation capacity and IM of the developed devices and only a few studies present the installed power and the conversion efficiency of the technologies. From this analysis, one can conclude that the systems that make use of vehicle mechanical energy have a higher conversion efficiency and energy generation capacity than the systems that make use of solar radiation. In terms of energy generation, hydraulic and electromechanical systems present higher capacities.

In terms of IMs, photovoltaic systems are mainly applied using IM 1, while TEG systems are applied under the road surface, using IMs 3 and 4. Piezoelectric systems are also applied using IM 3, while hydraulic and electromechanical systems can both be installed using methods 1 or 2, with their surface in direct contact with vehicle wheels, to maximise the energy input to the system.

In terms of development status, one can conclude that none of these devices are fully validated and available on the market; they are generally at TRL 3 or 4. The TNO system is on TRL 7, but this is for cycle path application and does not present any evidence of application on roads.

To determine fully whether any technology is viable, an economic analysis should also be performed. In such an analysis, the most important factor is the levelised cost of electricity, which determines the cost per watt produced, relating the total economic investment in a technology to the energy generated (EIA, 2014). However, no technology is fully developed and available on the market. So, no economic data of any product are yet available and such an analysis cannot be performed at this stage.

4. Conclusions

The concept of road pavement energy harvesting has become increasingly popular over the last few years. Unlike the case of wind energy, the present situation shows a wide variety of energy harvesting systems, at several stages of development, competing against each other to get an opportunity in the market. In the last 15 years or so, the research and development activity in road pavement energy harvesting has been developed more by companies than by universities, leading to a lack of scientific evidence being available on the developed technologies. The tests performed were not fully characterised in the literature, making very limited information available about the experimental tests and results obtained. It is clear that none of the developed technologies have been fully developed and validated, as none of them have entered the market with a finished and certified product (with the exception of ASC, which is an energy harvesting system, but not to generate electrical energy). In the road pavement energy harvesting field, most of the technologies are at a laboratory and prototyping validation stage.

Comparing the technologies that make use of solar energy as their energy source with the technologies that make use of vehicle mechanical energy, the former is at a more advanced developmental stage, as it makes use of more mature systems and technologies. However, presently, most research and development is being performed on the latter, mainly due to the higher potential that these systems present, in terms of energy conversion efficiency, energy generation and adaptability to road pavement conditions.

Comparing the technologies that make use of solar energy as an energy source, photovoltaic systems are the most efficient and mature. However, the implementation on road pavements is still a challenge, as glass has been used on the photovoltaic cells, causing difficulties for vehicle adherence, which is essential to guarantee rolling capacity and safety conditions. Systems that make use of TEGs are easier to install on the road pavement; however, efficiency is considerably reduced.

Comparing the technologies that make use of vehicle mechanical energy as an energy source, piezoelectric technology was the first to get the attention of researchers. However, due to its lower energy conversion efficiency, the developments with this technology have decreased in the recent years. On the other hand, there has been an increase in research and development of electromechanical systems that harvest vehicle mechanical energy and, using electromagnetic generators, generate electrical energy. These, together with hydraulic systems, have registered the highest energy generation values in experimental tests. Their installation is also simpler than the installation of piezoelectric devices and they currently present a higher likelihood of success as an effective solution to transform vehicle mechanical energy into electrical energy effectively.

MEMS also present potential in this field since they have been successful in other applications. However, in the case of road pavement energy harvesting, they have been applied to harvesting pavement vibrations instead of directly harvesting vehicle mechanical energy. Pavement vibrations represent a small amount of the available energy, leading to a low level of energy generation. In the future, these systems should also be developed to harvest vehicle mechanical energy in order to maximise energy generation.

Acknowledgements

The present research work was carried out in the framework of project PAVENERGY – Pavement Energy Harvest Solutions (PTDC/ECM-TRA/3423/2014) and also project EMSURE – Energy and Mobility for Sustainable Regions (CENTRO-07-0224-FEDER-002004). The author Francisco Duarte is also grateful to the Portuguese Foundation of Science and Technology for the financial support provided to this study through Grant SFRH/BD/95018/2013.

REFERENCES

Abramovich H, Milgrom C, Harash E, Azulay L and Amit U (2010a) *Multi-Layer Modular Energy Harvesting Apparatus, System And Method*. US Patent US20100045111 A1, Feb.

Abramovich H, Harash E, Milgrom C *et al.* (2010b) *Modular Piezoelectric Generators*. International Patent Application PCT/IL2009/000365, Patent WO2010116348 A1, Oct.

Abramovich H, Milgrom C, Harash E *et al.* (2012) *Piezoelectric-Based Weight in Motion System and Method for Moving Vehicles*. International Patent Application PCT/IL2011/000741, Patent WO2012038955 A1, Mar.

Andriopoulou S (2012) *A Review on Energy Harvesting from Roads*. KTH, Stockholm, Sweden.

Angel R and Gomez J (1997) *Road Vehicle-Actuated Air Compressor*. US Patent US5634774, Jun.

Arroyo E and Badel A (2011) Electromagnetic vibration energy harvesting device optimization by synchronous energy extraction. *Sensors and Actuators A: Physical* **171(2)**: 266–273.

Baugher JP (2013) *Energy Harvesting with a Mico-Electro-Mechanical System (MEMS)*. US Patent US20130193930 A1, Aug.

Beeby SP, Tudor MJ and White NM (2006) Energy harvesting vibration sources for microsystems applications. *Journal of Measurement Science and Technology* **17(12)**: R175.

Beeby SP, Torah RN, Tudor MJ *et al.* (2007) A micro electromagnetic generator for vibration energy harvesting. *Journal of Micromechanics and Microengineering* **17(7)**: 1257.

Bobes-Jesus V, Pascual-Muñoz P and Castro-Fresno D (2013) Asphalt solar collectors: a literature review. *Applied Energy* **102**: 962–970.

Bowen L and Near C (2000) *Low Voltage Piezoelectric Actuator*. US Patent 6,111,818, Aug.

Brusaw S and Brusaw J (2014) *Solar Roadway Panel*. Design Patent USD712822S, Sep.

Buhaug H and Urdal H (2013) An urbanization bomb? Population growth and social disorder in cities. *Global Environmental Change* **23(1)**: 1–10.

Calvo R and Calvo J (2004) *Highway Electric Power Generator*. US Patent US6767161 B1, Jul.

Chang H and Lee C (2011) *On-Road Energy Conversion and Vibration Absorber Apparatus*. US Patent US20110215593 A1, Sep.

Chen R (2009) *Vehicular Movement Electricity Converter Embedded Within a Road Bump*. US Patent US20090315334 A1, Dec.

Davis CR (2007) *Roadway Power Generating System*. US Patent US20070181372 A1, Aug.

Daya A (2010) *Road Vehicle Actuated Energy Device*. US Patent US7714456 B1, May.

Duarte F and Casimiro F (2011) *Pavement Module for Generating Electric Energy from the Movement of People and Vehicles*. International Patent Application PCT/IB2011/052164, Patent WO2011145057 A2, Nov.

Duarte F and Casimiro F (2013) *Electromechanical System for Electric Energy Generation and Storage Using a Surface Motion*. International Patent Application PCT/IB2013/050616, Patent WO2013114253 A1, Aug.

Duarte F, Casimiro F, Correia D, Mendes R and Ferreira A (2013a) Waynergy people: a new pavement energy harvest system. *Proceedings of the Institution of Civil Engineers – Municipal Engineer* **166(4)**: 250–256, http://dx.doi.org/10.1680/muen.12.00049.

Duarte F, Casimiro F, Correia D, Mendes R and Ferreira A (2013b) A new pavement energy harvest system. *Proceedings of the International Renewable and Sustainable Energy Conference, Ouarzazate, Morocco*. IEEE – Institute of Electrical and Electronics Engineers, New York, NY, USA (CD-ROM).

Duarte F, Champalimaud J and Ferreira A (2014) Waynergy vehicles: an innovative pavement energy harvest system. *Proceedings of the 2nd International Congress on Energy Efficiency and Energy Related Materials, Oludeniz, Turkey. Springer*, London, UK (CD-ROM).

Duarte F, Champalimaud J and Ferreira A (2016) Waynergy vehicles: an innovative pavement energy harvest system. *Proceedings of the Institution of Civil Engineers – Municipal Engineer* **169(1)**: 13–18, http://dx.doi.org/10.1680/muen.14.00021.

EIA (U.S. Energy Information Administration) (2014) *Levelized Cost and Levelized Avoided Cost of New Generation Resources in the Annual Energy Outlook 2014*. U.S. Energy Information Administration, Washington DC, USA. See http://www.eia.gov/forecasts/aeo/pdf/electricity_generation.pdf (accessed 05/09/2015).

Elliott SJ and Zilletti M (2014) Scaling of electromagnetic transducers for shunt damping and energy harvesting. *Journal of Sound and Vibration* **333(8)**: 2185–2195.

EnergyIntelligence (2014) http://www.energyintel.us/ (accessed 08/12/2014).

Galich T (2002) *Force Stand for Electrical Energy Producing Platform*. US Patent US6376925 B1, Apr.

Gao Q, Huang Y, Li M, Liu Y and Yan YY (2010) Experimental study of slab solar collection on the hydronic system of road. *Solar Energy* **84(12)**: 2096–2102.

Genziko (2014) http://www.genziko.com/ (accessed 10/12/2014).

Harb A (2010) Energy harvesting: state-of-the-art. *Renewable Energy* **36(10)**: 2641–2654.

Hasebe M, Kamikawa Y and Meiarashi S (2006) Thermoelectric generators using solar thermal energy in heated road pavement. In *Proceedings ICT '06 – 25th International Conference on Thermoelectrics (ICT), Vienna, Austria*. IEEE – Institute of Electrical and Electronics Engineers, New York, NY, USA, pp. 697–700.

Hendrickson B (2010a) *Adaptive, Low-Impact Vehicle Energy Harvester*. US Patent US20100192561 A1, Aug.

Hendrickson B (2010b) *Vehicle Energy Harvesting Roadway*. International Patent Application PCT/US2010/020676, Patent WO2010081113 A1, Jul.

Hendrowati W, Guntur HL and Sutantra IN (2012) Design, modelling and analysis of implementing a multilayer piezoelectric vibration energy harvesting mechanism in the vehicle suspension. *Engineering* **4(11)**: 728–738.

HES (Highway Energy Services) (2015) http://www.hughesresearch.co.uk/ (accessed 05/01/2015).

Hill D, Agarwal A and Tong N (2014) *Assessment of Piezoelectric Materials for Roadway Energy Harvesting*. DNV KEMA Energy and Sustainability, Sacramento, CA, USA. See http://www.energy.ca.gov/2013publications/CEC-500-2013-007/CEC-500-2013-007.pdf (accessed 05/09/2015).

Horianopoulos D and Horianopoulos S (2007) *Traffic-Actuated Electrical Generator Apparatus*. International Patent Application PCT/CA2006/001710, Patent WO2007045087 A1, Apr.

Houghton L (2011) *Hydraulic Powermat*. UK Patent Application GB2476826 A, Jul.

Hughes P, Hughes AM, Hughes R and Hughes AP (2004) *Apparatus For Converting Kinetic Energy*. International Patent Application PCT/GB2004/000364, Patent WO2004067850 A1, Aug.

Hughes P, Hughes AM, Hughes R and Hughes AP (2009) *Improvements in and Relating to Apparatus for Converting Kinetic Energy*. International Patent Application PCT/GB2009/050145, Patent WO2009101448 A1, Aug.

Hunter J (2006) *Vehicle Actuated Road Imbedded Magneto Generator*. US Patent US7102244 B2, Sep.

ICAX (2014) http://www.icax.co.uk (accessed 24/11/2014).

IEA (International Energy Agency) (2012) *Technology Roadmap: Fuel Economy of Road Vehicles*. International Energy Agency, Paris, France. See http://www.iea.org/publications/freepublications/publication/Fuel_Economy_2012_WEB.pdf.

IEA (2013) *Key World Energy Statistics 2013*. International Energy Agency, Paris, France. See http://www.iea.org/publications/freepublications/publication/KeyWorld2013.pdf (accessed 05/09/2015).

Innowattech (2014) http://www.innowattech.co.il/ (accessed 21/07/2014).

Jang JS (2011) *Electrical Generator Apparatus, Particularly for Use on a Vehicle Roadway*. US Patent US20110187125 A1, Aug.

Kang-Won W and Correia AJ (2010) *A Pilot Study for Investigation of Novel Methods to Harvest Solar Energy from Asphalt Pavements*. Korea Institute of Construction Technology (KICT), Goyang City, South Korea.

Kazmierski T and Beeby S (eds.) (2009) *Energy Harvesting Systems – Principles, Modeling and Applications*. Springer, New York, NY, USA.

Kelk (2015) http://www.kelk.co.jp/english/index.html (accessed 02/02/2015).

Kenney T (2004) *System and Method for Electrical Power Generation Utilizing Vehicle Traffic on Roadways*. US Patent US20040130158 A1, Jul.

Khaligh A and Onar OC (2010) *Energy Harvesting: Solar, Wind, and Ocean Energy Conversion Systems*. CRC Press Inc, Boca Raton, FL,USA.

Kinergy Power (2014) http://www.kinergypower.com/index.shtml (accessed 29/12/2014).

Klein G, Tsikhotsky E, Abramovich H and Milgrom C (2012) *Modular Piezoelectric Generators with a Mechanical Force Multiplier*. International Patent Application PCT/IL2012/050369, Patent WO2013038415 A1, Mar.

Loomans M, Oversloot H, de Bondt A, Jansen R and van Rij H (2003) Design tool for the thermal energy potential of asphalt pavements. *Proceedings of Building Simulation '03 – the 8th International Building Performance Simulation Association International Conference, Eindhoven, the Netherlands*, pp. 745–752.

Lund JW (2000) Pavement snow melting. *Geo-Heat Center Quarterly Bulletin* **21(2)**: 12–19.

Lundgren R (1984) *Device for Generating Electricity by Pedestrian and Vehicular Traffic*. US Patent US4434374, Feb.

Luttrull J (2005) *Roadway Generating Electrical Power by Incorporating Piezoelectric Materials*. US Patent US20050127677 A1, Jun.

Mallick RB, Chen BL and Bhowmick S (2009) Reduction of urban heat island effect through harvest of heat energy from asphalt pavements. In *Proceedings of the 2nd International Conference on Countermeasures to Urban Heat Islands Effect, September, Berkeley, CA, USA* (Melvin P (ed.)). Lawrence Berkeley National Laboratory, Berkeley, CA, USA, pp. 1–20.

Mankins JC (1995) *Technology Readiness Levels. White Paper*. National Aeronautics and Space Administration (NASA), Washington, DC, USA.

Mansfield R, Shani N and Shani D (2012) *Method and System for Energy Harvesting, Recapture and Control*. International Patent Application PCT/US2012/000033, Patent WO2012099706 A2, Jul.

Martinez S (1980) *Highway Turbine*. US Patent US4238687, Dec.

Mitcheson PD, Green TC, Yeatman EM and Holmes AS (2004) Architectures for vibration-driven micropower generators. *Journal of Microelectromechanical Systems* **13(3)**: 429–440.

Morita K and Tago M (2000) Operational characteristics of the Gaia snow-melting system in Ninohe, Iwate, Japan. In *Proceedings of World Geothermal Congress 2000*. Oregon Institute of Technology, Klamath Falls, OR, USA, pp. 3511–3516.

Morley B, Roberts E, Dean L, Debenham M and Hammond J (2005) *Rollers Set in Road to Extract Energy from Vehicles*. UK Patent Application GB2408074 A, May.

Munaz A, Lee BC and Chung GS (2013) A study of an electromagnetic energy harvester using multi-pole magnet. *Sensors and Actuators A: Physical* **201(2013)**: 134–140.

Nakatsu S (1994) *Road Apparatus*. UK Patent Application GB2290115 A, Dec.

Near C (2013) *Power Generator*. US Patent US20130207520 A1, Aug.

NewEnergyTechnologies (2015) http://www.newenergytechnologiesinc.com/ (accessed 04/02/2015).

Parr A (2011) *Hydraulics and Pneumatics: a Technician's and Engineer's Guide*. Elsevier, Oxford, UK.

Peralta M, Costa-Krämer JL, Medina E and Donoso A (2014) Analysis and fabrication steps for a 3D-pyramidal high density coil electromagnetic micro-generator for energy harvesting applications. *Sensors and Actuators A: Physical* **205(2014)**: 103–110.

Pirisi A (2012) *System for Converting Potential or Kinetic Energy of a Body Weighting Upon or Travelling Over a Support or Transit Plane into Useful Energy*. US Patent US20120248788A1, Oct.

Pirisi A, Grimaccia F, Mussetta M and Zich RE (2012a) Novel speed bumps design and optimization for vehicles' energy recovery in smart cities. *Energies* **5(11)**: 4624–4642.

Pirisi A, Grimaccia F, Mussetta M and Zich RE (2012b) An evolutionary optimized device for energy harvesting from traffic. In *2012 IEEE Congress on Evolutionary Computation (CEC)*. IEEE – Institute of Electrical and Electronics Engineers, New York, NY, USA.

Pirisi A, Mussetta M, Grimaccia F and Zich RE (2013) Novel speed-bump design and optimization for energy harvesting from traffic. *IEEE Transactions on Intelligent Transportation Systems* **14(4)**: 1983–1991.

Priya S and Inman DJ (eds.) (2009) *Energy Harvesting Technologies*, vol. 21. Springer, New York, NY, USA.

RES (Road Energy Systems) (2007) *Energy from Asphalt – Asphalt Solar Collectors for Heating and Cooling Buildings and Roads*. Road Energy Systems, Avenhom, the Netherlands. See http://www.ooms.nl/en/7/301/road_energy_system.aspx (accessed 05/09/2015).

Saha CR (2011) Modelling theory and applications of the electromagnetic vibrational generator. In *Sustainable Energy Harvesting Technologies – Past, Present and Future* (Tan YK (ed.)). InTech, Rijeka, Croatia, pp. 55–108. See http://www.intechopen.com/books/sustainable-energy-harvestingtechnologies-past-present-and-future/modelling-theory-and-applications-of-the-electromagnetic-vibrationalgenerator (accessed 05/09/2015).

Schreier M, Roschewsky N, Dobler E et al. (2013) Current heating induced spin seebeck effect. *Applied Physics Letters* **103(242404)**: 1–5.

Smith R (1979) *Road Vehicle-Actuated Air Compressor and System Therefore*. US Patent US4173431, Nov.

SolaRoad (2014) http://www.solaroad.nl/en/ (accessed 18/11/2014).

SR (Solar Roadways) (2015) http://www.solarroadways.com/main.html (accessed 06/02/2015).

Stephen NG (2006) On energy harvesting from ambient vibration. *Journal of Sound and Vibration* **293(1)**: 409–425.

Sullivan C, Bondt A, Jansen R and Verweijmeren H (2007) *Innovation in the Production and Commercial Use of Energy Extracted from Asphalt Pavements*. Ooms International Holding bv, Chelford, Cheshire, UK. See http://www.materialedge.co.uk/docs/Energy%20from%20Asphalt%20paper%2020%2011%2006.pdf (accessed 05/09/2015).

TNO (2014) https://www.tno.nl/index.cfm/ (accessed 30/05/2014).

Tu Y, Li J and Guan C (2010) Heat transfer analysis of asphalt concrete pavement based on snow melting. In *2010 International Conference on Electrical and Control Engineering (ICECE), Wuhan, China*. IEEE – Institute of Electrical and Electronics Engineers, New York, NY, USA, pp. 3795–3798.

Underground Power (2014) http://www.upgen.it/ (accessed 18/12/2014).

Valon F (2007) *Highway's Electrogenerators*. US Patent US20070246940 A1, Oct.

Waydip (2015) http://www.waydip.com/ (accessed 20/02/2015).

Wischke M, Masur M, Kroer M and Woias P (2011) Vibration harvesting in traffic tunnels to power wireless sensor nodes. *Smart Materials and Structures* **20(8)**: 1–8.

Wu G and Yu X (2012) Thermal energy harvesting across pavement structures. *Proceedings of the Transportation Research Board (TRB) 91st Annual Meeting*. Transportation Research Board, Washington, DC, USA.

Wu G and Yu X (2013) Computer-aided design of thermal energy harvesting system across pavement structure. *International Journal of Pavement Research and Technology* **6(2)**: 73–79.

Xiang HJ, Wang JJ, Shi ZF and Zhang ZW (2013) Theoretical analysis of piezoelectric energy harvesting from traffic induced deformation of pavements. *Smart Materials and Structures* **22(9)**: 095024.

Xiong H, Wang L, Wang D and Druta C (2012) Piezoelectric energy harvesting from traffic induced deformation of pavements. *International Journal of Pavement Research and Technology* **5(5)**: 333–337.

Yildiz F (2009) Potential ambient energy-harvesting sources and techniques. *The Journal of Technology Studies* **35(35)**: 40–48.

Zhao H, Yu J and Ling J (2010) Finite element analysis of cymbal piezoelectric transducers for harvesting energy from asphalt pavement. *Journal of the Ceramic Society of Japan* **118(1382)**: 909–915.

Zhao H, Qin L, Tao Y and Ling J (2013) Study on structure of PZT piles based transducer for harvesting energy from asphalt pavement. *Proceedings of the International Journal of Pavements Conference, São Paulo, Brazil*. IJP, São Paulo, Brazil (CD-ROM).

Zorlu Ö and Külah H (2013) A MEMS-based energy harvester for generating energy from non-resonant environmental vibrations. *Sensors and Actuators A: Physical* **202**: 124–134.

ICE Themes Geothermal Energy, Heat Exchange Systems and Energy Piles

Craig and Gavin
ISBN 978-0-7277-6398-3
https://doi.org/10.1680/gehesep.63983.079
ICE Publishing: All rights reserved

Chapter 5
Uncertainties in the design of ground heat exchangers

Olga Mikhaylova BEng, MEng
Department of Infrastructure Engineering, The University of Melbourne, Parkville, Australia
(corresponding author: narsilio@unimelb.edu.au)

Ian W. Johnston BSc, PhD
Department of Infrastructure Engineering, The University of Melbourne, Parkville, Australia

Guillermo A. Narsilio CEng, MSc, PhD
Department of Infrastructure Engineering, The University of Melbourne, Parkville, Australia

Ground source heat pump (GSHP) systems have the potential to reduce the carbon dioxide footprint of buildings by providing them with sustainable heating and cooling energy from the ground at shallow depths. The financial viability of these systems depends on the adequate sizing of the ground heat exchangers (GHEs), which generally contribute the most to the overall capital costs of GSHP systems. Consideration of the uncertainties in the design of GHEs has the potential to improve GHE design by increasing the confidence in sizing GHEs for particular applications. This chapter proposes a methodology for the estimation of the uncertainty in the design length of GHEs by considering the uncertainties involved with the selection of design parameters. By using the proposed methodology, a case study is presented where borehole GHEs are sized following a commonly applied design process recommended in the American Society of Heating, Refrigerating, and Air-Conditioning Engineers (Ashrae) handbook. The uncertainty in the resulting length of GHEs is estimated, and the sensitivity of this uncertainty to the uncertainties in individual design parameters is discussed. In addition, measures to reduce the length uncertainty are considered.

Notation

A	borefield geometrical aspect ratio
B	distance between boreholes: m
C_p	ground heat exchanger (GHE) fluid (water) thermal heat capacity: J/(kg K)
F	correlation function
G	G function representing the cylindrical heat source solution
H	borehole depth: m
h_{conv}	internal convection coefficient: W/(m² K)

k	ground thermal conductivity: W/(m K)
k_{grout}	grout thermal conductivity: W/(m K)
k_{pipe}	U loop pipe thermal conductivity: W/(m K)
L	design length of GHE: m
L_B	baseline length of GHE: m
L_u	centre-to-centre distance between pipes of U loop: m
m_{fls}	fluid mass flow rate per kilowatt of peak hourly ground load: kg/(s kW)
N_B	number of borehole GHEs
q_{hc}	peak hourly ground load for cooling: kW
q_{hh}	peak hourly ground load for heating: kW
q_{mc}	monthly ground load for cooling: kW
q_{mh}	monthly ground load for heating: kW
q_y	yearly average ground load: kW
R_{10y}	effective thermal resistance of the ground to 10-year ground load: m K/W
R_{1m}	effective thermal resistance of the ground to 1-month ground load: m K/W
R_{6h}	effective thermal resistance of the ground to 6-h ground load: m K/W
R_b	effective thermal resistance of the borehole: m K/W
R_{conv}	convection resistance inside each tube of the U loop: m K/W
R_g	grout resistance: m K/W
R_p	convection resistance for each tube of the U loop: m K/W
r_{bore}	borehole radius: m
$r_{p,in}$	U loop inner radius: m
$r_{p,ext}$	U loop outer radius: m
T_g	undisturbed ground temperature: °C
T_{inHP}	heat pump entering water temperature: °C
T_m	mean GHE fluid temperature: °C
T_{outHP}	heat pump outlet water temperature: °C
T_p	temperature penalty: °C
t	time
t_s	characteristic time: years
α	ground thermal diffusivity: m^2/d
ΔL	length uncertainty range: m
ΔL_{80}	length uncertainty range calculated with 80% probability: m

Introduction

Ground source heat pump (GSHP) systems use the ground as a heat source and sink for sustainable heating and cooling of buildings (Brandl, 2006). The technology involves using ground loops or ground heat exchangers (GHEs) and heat pumps and has the potential to contribute to the reduction of greenhouse gas emissions from buildings (Johnston et al., 2011). The high installation cost of a GSHP system can potentially discourage the adoption of this system in favour of a cheaper and more traditional, but less environmentally friendly, heating and air-conditioning system. Since the capital costs of GSHP systems are largely controlled by the installation costs of GHEs, it is important that the lengths of GHEs are determined as accurately as possible so that costs are not further increased by unnecessary oversizing.

Of course, it is also important to ensure that the GHEs are not undersized, so that the system is unable to fulfil its design requirements.

The sizing of GHEs is a relatively complex process and involves the prediction of building thermal demands, the selection of heat pumps and other mechanical equipment and the consideration of GHE-ground thermal energy interactions over the lifetime of the system. For large GSHP systems, it is relatively common to undertake a thermal response test (TRT) to determine ground thermal properties at a particular site. A TRT is a technique to measure the ground thermal conductivity and borehole thermal resistance in the appropriate position by applying constant power to the fluid circulating within a GHE and measuring the mean fluid temperature in the GHE over time (Spitler and Gehlin, 2015). While there are many relatively sophisticated design tools available to assist with the design of GHEs (Yang *et al.*, 2010), many GSHP systems are designed on the basis of simplified design guidelines such as presented by Ashrae (2007) and IGSHPA (2009).

Irrespective of the design tool used, a series of design parameters must be chosen as input to the design process. With a deterministic design, each parameter is given a value and a single GHE length is calculated. However, while such a process is based on the 'best' estimate of these values, there is always some uncertainty about what these values may be. This uncertainty can lead to the design process giving overdesigned or underdesigned GHE lengths. Bernier (2002) considered vertical GHE design length uncertainties when the Ashrae design approach is followed. However, the focus of that study was the methodology of the estimation, not the actual magnitude of the resulting length uncertainty.

This chapter presents a methodology for the estimation of the uncertainty in the GHE design length when a particular set of design recommendations is followed. The proposed method also considers uncertainties in design parameters and evaluates their individual contributions to the overall GHE design length uncertainty. Based on the proposed methodology, a case study is presented in which the commonly used Ashrae handbook design recommendations (Ashrae, 2007) are used to size borehole GHEs for a GSHP system. This example demonstrates the expected uncertainty in the GHE design length for the system when a simplified design method is followed. In addition, based on a sensitivity analysis, measures to reduce the design length uncertainty, including a TRT, are discussed.

Methodology

GHE design length uncertainty estimation

In this chapter, the common Ashrae design guideline (Ashrae, 2007), reorganised by Philippe *et al.* (2010) to a GHE sizing spreadsheet, is used to size GHEs. In this approach, the required length of GHEs is calculated by employing the following equation

$$L = \frac{q_h R_b + q_y R_{10y} + q_m R_{1m} + q_h R_{6h}}{T_m - (T_g + T_p)} \quad (1)$$

where q_y, q_m and q_h are the yearly average, highest monthly and peak hourly ground loads correspondingly; R_b is the effective thermal resistance of the borehole; R_{10y}, R_{1m} and R_{6h} are the effective ground thermal resistances to 10-year, 1-month and 6-h ground loads respectively; T_m and T_g are the mean GHE fluid and undisturbed ground temperatures correspondingly and T_p is the temperature penalty. An overview of the calculations of the parameters in Equation 1 is given in Table 1. For more detailed explanation of the calculation procedure, see Philippe et al. (2010).

Following this guideline, the GHEs are sized separately for heating and cooling and the longest required length is adopted as a design length. More detailed information about the Ashrae method and GHE sizing spreadsheet can be found in Ashrae (2007) and Philippe et al. (2010).

The proposed methodology aims to quantify the uncertainty in the design length of GHEs by considering the predictions of a particular design process. The GHE design length is a function of design parameters, including design thermal loads, ground and grout thermal properties and GHE geometry, all of which are not deterministic in nature and have some degree of uncertainty associated with them. Hence, uncertainty in the GHE design length can be quantified by considering the uncertainties in design parameters by using uncertainty propagation methods. As a result, the required length is expressed probabilistically in the form of a probability density function (pdf).

Along with the uncertainty analysis, a sensitivity analysis is proposed to determine the sensitivity of a length to uncertainties in individual design parameters. The sensitivity analysis may be able to reduce the scope of uncertainty analysis if some particular parameter is found to be insignificant.

In general, two types of methods are available to solve uncertainty propagation problems: the analytical methods of moments and the numerical probabilistic (Monte Carlo) methods (Haldar and Mahadevan, 2000). The methods of moments use the principles of probability theory to analytically estimate the uncertainty of a function based on uncertainties in the variables of this function. These methods utilise an explicitly expressed relationship between functions and their variables and can become mathematically intensive if many variables are involved (Haldar and Mahadevan, 2000).

Monte Carlo simulation methods are based on numerical experiments where a large number of random input parameters are generated to find the outcomes of a function, such as Equation 1. The Monte Carlo methods can handle large and complex implicit functions, so they are widely used in uncertainty analyses of engineering systems (Modarres, 2006). Considering the large number of input parameters in sizing GHEs, the Monte Carlo method is proposed for a sensitivity analysis and an estimation of the GHE length uncertainty. In this chapter, the commercial @Risk software was used to perform Monte Carlo simulations. This software utilises a built-in Mersenne twister to generate random numbers (Matsumoto and Nishimura, 1998).

Table 1 Calculation of the parameters in Equation 1

Parameter	Equations[a]
Effective thermal resistance of the borehole R_b	$R_b = R_g + \dfrac{R_p + R_{conv}}{2}$ where $R_g = \dfrac{1}{4\pi k_{grout}} \left\{ \ln\left(\dfrac{r_{bore}}{r_{p,ext}}\right) + \ln\left(\dfrac{r_{bore}}{L_u}\right) + \dfrac{k_{grout} - k}{k_{grout} + k} \ln\left[\dfrac{r_{bore}^4}{r_{bore}^4 - (L_u/2)^4}\right] \right\}$ is the grout resistance $R_p = \dfrac{\ln(r_{p,ext}/r_{p,in})}{2\pi k_{pipe}}$ is the convection resistance for each tube of the U loop $R_{conv} = \dfrac{1}{2\pi r_{p,in} h_{conv}}$ is the convection resistance inside each tube of the U loop
Effective ground thermal resistance to 10-year ground load R_{10y}	$R_{10y} = \dfrac{1}{k}\left[G\left(\dfrac{at_{10y+1m+6h}}{r_{bore}^2}\right) - G\left(\dfrac{at_{1m+6h}}{r_{bore}^2}\right) \right]$ where G is the G function representing the cylindrical heat source solution and t is the time
Effective ground thermal resistance to 1-month ground load R_{1m}	$R_{1m} = \dfrac{1}{k}\left[G\left(\dfrac{at_{1m+6h}}{r_{bore}^2}\right) - G\left(\dfrac{at_{6h}}{r_{bore}^2}\right) \right]$

Table 1 Continued

Parameter	Equations[a]
Effective ground thermal resistance to 6-h ground load R_{6h}	$R_{6h} = \dfrac{1}{k} G\left(\dfrac{\alpha t_{6h}}{r_{bore}^2}\right)$
Mean GHE fluid temperature T_m	$T_m = \dfrac{T_{inHP} + T_{outHP}}{2}$ where T_{inHP} is the heat pump inlet temperature defined by the designer and $T_{outHP} = T_{inHP} \pm (1000/m_{fls} \cdot C_p)$ is the heat pump outlet temperature
Temperature penalty T_p	$T_p = \dfrac{q_y}{2\pi kL} F\left(\dfrac{t_{10y}}{t_s}, \dfrac{B}{H}, NB, A\right)$ where F is the correlation function (see Philippe et al., 2010), $t_s = H^2/9\alpha$ is the characteristic time, H is the borehole depth and A is the borefield geometrical aspect ratio (see Philippe et al., 2010)

[a] For the definitions of intermediate parameters involved in the equations, see the Notation and Philippe et al. (2010)

Quantification of design parameter uncertainty

In general, on the basis of how they are determined, there are three categories of design parameters as follows.

- *Parameters determined by site conditions* are defined by the conditions of a project site (e.g. geological conditions, groundwater level and undisturbed ground temperature). The uncertainties in such design parameters are associated with the lack of knowledge about their exact values and with their natural variations.
- *Parameters determined by design decisions* are selected during the design process and expected to be achieved when the system is installed (e.g. geometry of the boreholes, the location of ground loops within a borehole, grout thermal conductivity). Due to imperfections of the installation process and natural variability inherent in design specifications, the design values of such parameters might be different from their as-built values. Since the actual, as-built parameters influence the performance of an installed system, the uncertainties in such design parameters have to be considered in the evaluation of the uncertainty in the required length of GHEs.
- *Parameters determined by site conditions and design decisions* are parameters that are influenced by both the preceding factors. The building thermal load is one such parameter. On the one hand, the climatic conditions and their natural variability are largely controlled by site conditions, whereas the response of the building and the variation in the activities it contains are related to design decisions made. All of these are accompanied by uncertainty.

An estimation of the uncertainties involved with any design parameter can be obtained by considering available information about an expected value of this parameter. The more information available about the parameter, the greater is the likelihood of predicting its value with a higher precision and accuracy. For some parameters, such as the thermal conductivity of the plastic ground loop pipes, the value is known very accurately. However, for most parameters, particularly those related to natural and human activities, the values cannot be predicted with a high degree of certainty. Therefore, it becomes necessary to consider a range of possible values, and this can be achieved through the use of pdfs. Based on the data available to predict a value of a design parameter, four levels of information are identified, with level 1 corresponding to when the least information is available about a parameter and level 4 when the most information is available. Table 2 shows a possible range of pdfs applicable depending on a level of information about parameters.

Case study

The new Elizabeth Blackburn School of Sciences located in Melbourne, Australia, is used as a case study to obtain building thermal demands and ground and climate conditions (Mikhaylova *et al.*, 2015). Note that the case study does not examine the GHEs installed in this particular building but follows the process of sizing GHEs if the Ashrae design guidelines are to be used. The parameters of the system under consideration are summarised in Table 3. The configuration of the borefield of GHEs and the cross-section of an individual GHE are shown in Figure 1. The aim of the design is to estimate the overall required length of GHEs for a given configuration of the borefield. All calculations are made for Melbourne climatic conditions.

Table 2 Design parameter uncertainty estimation

Level of information available about a parameter value		Description	Type of pdf/ deterministic value	Examples
Level 1	↑ Least information	Not enough or exact information is available to predict the expected true value of a parameter. The value is usually estimated based on general recommendations to be in a range between minimum and maximum.	Uniform distribution	Ground thermal conductivity when no TRT is performed on site
Level 2		A parameter is reasonably expected to achieve a certain value but the true value can deviate from this expected value due to unforeseen circumstances or measurement errors.	Triangular distribution	Ground thermal conductivity measured on site by a TRT
Level 3		There are data available about the typical expected distributions of the values of a parameter.	Actual evaluated distribution	Grout thermal conductivity when the expected distribution of its probable values is available from a supplier
Level 4	↓ Most information	The value of a parameter and its likely variation is well established and the variation is not significant in comparison with the possible variations in the values of other parameters.	Single, deterministic value	Thermal conductivity of high-density polyethylene (HDPE) pipes of U-loops

Table 3 Parameters of the GSHP system

Parameter	Value
Building thermal design parameters	
Peak thermal demands for heating/cooling	80/120 kW
Design air temperature for peak heating/cooling	2·1/35·7°C
Building balance point temperature for heating/cooling	16/20°C
Building operating hours	Weekdays: 7 am–6 pm (weekends and other time: off)
GSHP system	
Heat pump min/max entering water temperatures T_{inHP}	7/35°C
Coefficient of performance (COP) of heat pumps for heating and cooling	4
GHE fluid (water) thermal heat capacity C_p	4200 J/(kg K)
Borehole GHEs	
Number of borehole GHEs N_B/parallel borelines	28/4
Spacing between GHEs/borelines B	4/4 m
Number of U loops in one GHE	1
Borehole diameter $2r_{bore}$	114 mm
U loop inner diameter $2r_{p,in}$/outer diameter $2r_{p,ext}$	22/25 mm
U loop HDPE pipe thermal conductivity k_{pipe}	0·45 W/(m K)
Internal convection coefficient h_{conv}	1000 W/(m² K)

Figure 1 (a) A plan view of the field of borehole GHEs; (b) cross-section of a GHE

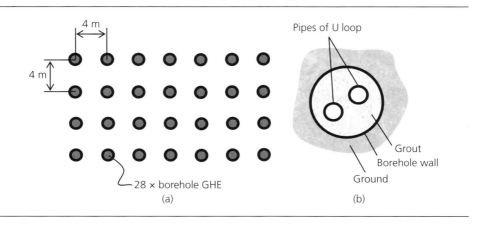

Design parameter uncertainties

In total, uncertainties in 12 design parameters are considered in this case study. Their baseline values and the uncertainties adopted are explained in the following sections and summarised in Table 4. All other design parameters are assumed to have insignificant uncertainties because their values are reasonably well known and are not expected to change during the installation of the system. Their values are taken as presented in Table 3.

Design parameters determined by site conditions

- Ground thermal conductivity k: A TRT is rarely conducted for small or medium-size GSHP projects because of the high cost of this full-scale on-site test. At the design stage, the ground thermal conductivity is normally estimated based on the known results of TRTs performed in nearby locations. When not much data are available, the ground thermal conductivity is assumed based on the recommended values for similar geological materials in the various design manuals and guidelines.
  As in the prototype building, Silurian mudstone is assumed to underlay the project site. For thermal conductivity of clay stones, Kavanaugh and Rafferty (1997) suggested a range of 1·8–3·0 W/(m K). When relying on these recommendations, the baseline value

Table 4 Uncertainties in the design parameters considered in the case study

Design parameter determined by	Baseline value	Estimated uncertainty[a]
Site conditions		
Ground thermal conductivity k: W/(m K)	2·4	Uniform pdf (1·8–3·0)
Ground thermal diffusivity a: m²/d	0·095	Uniform pdf (0·070–0·120)
Undisturbed ground temperature T_g: °C	18·25	Uniform pdf (17·0–19·5)
Design decisions		
Grout thermal conductivity k_{grout}: W/(m K)	2·05	Uniform pdf (1·4–2·7)
Centre-to-centre distance between pipes of U loop, L_u: m	0·047	Triangular pdf (0·025, 0·047, 0·089)
Distance between boreholes B: m	4·0	Triangular pdf (3·6, 4·0, 4·4)
Fluid mass flow rate per kilowatt of peak hourly ground load m_{fls}: kg/(s kW)	0·074	Triangular pdf (0·067, 0·074, 0·081)
Site conditions and design decisions		
Peak hourly ground load for cooling q_{hc}: kW	129	Triangular pdf (98, 143, 180)
Peak hourly ground load for heating q_{hh}: kW	−42[b]	Triangular pdf (−69, −48, −30)[b]
Monthly ground load for cooling q_{mc}: kW	15·5	Triangular pdf (4·4, 13·8, 25·1)
Monthly ground load for heating q_{mh}: kW	−8·1[b]	Triangular pdf (−12·5, −9·1, −6·1)[b]
Yearly average ground load q_y: kW	0·44	Triangular pdf (−2·81, 0·44, 4·17)[b]

[a] For uniform pdfs, figures in parentheses show minimum and the maximum values in the range; for triangular pdfs, figures in parentheses show minimum, mean and maximum values in the range
[b] Negative values indicate that ground energy is extracted from the ground

of k is taken as 2·4 W/(m K), the average of the suggested range. The pdf of this design parameter is assumed to be uniform in the above range.
- Ground thermal diffusivity α: Similar to the ground thermal conductivity, ground thermal diffusivity is usually assumed based on the values recommended by different design guidelines. For this study, the pdf of α is estimated to be uniform in the range of 0·070–0·120 m^2/d based on the thermal diffusivity of sedimentary rocks recommended by Kavanaugh and Rafferty (1997). The baseline value of α is taken as the average of this range, 0·095 m^2/d.
- Undisturbed ground temperature T_g. Ground temperatures are rarely measured in a particular project site prior to the design of GHEs unless the project is particularly large. When GHEs are designed, T_g is usually assumed based on previous experience and the site annual mean ambient temperature, leading to the uncertainty in T_g. For the Melbourne metropolitan region, the ground temperature has been reported in the range of 17·0–19·5°C (Colls, 2013; Mikhaylova *et al.*, 2015; Wang *et al.*, 2012). For this case study, the baseline value of T_g is taken as 18·25°C, the average of the reported range. The pdf of T_g is assumed to be uniform in the range of 17·0–19·5°C.

Design parameters determined by design decisions

- Grout thermal conductivity k_{grout}: In the original design of the GHEs for the case study building, the grout was specified as a silica sand-rich cement-based grout. There were no exact specifications for the grout mixture given, so the expected k_{grout} cannot be determined accurately. According to IGSHPA (2000), the k_{grout} of such grouts can vary from 1·4 to 2·7 W/(m K). The baseline value of k_{grout} is taken as average of these values, 2·05 W/(m K). The pdf of this parameter is assumed to be uniform in the range of 1·4–2·7 W/(m K).
- Centre-to-centre distance between pipes of U loop L_u: Since during the installation of the U loops of GHEs, pipe leg spacers are unlikely to be used, the legs of U loops can deviate from their designed positions. Even though the positions of U loops may not necessarily be in the centres of boreholes, the Ashrae design method assumes only a centred position of U loops inside GHEs. In this approach, only a U loop leg distance can be specified. In terms of pipe spacing, the legs of U loops can be somewhere between touching each other in the centre of the borehole or diametrically opposite, touching the borehole wall. The positions of the U loop legs are likely to vary along the depths of GHEs as well. The distance between U loop legs along the depth of GHEs is designed to be 47 mm, but can vary between 25 and 89 mm. Hence, the value of this parameter is estimated to have a triangular pdf with the min/max = 25/89 mm and the mean = 47 mm.
- Distance between boreholes B: The expected design value of the borehole spacing is 4 m. The spacing between boreholes can be changed during the installation of GHEs because of the deviations of boreholes from their vertical design alignments due to drilling tolerances and on-site repositioning of the boreholes because of unforeseen conditions. Hence, the borehole spacing can be considered with uncertainty by following a triangular distribution with the mean equal to the baseline (expected) value of 4 m. No research on the deviations of the spacing along the depths of adjacent GHEs from their design values has been found by the authors, so the variation of this parameter is assumed to be ±10% of the design spacing.

- Fluid mass flow rate per kilowatt of peak hourly ground load m_{fls}: The system is designed for 0·074 kg/(s kW) of water circulating within the GHEs. However, when the system is installed, the water flow rate can deviate from the design value because of a number of factors, including incorrectly calculated pipe resistances and unexpected modifications of the pipework. It is assumed that the pdf of the water flow rate is a triangular distribution with min/max = 0·067/0·081 kg/(s kW) and the mean equal to its design value.

Design parameters determined by site conditions and design decisions

Ground thermal loads are mainly determined by building thermal demands, which are defined by the ambient air temperatures, building occupant behaviour and the thermal performance of building construction materials. The predicted building thermal demands can be different from the actual demands due to modelling assumptions and error, changes in the anticipated building occupant behaviour and natural variations in ambient air temperatures from year to year. These probable differences can lead to the differences in the predicted and actual ground thermal loads, which explain the uncertainties in these design parameters.

For the case study building, the maximum building energy demands for heating and cooling are estimated to be 80 and 120 kW respectively at the design minimum and maximum ambient air temperatures. There have been reports that predicted building demands can be considerably different from actual demands (for example Garber *et al.*, 2013). For this study, the uncertainties in the peak building loads are assumed to follow simple triangular distributions in the range of ±20% of their expected values with the means equal to these values. These uncertainties are assumed to cover all possible reasons for the uncertainty except the possible variations in the ambient air temperatures, which are taken into account separately.

The baseline values of the ground thermal loads are calculated by using the predicted deterministic building peak thermal loads. The hourly ambient air temperatures are assumed to follow the standard annual ambient air temperature set, which is used in design of conventional heating, ventilation and air-conditioning systems. Knowing the maximum building energy demands for heating and cooling, building energy demands at other ambient air temperatures are calculated by linear interpolation. Having the design building energy demands at each hour of a standard design year, the ground loads required for the GHE sizing (Equation 1) are estimated on the assumption that the coefficients of performance of heat pumps are constant for simplicity (Table 3).

To estimate the uncertainties in the ground loads, the uncertainty in the building peak loads discussed earlier are superimposed on the possible variations in the ambient air temperature from year to year. For this purpose, the 42-year record of half-hourly ambient air temperatures, from 1970 to 2012, for a Melbourne weather station is considered. By using the Monte Carlo method, ground thermal loads were calculated for different combinations of the peak loads and ambient air temperatures of a particular year from the 42-year temperature set. The results obtained were analysed and triangular pdfs were established for each of the ground thermal load parameters. The baseline values and estimated uncertainties in the ground thermal loads are summarised in Table 4 and Figure 2.

If all design parameters from Table 4 are taken as their baseline values, the baseline required length of GHEs, according to the Ashrae handbook, is $L_B = 1888$ m.

Figure 2 Estimated variations in ground loads. (a) Peak hourly ground load in cooling q_{hc}; (b) peak hourly ground load in heating q_{hh}; (c) monthly ground load in cooling q_{mc}; (d) monthly ground load in heating q_{mh}; (e) yearly average ground load q_y. [a]Negative values indicate that ground energy is extracted from the ground

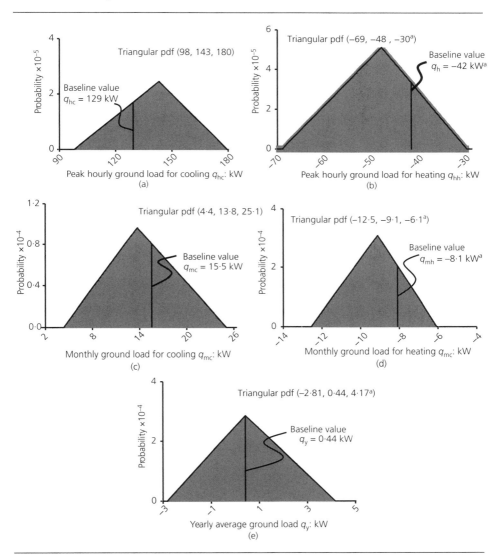

Sensitivity analysis

A sensitivity analysis was performed to identify the parameters whose uncertainties are the most and the least influential on the uncertainty in the required length of GHEs. By using Monte Carlo numerical simulations, the influence of possible variations in each parameter from Table 4 on the *mean* of the required length was calculated within the estimated ranges of possible values of this parameter. The tornado diagram in Figure 3 shows the results of the sensitivity analysis for the ten most influential parameters. From all 12 parameters considered, the mass flow rates of water, m_{fls}, and the distance between boreholes, B, have a minor influence on the length uncertainty and are not shown in the same figure.

A *length uncertainty range* ΔL is proposed to quantify the length uncertainty caused by each design parameter. ΔL is the difference between the maximum and the minimum mean values of the required length due to the variation in a specific design parameter. ΔL is expressed in metres and as a percentage of the baseline value of the length (Figure 3).

Ground thermal loads

In the case considered here, the peak building demand for cooling is significantly higher than the peak building demand for heating. Not surprisingly, the largest GHE design length uncertainty contributor is the peak hourly ground load for cooling, q_{hc}, with $\Delta L(q_{hc}) = 760$ m or 40·3%. Another ground thermal load parameter that significantly influences the design length uncertainty is the yearly average ground load q_y, which is the third most influential parameter, with $\Delta L(q_y) = 397$ m or 21·0%. Clearly, q_{hc} and q_y collectively have the greatest influence on the GHE design length. The monthly ground load in cooling q_{mc} has a moderate effect on the length uncertainty, with $\Delta L(q_{mc}) = 122$ m or 6·5%.

Another two parameters related to ground loads – monthly in heating, q_{mh}, and peak hourly in heating, q_{hh}, – have ΔL less than 1%. The influence of these design parameters on the length uncertainty can be considered insignificant, and they could be excluded from the GHE length uncertainty analysis.

Figure 3 Tornado diagram showing the sensitivity of the mean of the required length of GHEs to the uncertainties in the design parameters when no TRT is performed on site

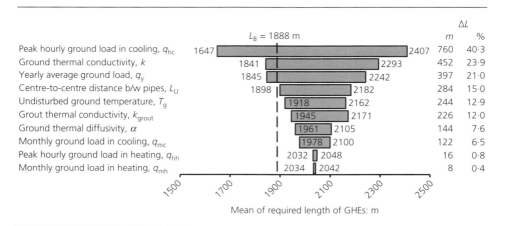

Ground thermal properties

The ground thermal conductivity k is the second most influential parameter with respect to the uncertainty in the required length (Figure 3). The variation in this parameter can influence the mean of the GHE length within $\Delta L(k) = 452$ m or 23·9%. Other ground thermal properties, the undisturbed ground temperature T_g and the ground thermal diffusivity α, have a moderate effect on the length uncertainty, with $\Delta L(T_g) = 244$ m or 12·9% and $\Delta L(\alpha) = 144$ m or 7·6%. The ground thermal property parameters collectively may have a large influence on the GHE length uncertainty.

Borehole characteristics

The centre-to-centre distance between pipes of a U loop, L_u, and the grout thermal conductivity k_{grout} are the two borehole characteristics that are among the ten most influential parameters with respect to the length uncertainty (Figure 3). They both substantially contribute to the length uncertainty, with $\Delta L(L_u) = 284$ m or 15·0% and $\Delta L(k_{grout}) = 226$ m or 12·0%.

Design length uncertainty

The pdf of the GHE design length was estimated by using the Monte Carlo numerical simulation method by considering the uncertainties in all design parameters presented in Table 4 all at once. The results of the estimations are shown in Figure 4.

As similarly done in the sensitivity analysis earlier, the length uncertainty range is proposed to quantify the length uncertainty obtained. Based on the pdf estimated, the length uncertainty range can be calculated with a certain degree of probability. For example the length uncertainty range calculated with 80% probability, ΔL_{80}, is the difference between the lengths calculated at the 90th and 10th percentiles. If all potential uncertainties are taken into account, $\Delta L_{80} = 842$ m or 45% with respect to the baseline length $L_B = 1888$ m (Figure 4). The estimated length uncertainty seems to be relatively high. Some practical measures of the length uncertainty reduction are discussed in the next section.

Figure 4 Estimated uncertainty in the required length of GHEs when no TRT is performed on site

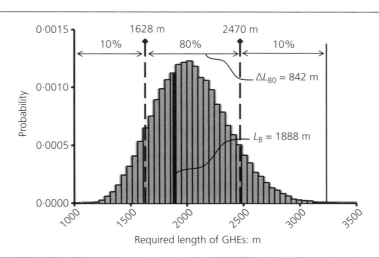

It should be noted that the uncertainty in the design tool is not considered in this chapter. Previous research has shown that the Ashrae design tool can undersize or oversize the GHEs (Cullin et al., 2015; Fossa and Rolando, 2015). A preliminary study undertaken by the authors indicates that if the Ashrae design tool uncertainty is considered, it can potentially become one of the dominant uncertainty contributors and could significantly increase the length uncertainty. More research is needed to properly quantify the design tool uncertainty to include it in the length uncertainty analysis.

Measures for design length uncertainty reduction

Thermal response test

One obvious way to reduce the uncertainty in the GHE length is to conduct a TRT on the site prior to the design of GHEs. This can substantially reduce uncertainties in k and T_g, and the methodology presented here can be used to quantify the cost-effectiveness of doing so. For the case study considered here, it is estimated that a TRT performed would reduce the uncertainties in k and T_g in terms of pdfs (e.g. from uniform to triangular) and range of values because these parameters are measured during a TRT (Zervantonakis and Reuss, 2006). The new estimated uncertainties in these two parameters are summarised in Table 5. The results of the sensitivity analysis when a TRT is performed on site are presented in Figure 5.

From Figure 5, a TRT substantially reduces the influence of k and T_g on the length uncertainty. This is demonstrated by the reduction in $\Delta L(k)$ from 23·9 to 6·9% and in $\Delta L(T_g)$ from 12·9 to 4·3% when the length uncertainty ranges are calculated with and without a TRT. As a result, the overall uncertainty in the required length is reduced, which is shown by the reduction in ΔL_{80} from 842 to 731 m or from 45 to 39% (Figures 4 and 6). Although there is a 6% reduction in the overall length uncertainty, the uncertainty is still fairly high. Hence, even though the additional site investigations would allow some reduction in the design GHE length uncertainty, further length uncertainty reduction actions need to be developed.

With the estimated pdf of the required length, a design length of GHEs can be selected with a certain degree of probability. For example if the length were decided to be selected with a probability of 90% of not undersizing the GHEs, 2470 or 2383 m would be adopted as the GHE design length for the cases of a TRT not performed or performed respectively (Figures 4 and 6). Hence, if the ground thermal properties were measured on site, the design length could be reduced by 87 m. It is only through the proposed methodology that the length reduction

Table 5 Uncertainty in ground thermal design parameters after a TRT

Design parameter	Baseline value	Estimated uncertainty[a]
Ground thermal conductivity k: W/(m K)	2·4	Triangular pdf (2·16, 2·4, 2·64)
Undisturbed ground temperature T_g: °C	18·25	Triangular pdf (17·75, 18·25, 18·75)

[a]For uniform pdfs, figures in parentheses show minimum and maximum values in the range; for triangular pdfs, figures in parentheses show minimum, mean and maximum values in the range

Figure 5 Tornado diagram showing the sensitivity of the mean of the required length of GHEs to the uncertainties in the design parameters when a TRT is performed on site

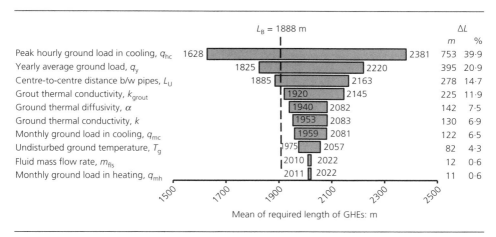

Figure 6 Estimated uncertainty in the required length of GHEs when a TRT is performed on site

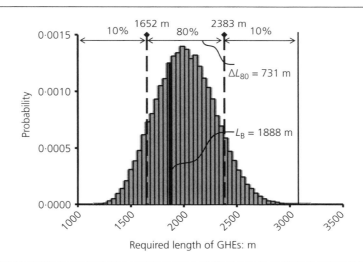

value (87 m) becomes available for the economic analysis of conducting a TRT. Note that this moderate reduction in the design length is estimated for the same expected (mean) values of k and T_g assumed initially (Table 4) and obtained during a TRT (Table 5). If the values of the parameters measured by the TRT were different from their initially assumed values, the resultant design length difference could have been larger.

Typically, in Australia, 1 m of installed GHE costs around A$ 100 (A$ 1 is approximately equal to US$ 0·8). Then, a potential capital saving from conducting a TRT is 87 m × A$ 100 = A$ 8700. A TRT test typically costs around A$ 2000 which makes the TRT financially beneficial for the project since it helps to save around 4% of the GHE cost. This is a preliminary assessment which is based on the design length calculated with 90% probability. Further investigations are needed to evaluate what is the optimum probability threshold for the selection of GHE design lengths.

Ground thermal loads

From Figures 3 and 5, q_{hc} and q_y largely affect the design length uncertainty. These design parameters, as all other ground loads, are controlled by building thermal demands which are predicted at the design stage. The likely error in these predictions will cause ground thermal load uncertainties. More consideration should be given to the estimation of the building thermal demands which determine the ground loads.

Another potential source of error is the methodology of the estimation of peak hourly ground loads that is not clearly defined in the Ashrae design guidelines. As a result, different studies interpret the peak loads differently. For example, Kavanaugh and Rafferty (1997) suggested maximum 4-h average peak loads be adopted as design ground peak loads, whereas Philippe *et al*. (2010) and Fossa (2011) proposed the use of maximum 6-hr average peak loads for this purpose. This error is related to the uncertainty in the Ashrae design tool and not discussed in detail here. Further research is needed to consider whether such an emphasis should be given to the peak loads in the length design formula (Equation 1) and how they should be better estimated based on rational engineering principles. Experimental studies of ground thermal reactions to short peak loads can be particularly useful for these investigations.

Borehole characteristics

Several measures can potentially reduce the GHE length uncertainty caused by the uncertainty in the grout thermal conductivity. Research on expected thermal conductivities of different grouts should be undertaken to better specify thermal properties of different grouts. The grout mixture should be well defined at a design stage, and a site control should be in place to ensure the design specifications are followed on site. These can potentially reduce variations in the grout properties over the depths of GHEs and between several GHEs.

In terms of distances between pipes of U loops, spacers to separate the legs of U loops can be of considerable assistance to ensure that the U loop pipes are installed at the design positions.

Conclusions

The chapter proposes a methodology for the estimation of GHE design length uncertainty based on uncertainties in design parameters. The reasons behind the design parameter uncertainties are discussed, and a practical approach to the estimation of these uncertainties is outlined. The proposed methodology is used to evaluate the likely GHE design length uncertainty for a case study when GHEs are sized by following the Ashrae design handbook recommendations. The most and least influential design parameters on the GHE length uncertainty are evaluated, and some measures for length uncertainty reduction are discussed.

In the case study considered, the uncertainty in the GHE design length appears to be high and needs to be reduced. For this cooling-dominant building, the most influential design parameter on the GHE length uncertainty is the peak hourly ground load for cooling. More effort should be taken at a design stage to estimate building thermal demands, which largely determine the peak ground loads. Another way to reduce uncertainty in the GHE length is to perform a site investigation prior to design. If a TRT is conducted on the site, the overall length uncertainty can be reduced and the TRT can be economically justified at least for the case study considered. Proper consideration of borehole characteristics at the design stage may also lead to length uncertainty reduction.

The case study does not include the uncertainty in the particular design tool used to size GHEs, which may substantially contribute to the length uncertainty. Further research is needed to develop a methodology to include this uncertainty into the GHE length uncertainty evaluation.

Acknowledgements

The authors would like to acknowledge the support provided by the Sustainable Energy Pilot Demonstration (SEPD) Program funded by the Department of Economic Development, Jobs, Transport and Resources of the Government of Victoria; and the Australian Research Council (FT140100227) which pays for G. Narsilio's salary.

REFERENCES

Ashrae (American Society of Heating, Refrigerating, and Air-Conditioning Engineers) (2007) Geothermal energy. *ASHRAE Handbook: HVAC Applications*. Ashrae, Atlanta, GA, USA.

Bernier M (2002) Uncertainty in the design length calculation for vertical ground heat exchangers. *ASHRAE Transactions* **108(1)**: 939–944.

Brandl H (2006) Energy foundations and other thermo-active ground structures. *Géotechnique* **56(2)**: 81–122, http://dx.doi.org/10.1680/geot.2006.56.2.81.

Colls S (2013) *Ground Heat Exchanger Design for Direct Geothermal Energy Systems*. PhD thesis. The University of Melbourne, Parkville, Australia.

Cullin JR, Spitler JD, Montagud C *et al*. (2015) Validation of vertical ground heat exchanger design methodologies. *Science and Technology for the Built Environment* **21(2)**: 137–149, http://dx.doi.org/10.1080/10789669.2014.974478.

Fossa M (2011) The temperature penalty approach to the design of borehole heat exchangers for heat pump applications. *Energy and Buildings* **43(6)**: 1473–1479, http://dx.doi.org/10.1016/j.enbuild.2011.02.020.

Fossa M and Rolando D (2015) Improving the Ashrae method for vertical geothermal borefield design. *Energy and Buildings* **93**: 315–323, http://dx.doi.org/10.1016/j.enbuild.2015.02.008.

Garber D, Choudhary R and Soga K (2013) Risk based lifetime costs assessment of a ground source heat pump (GSHP) system design: methodology and case study. *Building and Environment* **60**: 66–80, http://dx.doi.org/10.1016/j.buildenv.2012.11.011.

Haldar A and Mahadevan S (2000) *Probability, Reliability, and Statistical Methods in Engineering Design*. Wiley, New York, NY, USA.

IGSHPA (International Ground Source Heat Pump Association) (2000) *Grouting for Vertical Geothermal Heat Pump Systems: Engineering Design and Field Procedures Manual*. Oklahoma State University, Stillwater, OK, USA.

IGSHPA (2009) *Ground Source Heat Pump Residential and Light Commercial Design and Installation Guide*. Oklahoma State University, Stillwater, OK, USA.

Johnston IW, Narsilio GA and Colls S (2011) Emerging geothermal energy technologies. *KSCE Journal of Civil Engineering* **15(4)**: 643–653, http://dx.doi.org/10.1007/s12205-011-0005-7.

Kavanaugh SP and Rafferty KD (1997) *Ground-source Heat Pumps: Design of Geothermal Systems for Commercial and Institutional Buildings*. ASHRAE, Atlanta, GA, USA.

Matsumoto M and Nishimura T (1998) Mersenne twister: a 623-dimensionally equidistributed uniform pseudo-random number generator. *ACM Transactions on Modeling and Computer Simulation* **8(1)**: 3–30, http://dx.doi.org/10.1145/272991.272995.

Mikhaylova O, Johnston IW, Narsilio GA *et al.* (2015) Performance of borehole ground heat exchangers under thermal loads from a school building: full-scale experiment in Melbourne, Australia. *Proceedings World Geothermal Congress 2015*. Melbourne, Australia, pp. 19–25.

Modarres M (2006) *Risk Analysis in Engineering: Techniques, Tools, and Trends*. CRC Press, Boca Raton, FL, USA.

Philippe M, Bernier M and Marchio D (2010) Sizing calculation spreadsheet vertical geothermal borefields. *ASHRAE Journal* **52(7)**: 20–28.

Spitler JD and Gehlin SEA (2015) Thermal response testing for ground source heat pump systems – an historical review. *Renewable and Sustainable Energy Reviews* **50**: 1125–1137, http://dx.doi.org/10.1016/j.rser.2015.05.061.

Wang B, Bouazza A, Barry-macaulay D *et al.* (2012) Field and laboratory investigation of a heat exchanger pile. *Proceedings of the Geo-Congress*, American Society of Civil Engineers, Reston, VA, USA, pp. 4396–4405.

Yang H, Cui P and Fang Z (2010) Vertical-borehole ground-coupled heat pumps: a review of models and systems. *Applied Energy* **87(1)**: 16–27, http://dx.doi.org/10.1016/j.apenergy.2009.04.038.

Zervantonakis I and Reuss M (2006) Quality requirements of a thermal response t. *10th International Conference on Thermal Storage – Ecostock 2006: Thermal Energy Storage Here and Now*, pp. 435–441.

Craig and Gavin
ISBN 978-0-7277-6398-3
https://doi.org/10.1680/gehesep.63983.099
ICE Publishing: All rights reserved

Chapter 6
The role of ground conditions on energy tunnels' heat exchange

Alice Di Donna PhD
Dipartimento di Ingegneria Strutturale, Edile e Geotecnica, Politecnico di Torino, Turin, Italy
(corresponding author: alice.didonna@polito.it)

Marco Barla PhD
Confirmed Assistant Professor, Dipartimento di Ingegneria Strutturale, Edile e Geotecnica, Politecnico di Torino, Turin, Italy

Geotechnical structures are increasingly employed, in Europe as all around the world, to exchange heat with the ground and supply thermal energy for heating and cooling of buildings and de-icing of infrastructure. Most current practical applications are related to energy piles and retaining walls, but some examples of energy tunnels have been proposed recently. With respect to building foundations, tunnels involve a larger volume of ground and surface for heat exchange. When mechanised tunnelling is used, tunnel lining segments can be equipped and optimised for heat exchange by including hydraulic circuits into the concrete. This chapter deals with the use of a tunnel segmental lining for exploiting heat. Thermo-hydro finite-element analyses able to capture the key aspects of the problem are performed to investigate the heat exchange of the system under different underground scenarios. In particular, the influence of ground properties (hydraulic and thermal conductivities) and ground conditions (groundwater temperature and flow velocity) are investigated and discussed. Parametric analyses are employed to define design charts useful for the preliminary quantification of exploitable heat and the assessment of the applicability to different ground conditions.

Notation

A	effective pipe section area
c_w	water heat capacity
c_s	solid phase heat capacity
D	pipe diameter
∂_t	time derivative
$\mathbf{g}_i$	gravity vector
h	hydraulic head
$\mathbf{k}_{ij}$	intrinsic hydraulic conductivity tensor
k_x	intrinsic hydraulic conductivity along x
k_y	intrinsic hydraulic conductivity along y
k_z	intrinsic hydraulic conductivity along z
m	mass flow rate

99

n	porosity
p	pressure
Q	heat
$\mathbf{q}_{f,i}$	fluid flux vector
R	pipe radius
r_{hyd}	hydraulic radius
S	specific storage coefficient
s	pipe thickness
T	temperature
T_{wi}	inlet temperature
T_{wo}	outlet temperature
$v_{f,i}$	Darcy's velocity
$v_{p,\hat{z}}$	pipe water velocity
$v_{w,i}$	vector of water velocity
x, y, z	Cartesian coordinates
$\hat{z}$	axial pipe coordinate
α_L	longitudinal thermal dispersivity
α_T	transverse thermal dispersivity
β_w	water thermal expansion coefficient
Y_w	water compressibility
Y_s	solid compressibility
δ_{ij}	Kronecker delta
λ_{ij}	heat conduction and dispersion tensor
λ_s	solid phase thermal conductivity
λ_w	water thermal conductivity
μ	water dynamic viscosity
ρ_w	water density
ρ_s	solid phase density
∇	gradient
$\nabla \cdot$	divergence

Introduction

Underground geotechnical structures, such as deep and shallow foundations, diaphragm walls, tunnel linings and anchors, are increasingly employed, in Europe as all around the world, to exchange heat with the ground and supply thermal energy for heating and cooling of buildings and de-icing of infrastructure (Laloui and Di Donna, 2013). The thermal activation is achieved by installing absorber pipes in the geostructures, in which a circulating fluid extracts or injects heat from or into the ground. These systems belong to the category of low-enthalpy geothermal plants and are combined with heat pumps. Several practical applications of this technology are already operational, particularly in Austria, Germany, the UK and Switzerland (Adam, 2008; Bourne-Webb, 2013; Brandl, 2006; Pahud, 2013; Riederer et al., 2007; SIA, 2005). A recently growing interest in the application of this technology to tunnel linings has resulted in several scientific works devoted to investigating the feasibility and efficiency of such systems (Barla and Perino, 2014a, 2014b; Barla et al., 2016; Dupray et al., 2013; Franzius and Pralle, 2011; Lee et al., 2012; Markiewicz and Adam, 2009; Nicholson et al., 2013; Zhang et al., 2013).

With respect to building foundations, energy tunnels have the advantage of involving a larger volume of ground and surface for heat exchange. Nevertheless, the cases of real implementation of energy tunnels are limited, for the moment, to Austria and Germany (Franzius and Pralle, 2011; Markiewicz and Adam, 2009; Schneider and Moormann, 2010). This is probably because the initial costs of installation related to the effective energy advantages in the operational phase, which strongly depend on the specific site, are not yet completely assessed. To facilitate this, a preliminary design procedure is proposed in this chapter. It is based on design charts developed numerically and validated against available experimental data and can be used for preliminary assessment of the heat extracted or injected in different ground conditions.

The energy tunnel technological aspects will be briefly described in the following. Then this chapter will focus on the parametric numerical analyses performed to develop the preliminary design procedure.

Energy tunnels

A schematic representation of the hydraulic circuit that can be installed in a tunnel segmental lining to allow for heat exchange is presented in Figure 1. The main conduit connects the ground heat exchanger, installed into the lining, with the heat pump. The pipe circuit embedded in each lining segment is linked to those of the adjacent ones by hydraulic connections to form lining ring circuits. Each ring is usually made of five to seven segments. Two or more rings can be hydraulically connected in parallel, forming a secondary circuit, and each secondary circuit is then connected to the main conduit. Connecting in parallel two or more rings allows one to isolate one of them in case of malfunctioning without compromising the

Figure 1 Schematic representation of a tunnel segmental lining equipped with the ground heat exchanger system (Barla and Perino, 2014a)

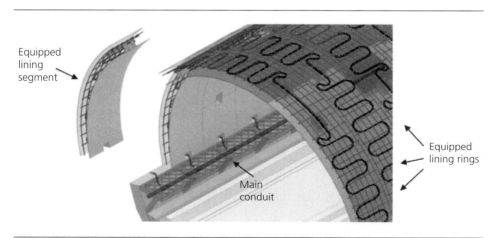

use of the entire plant and reduces the number of connections on the main conduit and the consequent significant head losses. A proper design of an energy tunnel should thus include a number of different aspects, among which the most important are

- the design of the main plant, including the heat pump and the main hydraulic conduit
- the dimensioning of the secondary plants – that is, the secondary conduits which link two or more rings in parallel and are connected to the main conduit
- the dimension, position, and number of the pipes to be installed in each tunnel lining segment
- the operational fluid temperature and pressure.

All these aspects being fixed, the efficiency of the system strongly depends on the heat exchange potential of the specific case – that is, on the specific site conditions. The most influential factors are the thermal and hydraulic properties of the subsoil and the presence of groundwater flow. In the following, the efficiency of an energy tunnel working under different ground conditions and properties is investigated through numerical analyses, assuming a fixed geometry and pipe configuration, typical of real tunnel applications.

Set-up of the numerical model

The preliminary assessment of the heat extracted or injected in different ground conditions is a key aspect of evaluating the feasibility of energy tunnels for district heating and cooling. To provide hints and assist city planners, a parametric numerical study was performed by adopting the thermo-hydro mathematical formulation implemented in the finite-element software Feflow (Diersch, 2009), which proved to be effective in this respect (Barla *et al.*, 2016).

Mathematical formulation

The thermo-hydro problem is governed by the following equations, written in the Eulerian coordinate system for a saturated medium composed of a solid skeleton and a liquid (water) phase.

- The mass conservation equation

$$S\partial_t p - n\beta_w \partial_t T + \nabla \cdot (nv_{w,i}) - nv_{w,i}\beta_w \nabla T = 0 \qquad (1)$$

where ∂_t, $\nabla \cdot$ and ∇ denote the time derivative, the divergence and the gradient operators respectively; $S = [nY_w + (1-n)Y_s]$ is the specific storage coefficient; n is the porosity; Y_w and Y_s are the water and solid compressibilities respectively; p is the pressure; β_w is the water thermal expansion coefficient; T is the temperature; and $v_{w,i}$ is the vector of water velocity with respect to the solid skeleton

- Darcy's velocity ($v_{f,i}$) law

$$v_{f,i} = nv_{w,i} = -\frac{\boldsymbol{k}_{ij}\rho_w g_i}{\mu}\nabla h = -\boldsymbol{k}_{ij}\nabla h \qquad (2)$$

where k_{ij} is the intrinsic hydraulic conductivity tensor (expressed in square metres), ρ_w is the water density, g_i is the gravity vector, μ is the water dynamic viscosity and h is the hydraulic head defined as

$$h = \frac{p}{\rho_w g_i} + z \qquad (3)$$

where z is the vertical coordinate

- The energy conservation equation

$$[n\rho_w c_w + (1-n)\rho_s c_s]\partial_t T + n\rho_w c_w v_{w,i}\nabla T - \nabla \cdot (\lambda_{ij}\nabla T) = 0 \qquad (4)$$

where c_w and c_s are the water and solid phase heat capacities and ρ_s is the solid phase density. The term λ_{ij} includes the heat conduction and the dispersion components as

$$\lambda_{ij} = [n\lambda_w + (1-n)\lambda_s]\delta_{ij} + \rho_w c_w \left[\alpha_T \sqrt{q_{f,i} q_{f,j}}\delta_{ij} + (\alpha_L - \alpha_T)\frac{q_{f,i} q_{f,j}}{\sqrt{q_{f,i} q_{f,j}}}\right] \qquad (5)$$

where λ_w and λ_s are the water and solid phase thermal conductivities, δ_{ij} is the Kronecker delta, α_T and α_L are the transverse and longitudinal thermo dispersivities respectively and $q_{f,i}$ is the fluid flux along direction i.

One-dimensional discrete features elements provided in Feflow were used to simulate the absorber pipes installed in the tunnel lining. The use of these elements to simulate pipes in geothermal systems has been validated and showed good agreement with confirmed analytical solutions (Diersch, 2009). The mass and energy conservation equations described above (Equations 1 and 4) are satisfied also for these elements, while the fluid flow inside them is described by the Hagen-Poiseuille law. Accordingly, fluid particles are assumed to move in pure translation with constant velocity, as occurs in circular tubes. The water velocity $v_{p,\hat{z}}$ along the axial direction $\hat{z}$ of the 1D elements obtained by integration of the equilibrium equation, and averaged over the circular section of a tube, is

$$v_{p,\hat{z}} = -\frac{r_{hyd}^2}{2\mu}\left(\frac{dp}{d\hat{z}} - \rho g\right) \qquad (6)$$

where r_{hyd} is the hydraulic radius defined as the ratio between the flow area and the wetted perimeter – that is, for a circular pipe of radius R

$$r_{hyd} = \frac{R}{2} \qquad (7)$$

Finite-element model geometry and characteristics

The three-dimensional finite-element model used for the purpose of this chapter is based on the analyses performed by Barla et al. (2016) and assumes that the energy tunnel hydraulic plant is composed of a series of secondary circuits made of five rings of segmental lining connected in parallel. Each block of five rings is then connected to the main inlet and outlet conduits. With reference to the case study of the south extension of Metro Torino line 1, Barla *et al.* (2016) performed an optimisation analysis using different numerical models and concluded that a 3D model of one ring of tunnel lining would properly reproduce the thermo-hydro behaviour of the entire tunnel, reducing the complexity of the model with respect to solutions involving longer portions of tunnel. This assumption was reasonable in the case mentioned above, because the groundwater flow is perpendicular to the tunnel axis, so adjacent rings do not influence each other from a thermal point of view (Barla *et al.*, 2016). Accordingly, the model used in this chapter makes the same assumptions and is composed of 648 999 nodes, 1 250 928 triangular prismatic six-node elements and 2465 1D elements to simulate the pipes. The model is 77·85 m high, 120 m large and 1·4 m wide. The centre of the tunnel is 21·5 m from the ground surface, the excavation has a diameter of 6·8 m and the lining is 30 cm thick. The vertical cross-section and a perspective view of the finite-element mesh are shown in Figure 2. To consider the influence of the air inside the tunnel, a 30-cm-thick limit layer was added inside the concrete lining – that is, a layer where the temperature varies from that of the tunnel lining to that measured inside the tunnel. The latter is known according to the monitoring data acquired in the Metro Torino tunnel from June 2010 to May 2011, as described by Barla *et al.* (2016), and is imposed on the internal nodes of the air layer as a Cauchy boundary condition. One-dimensional discrete feature elements are used to simulate the pipes in the tunnel segmental lining. They are located close to the external boundary of the lining (at 10 cm from the interface with the ground) and with a spacing of 30 cm. The pipes have an external diameter of

Figure 2 Geometry and dimension of the 3D model of the Metro Torino line tunnel

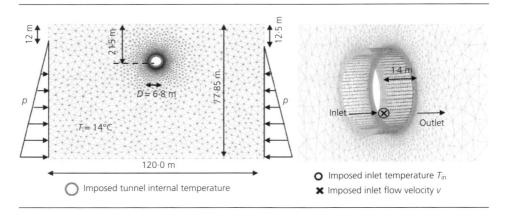

Table 1 Pipe properties

External diameter D: mm	25
Tube thickness s: mm	2·3
Effective section A: mm^2	326·8
Hydraulic radius r_{hyd}: mm	5·1

25 mm and are 2·3 mm thick (Table 1). They are represented in Figure 3, where the inner pipe surplus material is required to exchange heat between the 1D element (represented at a single node) and the pipe wall. The material parameters of the inner pipe surplus, tubes, concrete lining and limit layer of the air inside the tunnel are listed in Table 2. The initial temperature is equally fixed for the whole domain, while the position of the groundwater table is obtained by initialising the model with the values of hydraulic head on the left and right sides of the mesh and running the model to equilibrium. The operational conditions of the geothermal system are simulated by circulating fluid in the pipes at given inlet temperature and velocity. According to Barla *et al.* (2016), the inlet temperatures are assumed equal to 4 and 28°C for winter (heating mode) and summer (cooling mode) respectively (Dirichlet boundary conditions). The imposed inlet fluid velocity (Neumann boundary condition) is equal to 0·4 m/s, as resulting from the optimisation process presented by Barla *et al.* (2016).

Figure 3 Representation of the heat exchanger pipes in the tunnel lining

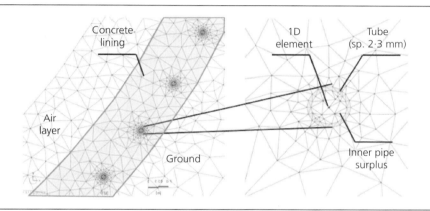

Table 2 Material properties

Property	Pipe surplus	Pipes	Concrete	Internal air
Horizontal hydraulic conductivity $k_x = k_z$: m/s	10^{-4}	10^{-16}	10^{-16}	10^{-2}
Vertical hydraulic conductivity k_y: m/s	10^{-4}	10^{-16}	10^{-16}	10^{-2}
Specific storage coefficient S: m^{-1}	10^{-4}	10^{-4}	10^{-4}	10^{-4}
Porosity n	0·0	0·0	0·0	1·0
Heat capacity of the fluid, $\rho_w c_w$: MJ/(m^3 K)	—	—	—	10^{-3}
Heat capacity of the solid, $\rho_s c_s$: MJ/(m^3 K)	10^{-18}	2·16	2·19	—
Thermal conductivity of the fluid, λ_w: W/(m K)	—	—	—	0·53
Thermal conductivity of the solid, λ_s: W/(m K)	1000	0·38	2·3	—
Longitudinal dispersivity a_L: m	—	—	—	5
Transverse dispersivity a_T: m	—	—	—	0·5

Parametric numerical analyses for various ground conditions

The model described previously was run at first by imposing the subsoil conditions pertinent to the south extension of Metro Torino line 1, which is considered as a reference case. Later, different subsoil conditions were tested by running parametric analyses varying the subsoil initial temperature, the groundwater flow velocity or the ground thermal conductivity. The model geometry, the inlet temperature and the inflow fluid velocity were held true for all the numerical simulations performed. Only the case of groundwater flow perpendicular to the tunnel axis was considered.

For the reference case, the initial soil temperature is equal to 14°C, the groundwater flow velocity is 1·5 m/d and the soil particles' thermal conductivity is 2·8 W/mK. The groundwater flow velocity was obtained by imposing a hydraulic head gradient equal to 0·49 m between the two lateral boundaries of the numerical model and applying Darcy's law (Equation 2). The soil thermo-hydro properties are listed in Table 3 (Barla and Barla, 2012; Barla et al., 2013; Bottino and Civita, 1986; Civita and Pizzo, 2001).

The heat Q (expressed in watts) extractable during winter and injectable during summer is computed by using the following equation

$$Q = mc_w |T_{wo} - T_{wi}| \qquad (8)$$

where m is the mass flow rate expressed in kilograms per second, T_{wi} is the inlet (imposed) temperature of the pipe circuit and T_{wo} is the outlet temperature (result of the numerical simulation). The imposed inlet and computed outlet temperatures for the reference case (fluid velocity of 0·4 m/s) are presented in Figures 4(a) and 4(b), for 1 month simulations in winter and summer conditions respectively. The resulting difference in temperature is about 3°C in the winter mode and 4°C in the summer mode. The extractable heat during winter is 1·7 kW and the transferable heat during summer is 2·3 kW. The efficiency in summer is higher due to the assumed higher difference between the inlet and soil temperatures with respect to the winter

Table 3 Soil thermo-hydro properties for the reference case

Property	Ground
Horizontal hydraulic conductivity $k_x = k_z$: m/s	$4·15 \times 10^{-3}$
Vertical hydraulic conductivity k_y: m/s	$2·075 \times 10^{-4}$
Specific storage coefficient S: m^{-1}	10^{-4}
Porosity n	0·25
Heat capacity of the fluid, $\rho_w c_w$: MJ/(m^3 K)	4·2
Heat capacity of the solid, $\rho_s c_s$: MJ/(m^3 K)	2·0
Thermal conductivity of the fluid, λ_w: W/(m K)	0·65
Thermal conductivity of the solid, λ_s: W/(m K)	2·8
Longitudinal dispersivity a_L: m	3·1
Transverse dispersivity a_T: m	0·3

Figure 4 Imposed inlet and computed outlet temperatures for (a) winter conditions and (b) summer conditions

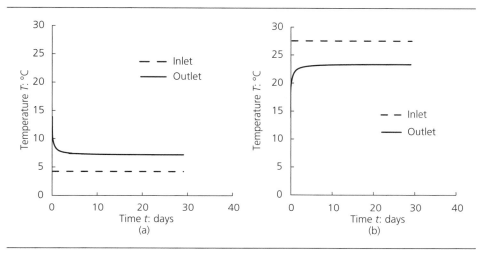

case. Normalising this result with respect to (*a*) the tunnel lining surface area and (*b*) the tunnel length, the results summarised in Table 4 are obtained. The resulting system is particularly favourable with respect to other case studies available in the literature (e.g. Franzius and Pralle, 2011). The main reason is the presence of the significant groundwater flow in the area, which allows for a continuous thermal recharge of the ground, significantly improving the heat extraction and injection efficiency.

To study the influence of the environmental conditions on the energy efficiency of the tunnel, numerical simulations were run with the same 3D finite-element model, varying the following three factors between reasonably realistic boundary values – that is, assuming the following.

- The ground temperature ranges between 8 and 18°C, typical of continental regions.
- The groundwater flow velocity varies between 0 and 2·0 m/d, imposed by applying different hydraulic head gradients between the two external boundaries of the numerical models. Consequently, the influence of hydraulic conductivity is indirectly taken into account, according to Darcy's law.
- The total thermal conductivity varies between 0·9 and 3·9 W/mK, which correspond to solid particle thermal conductivities of between 1 and 5 W/mK.

Table 4 Extracted/injected heat in winter and summer for the reference case

Season	Total extracted/injected power Q: kW	Extracted/injected power Q per square metre: W/m^2	Extracted/injected power Q per metre: W/m
Winter	1·67	52·76	1193·42
Summer	2·34	73·87	1670·81

All the other soil thermo-hydro properties are assumed to be equal to those measured in Torino (Table 3).

The effects of the three investigated parameters are discussed and quantified in the following sections.

Influence of ground temperature

Based on the results of the parametric analyses, Figures 5(a) and 5(b) illustrate the effect of ground temperature on the energy performance of the system in heating (winter) and cooling (summer) modes respectively. In winter, the efficiency increases linearly with the increase in ground temperature, as the difference between the inlet and ground temperatures increases. In summer, the opposite behaviour is shown. Irrespective of the groundwater flow and soil thermal conductivity, the efficiency of the system increases by about 25% of the initial extracted heat power (for $T = 8°C$) per degree Celsius of the soil temperature in winter and decreases by about 5% per degree Celsius in summer.

Influence of the groundwater flow velocity

Figures 6(a) and 6(b) illustrate the effect of groundwater flow velocity on the energy performance of the system for winter and summer modes respectively. As already mentioned, only the case with groundwater flow perpendicular to the tunnel axis was considered. In both cases the exchanged heat increases in a non-linear way when the groundwater flow velocity increases. For the same soil thermal conductivity, the increase in extracted/injected heat with

Figure 5 Effect of temperature on the energy performance (the numbers near the curves indicate groundwater flow velocities (gwf)): (a) winter mode (experimental data from Franzius and Pralle, 2011) and (b) summer mode (experimental data from Lee et al., 2012)

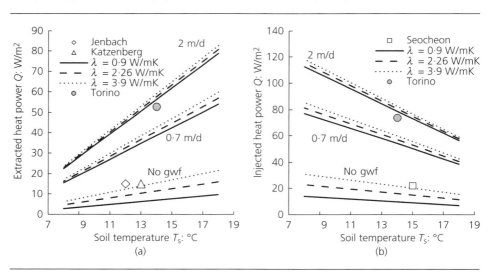

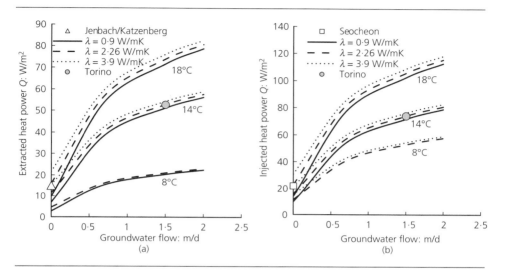

Figure 6 Effect of groundwater flow on the energy performance (the numbers near the curves indicate T's): (a) winter mode (experimental data from Franzius and Pralle, 2011) and (b) summer mode (experimental data from Lee et al., 2012)

increasing groundwater flow velocity with respect to the initial value (i.e. groundwater flow velocity = 0) is the same, irrespective of the ground temperature. Conversely, if for a soil thermal conductivity of 5 W/mK the extracted/injected heat for a groundwater flow velocity of 2 m/d is about four times that measured without groundwater flow, for a soil thermal conductivity of 1 W/mK it is about eight times. In other words, when the conductive component (thermal conductivity) of the heat exchange process becomes more important, the effect of the velocity of the groundwater flow (convection) becomes less significant.

Influence of thermal conductivity

Figures 7(a) and 7(b) illustrate the effect of soil thermal conductivity on the energy performance of the system for winter and summer modes respectively. In both cases the exchanged heat increases linearly with increasing thermal conductivity, but this effect is definitely more evident in the case of no water flow and it decreases with increasing groundwater flow velocity, irrespective of ground temperature. The extracted/injected heat increases by about 25% with respect to the initial value (i.e. for $\lambda = 0.9$ W/mK), per unit increment of thermal conductivity (W/mK) if no water flow is present and only by 1% in the case of 2 m/d groundwater flow velocity. This is in agreement with what was previously discussed on combination of convection and conduction on the heat exchange process.

Comparison with monitoring data

To provide validation and consistency with the results obtained, Figures 5, 6 and 7 show a comparison with the few monitoring data available in the literature for real applications of energy tunnels.

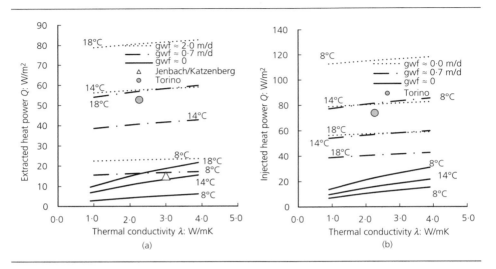

Figure 7 Effect of soil thermal conductivity on the energy performance (the numbers near the curves indicate T's): (a) winter mode (experimental data from Franzius and Pralle, 2011) and (b) summer mode (experimental data from Lee et al., 2012)

For the winter case, the monitoring data presented by Franzius and Pralle (2011) for the real energy tunnel of Jenbach and the German test field of Katzenberg were considered. According to these authors, the soil thermal conductivity on both sites was around 3 W/mK, while the ground temperatures were about 12 and 13°C for the Jenbach tunnel and the German field test respectively. In the case of the German field test, the soil hydraulic conductivity was relatively small, so it is reasonable to assume no groundwater flow. Conversely, at the Jenbach tunnel site, the groundwater flow was present but nearly parallel to the tunnel axis. However, the additional heat exchange due to the groundwater flow in the parallel direction to the tunnel axis and the related influence between the lining rings is not taken into account in the numerical modelling. Therefore, for the purpose of comparison, the groundwater flow is assumed null in Figure 6 for both case studies. The monitoring data measured by Franzius and Pralle (2011) indicate an extracted heat typically in the range of 10–20 W/m², which corresponds to the model prediction in a satisfactory way.

With reference to summer conditions, the data presented by Lee et al. (2012) for the Seocheon tunnel in South Korea were included in Figures 5(b), 6(b) and 7(b). These authors carried out a series of thermal response tests in the appropriate positions. They considered the groundwater to be static and the soil temperature to be equal to 15°C. The inlet temperature was around 30°C and the inlet fluid velocity was 0·1 m/s. Information concerning the soil thermal conductivity is not available. According to the measurements of Lee et al., the average injected heat was about 22 W/m², which satisfactorily agrees with the numerical results presented in this chapter.

Preliminary design charts

Based on the analyses presented so far, the heat exchange increases with increasing ground thermal conductivity and groundwater flow velocity, in both winter and summer. Additionally, when the system is used for heating (winter mode) the higher the ground temperature the higher the heat exchange, while when the system is used for cooling (summer mode) the opposite is true. Considering these findings, two design charts, one for heating (winter) mode and one for summer (cooling) mode, were developed and are presented in Figures 8 and 9 respectively. On the basis of the specific site conditions – that is, ground temperature, soil thermal conductivity and groundwater flow velocity – the charts can give the designer a reasonable idea of the heat that can potentially be exchanged from the ground in winter and injected during summer, expressed in watts per square metre of tunnel lining.

These charts were developed with the numerical model presented so far, so they are valid only for groundwater flow perpendicular to the tunnel axis and for the assumed geometry, inlet temperature and fluid velocity, which are reasonable and typical of energy tunnel applications. Nevertheless, the final results may differ for different assumptions. In particular, a groundwater flow along a direction not perpendicular to the tunnel axis would probably result in a reduced heat exchange for the same groundwater velocity. Moreover, in such cases the heat exchange of one ring is likely to be influenced by the adjacent ones. As a consequence, the design charts would provide optimistic results in such conditions. This aspect will be considered in future developments.

Figure 8 Design chart for winter mode (extracted heat in watts per square metre) (experimental data from Franzius and Pralle, 2011 and Lee et al., 2012)

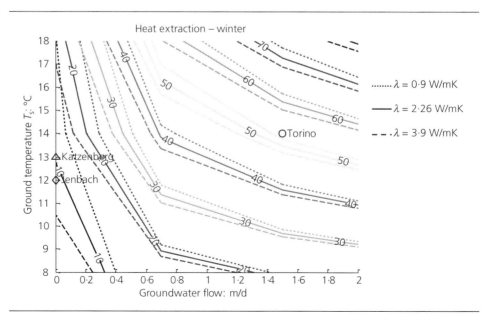

Figure 9 Design chart for summer mode (heat injection in watts per square metre) (experimental data from Franzius and Pralle, 2011, and Lee *et al.*, 2012)

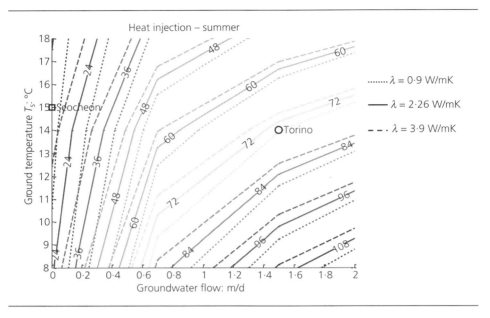

The winter chart (Figure 8) outlines that the most favourable condition for heat exchange is when the groundwater flow velocity and ground temperature are maximum. As previously discussed, the effect of soil thermal conductivity is more significant in the absence of groundwater flow, while it becomes almost negligible with a groundwater flow of up to 2 m/d. Reasonable values of heat extraction range between 10 and 70 W/m², with maximum values reached only in the presence of significant groundwater flow.

The summer chart (Figure 9) outlines that the most favourable condition for heat injection is when the groundwater flow velocity is maximum and the ground temperature is minimum. The heat exchange in summer is higher due to the assumed higher difference between inlet and soil temperatures with respect to the winter case, ranging between 10 and 100 W/m² of injected heat. It is to be mentioned that reference was made to the continental climate region. Charts may not be applicable for other climatic regions. Moreover, even if a higher groundwater flow velocity definitely improved the heat exchange potential, it results in soil temperature variations that propagate with the groundwater flow. This could threaten the underground thermal equilibrium and therefore deserves to be accurately considered.

The design charts also give a visual idea of the effect of soil thermal conductivity: for a fixed ground temperature a higher groundwater flow velocity is necessary to reach the same value of extracted/injected heat when the soil thermal conductivity decreases. The experimental data discussed so far (Franzius and Pralle, 2011; Lee *et al.*, 2012) were also plotted in these graphs for comparison and validation. In the case of Metro Torino line 1 (Barla *et al.*, 2016), the design charts would indicate a potential heat extraction of between 50 and 55 W/m^2 in winter and a potential heat injection of about 70 W/m^2 in summer (Figures 8 and 9). When accurate site data are available, these can also give an indication of the preferable location for the installation of a thermo-active tunnel.

Conclusions

Finite-element numerical analyses were performed to investigate and quantify the effect of groundwater flow velocity, soil temperature and thermal conductivity on the heat exchange potential of energy tunnels. The results show the following.

- In the absence of groundwater flow, when the heat exchange occurs essentially by conduction, the soil thermal conductivity plays a primary role and the exchanged heat is more than doubled, passing from 0·9 to 3·9 W/mK, for both summer and winter operational modes.
- When groundwater flow is present, the heat exchange results from a combination of conduction and convection, and the most influential factor becomes the intensity of groundwater flow. With an increase from 0 to 2 m/d, the exchanged heat improves by a factor of 3 to 8. However, in the case of high groundwater flow velocity, the risk of thermal pollution should be assessed.
- The heat exchange improves by about 7% per degree Celsius of ground temperature, being more favourable in hot regions in winter and in cold regions in summer.

The obtained results were shown to be in good agreement with monitoring data available in literature. To assist designers and city planners in evaluating the potential of heat exploitation from energy tunnels in relation to different site conditions, two design charts are presented. They can be used for a preliminary assessment of the potential energy exploitation in both summer and winter conditions, on the basis of the specific site conditions where the system has to be installed.

Acknowledgements

The authors are thankful to Professor Jan Franzius for the information provided on the case studies of Jenbach and Katzenberg, which permitted the comparisons shown in this chapter.

REFERENCES

Adam D (2008) Tunnels and foundations as energy sources – practical applications in Austria. In *5th International Symposium on Deep Foundations on Bored and Auger Piles (BAP V)* (Van Impe WF and Van Impe PO (eds)). CRC Press, Boca Raton, FL, USA pp. 337–342.

Barla M and Barla G (2012) Torino subsoil characterisation by combining site investigations and numerical modelling. *Geomechanics and Tunnelling* **5(3)**: 214–231, http://dx.doi.org/10.1680/geot.201200008.

Barla M and Perino A (2014a) Energy from geo-structures: a topic of growing interest. *Environmental Geotechnics* **2(1)**: 3–7, http://dx.doi.org/10.1680/envgeo.13.00106.

Barla M and Perino A (2014b) Geothermal heat front the Turin metro south extension tunnels. *Proceedings of the World Tunnel Congress 2014: Tunnels for a Better Life Iguaçu, Brazil.*

Barla G, Antolini F, Barla M et al. (2013) *Analisi e Verifica delle Condizioni di Esercizio in Sicurezza del Palazzo Uffici Provinciali di Corso Inghilterra 7 tenuto conto del Centro*

Direzionale di Intesa Sanpaolo – Relazione sulle Prove di Emungimento e di Immissione. Politecnico di Torino Turin, Italy Report No. PTO03 (in Italian).

Barla M, Di Donna A and Perino A (2016) Application of energy tunnels to an urban environment. *Geothermics*, http://dx.doi.org/10.1016/j.geothermics.2016.01.014.

Bottino G and Civita M (1986) Engineering geological features and mapping of subsurface in the metropolitan area of Turin, North Italy. *5th International IAEG Congress, Buenos Aires, Argentina*, pp. 1741–1753.

Bourne-Webb PJ (2013) Observed response of energy geostructures. In *Energy Geostructures: Innovation in Underground Engineering* (Laloui L and Di Donna A (eds)). ISTE, London, UK and Wiley, Hoboken, NJ, USA, pp. 45–78.

Brandl H (2006) Energy foundations and other thermo-active ground structures. *Géotechnique* **56(2)**: 81–122, http://dx.doi.org/10.1680/geot.2006.56.2.81.

Civita M and Pizzo S (2001) L'evoluzione spazio-temporale del livello piezometrico dell'acquiferolibero nel sottosuolo di Torino. *GEAM* **38(4)**: 271–287 (in Italian).

Diersch HJG (2009) *DHI Wasy Software – Feflow 6.1 – Finite Element Subsurface Flow & Transport Simulation System: Reference Manual.* DHI Wasy, Berlin, Germany. See http://www.mikepoweredbydhi.com/products/feflow.

Dupray F, Mimouni T and Laloui L (2013) Alternative uses of heat-exchanger geostructures. In *Energy Geostructures: Innovation in Underground Engineering* (Laloui L and Di Donna A (eds)). ISTE London, UK and Wiley, Hoboken, NJ, USA, pp. 119–138.

Franzius JN and Pralle N (2011) Turning segmental tunnels into sources of renewable energy. *Proceedings of the ICE – Civil Engineering* **164(1)**: 35–40, http://dx.doi.org/10.1680/cien.2011.164.1.35.

Laloui L and Di Donna A (2013) *Energy Geostructures: Innovation in Underground Engineering.* ISTE, London, UK and Wiley, Hoboken, NJ, USA.

Lee C, Park S, Won J *et al.* (2012) Evaluation of thermal performance of energy textile installed in Tunnel. *Renewable Energy* **42**: 11–22, http://dx.doi.org/10.1016/j.renene.2011.09.031.

Markiewicz R and Adam D (2009) Energy from earth-coupled structures, foundations, tunnels and sewers. *Géotechnique* **59(3)**: 229–236, http://dx.doi.org/10.1680/geot.2009.59.3.229.

Nicholson DP, Chen Q, Pillai A and Chendorain M (2013) Developments in thermal piles and thermal tunnel lining for city scale GSHP systems. Thirty-Eighth Workshop on Geothermal Reservoir Engineering, Stanford University, CA, USA.

Pahud D (2013) A case study: the dock midfield of Zurich airport. In *Energy Geostructures: Innovation in Underground Engineering* (Laloui L and Di Donna A (eds)). ISTE, London, UK and Wiley, Hoboken, NJ, USA, pp. 281–295.

Riederer P, Evers G, Gourmez D *et al.* (2007) *COnception de FOndations GEothermiques*, Rapport d'étude n° ESE/ENR n° 07.044RS, Diffusion libre, CSTB, Département Energie Santé Environnement, Pôle ENergies Renouvelables. Paris, France(in French).

Schneider Ma and Moormann C (2010) GeoTU6 – a geothermal research project for tunnels. *Tunnel* **29(1)**: 14–21.

SIA (Société Suisse des Ingénieurs et des Architects) (2005) *SIA DO 190: Utilisation de la chaleur du sol par des ouvrages de fondation et de soutènement en béton – guide pour la conception, la réalisation et la maintenance* SIA, Zurich, Switzerland (in French).

Zhang G, Xia C, Sun M, Zou Y and Xiao S (2013) A new model and analytical solution for the heat conduction of tunnel lining ground heat exchangers. *Cold Regions Science and Technology* **88**: 59–66, http://dx.doi.org/10.1016/j.coldregions.2013.01.003.

Craig and Gavin
ISBN 978-0-7277-6398-3
https://doi.org/10.1680/gehesep.63983.115
ICE Publishing: All rights reserved

Chapter 7
Simulations of a photovoltaic-thermal ground source heat pump system

Kevin E. Varney MSc, PhD, MBCS
Research Student, University of Reading, Reading, UK

Maria M. Vahdati PhD, CEng, MIChemE
Lecturer, University of Reading, Reading, UK

A ground source heat pump assisted by an array of photovoltaic (PV)-thermal modules was studied in this work. Extracting heat from an array of PV modules should improve the performance of both the PV cells and the heat pump. A series of computer simulations compare the performance of a ground source heat pump with a short ground circuit, used to provide space heating and domestic hot water at a house in southern England. The results indicate that extracting heat from an array of PV-thermal modules would improve the performance of a ground source heat pump with an undersized ground loop. Nevertheless, open air thermal collectors could be more effective, especially during winter. In one model more electricity was saved in ohmic heating than was generated by cooling the PV cells. Cooling the PV modules was found to increase their electrical output up to 4%, but much of the extra electricity was consumed by the cooling pumps.

1. Introduction

There have been a number of studies into solar-assisted ground source heat pumps in the past. From five of these studies Varney and Vahdati (2013) concluded that the benefit of using solar thermal energy to improve the performance of a ground source heat pump was marginal. While the coefficient of performance (COP) was improved during certain months of the year, particularly in spring, the electricity saved in the heat pump's compressor was expended again by the additional circulation pumps required to pump glycol through the solar collectors. For example, while Bakker *et al.* (2005) improved the COP of their system from 2·60 to 2·66, Pahud and Lachal (2005) decreased the COP of theirs from 3·8 to 3·7.

Nevertheless, in a series of software simulations performed by Kjellsson *et al.* (2010), a solar collector was found to be useful to a ground source heat pump. With an adequately sized borehole, the solar collector provided about 15% of the source heat required. Moreover, when the borehole depth was reduced, the injection of solar heat was more beneficial. At 80% optimal borehole depth, solar heat injection enabled a 29% saving of electrical consumption, mostly in the form of ohmic heating.

The system described by Bakker et al. (2005) used photovoltaic (PV)-thermal modules to provide extra heat for the ground source heat pump. A PV-thermal-assisted ground source heat pump has the advantage that not only would the transferred heat usefully improve the performance of the heat pump, but the cooling effect on the PV-thermal panel would increase electricity generation.

In this study, software models were implemented to simulate a PV-thermal-assisted ground source heat pump. The model was based on a system at a test house in Oxfordshire, England, in which the heat pump was sized to provide 70% of the maximum expected heat demand. Figure 1 is a diagram of the test house's heating system. The ground circuit for collecting source heat for the heat pump was undersized, although a pair of wall-mounted ambient heat collectors (termed air panels) helped compensate. The models were undertaken to determine how well the addition of an array of PV-thermal modules could supplement the source heat to a ground source heat pump with a short ground circuit.

2. Model

The software tool chosen to model the PV-thermal/ground source heat pump system was TRNSYS version 17·0. The TRNSYS model shown in Figure 2 was simplified in two notable respects from the system shown in Figure 1: there is no bypass in the space heating circuit, and there is one series of six ground panels rather than two series of three.

Weather data for Bracknell, England was selected. Average daily solar radiation varied from 0·53 kWh/m^2 per day in December to 4·9 kWh/m^2 per day in July. Average monthly ambient temperatures varied from 4·4°C in February to 17·5°C in August (see Figure 3).

The house was modelled as a single zone, lumped capacity building (type 12c), which used the energy/degree-day concept. The overall loss coefficient was set at 220 W/K, while thermal capacitance was set at 75 000 kJ/K. The effectiveness C_{min} product for the space heating system was set at 225 W/K.

The heat pump was modelled as a type 927 water-to-water ground source heat pump. The brine was given the density and specific heat capacity of 35% mono-propylene glycol solution. The heat transfer fluid was set the same attributes as water. The rated heating capacity was set to 4 kW. The source flow rate was set at 0·3 l/s and the load flow rate at 0·2 l/s.

The underground pipe through which glycol flows from the heat pump to the manifold was modelled using a 952-auto type. However, the main part of the ground circuit was modelled using a type 997 horizontal ground heat exchanger, which was configured to represent six thermal collectors buried at a depth of 1 m, embedded in 200 mm of sand and covered over with 800 mm of clay soil. Each thermal collector comprised 18 2 m lengths of 40 mm high-density polyethylene pipe, based on a design by the Swedish manufacturer, IVT (IVT Industrier AB, 2005).

Figure 1 Heating system at the test house: NRV, non-return valve; PVT, photovoltaic-thermal

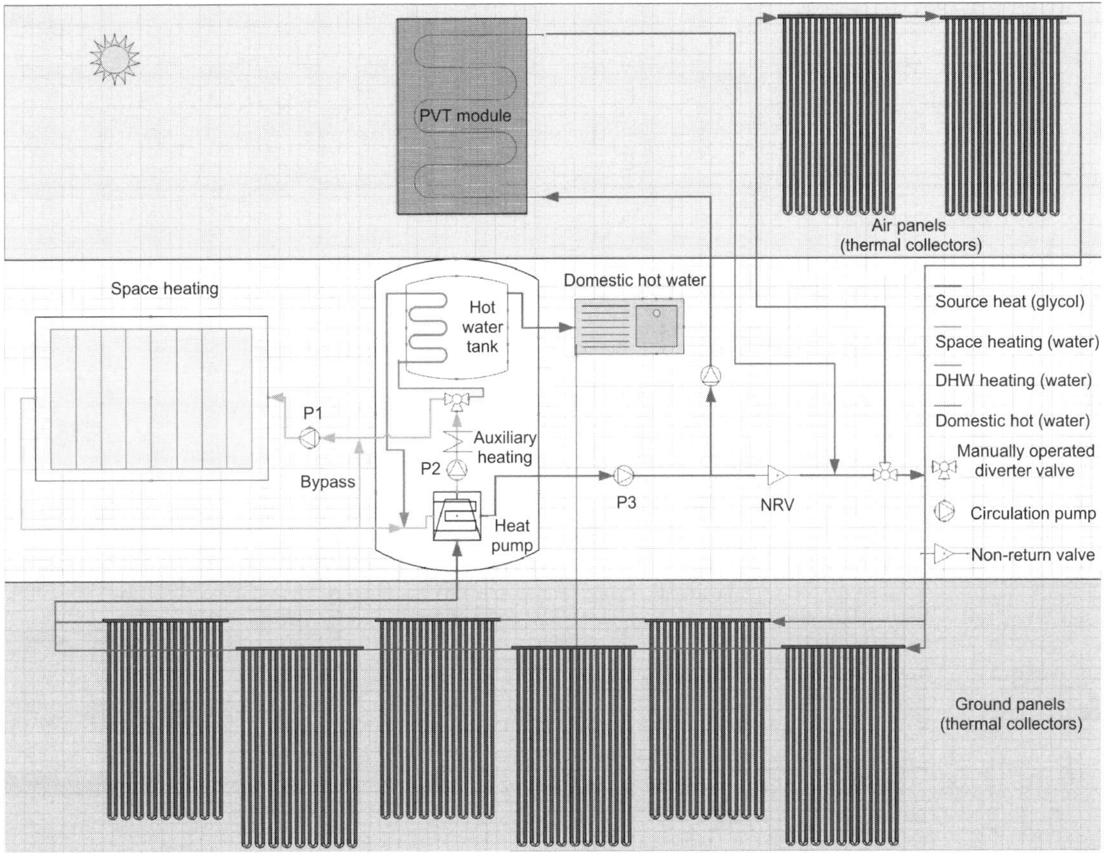

Figure 2 TRNSYS model of the test house: PV, photovoltaic; PVT, photovoltaic-thermal

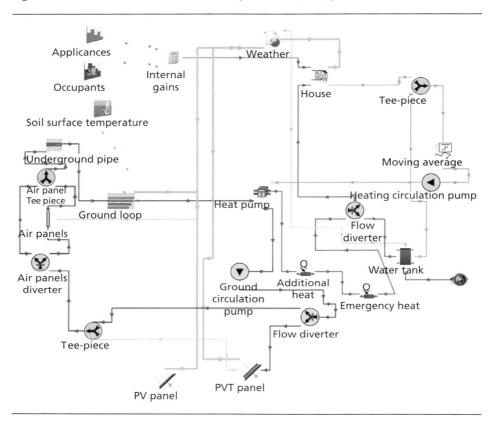

Figure 3 Ambient temperature and solar radiation at Bracknell

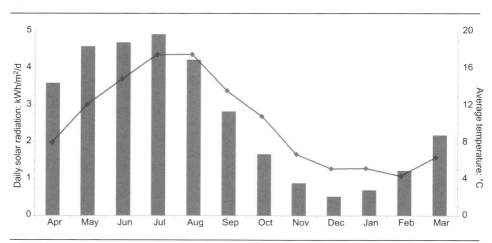

A type 560 component was configured to represent an array of 16 PV-thermal modules, with an area of 26·24 m². Each PV-thermal heat exchanger was modelled as 11 tubes of 6 mm dia. PV efficiency was set at 14·6% at standard test conditions. The efficiency modifier was set at −0·44%/K. A type 562 component was set to represent a similar sized array of Sharp ND series 240 W PV modules (Sharp Energy Solutions Europe, 2011), fixed on a south-facing roof with a 30° slope.

This study refers to air panels: these are ambient heat collectors attached to an outside wall, not in direct sunlight. They were modelled as a length of high-density polyethylene pipe equivalent to eight thermal collectors. The pipe length was set at 300 m with an inner diameter of 37 mm and an outer diameter of 40 mm. The loss coefficient was set at a constant 14 W/m²/K.

There were three user-defined types included in the model (not shown). The first controlled the operation of the heating system, including the heat pump and auxiliary heaters, the second controlled glycol flow through the PV-thermal heat exchanger, while the third controlled glycol flow through the air panels and the ground circuit.

The heat load was based on the house being inhabited by two occupants.

Five variants of the model were devised.

1. Control system: this system was designed to enable comparison with a ground source heat pump with a short ground circuit and no other sources of heat.
2. System with air panels: glycol passes through eight air panels before entering the ground circuit.
3. System with PV-thermal array (biased for electricity): glycol is pumped through the PV-thermal array when a 15 K difference between the glycol and PV surface is exceeded, when it may be assumed the sun is shining.
4. System with PV-thermal array (biased for source heat): glycol is pumped through the PV-thermal array when a 5 K difference between the glycol and PV surface is exceeded. In effect, this is a hybrid solar/ambient heat collector.
5. System with PV-thermal array (ground recharge): glycol pumped through PV surface exceeds glycol temperature by 15 K, even if the heat pump is not in operation.

3. Results

The simulations indicated a maximum space heating load of nearly 6 kW. The heat pump was not configured for cooling, but the room temperature never exceeded 23°C. The ground source heat pump controller unit used ohmic heating when necessary to maintain the indoor temperature at 19°C throughout the heating season.

The electrical outputs of the three systems with PV-thermal arrays are shown in Table 1.

The annual electrical output of an equivalent array of PV panels was 2936 kWh. The PV cooling pump was attributed a value of 25 W. The pressure loss imposed by the PV-thermal array was estimated at 24 kPa, which a 25 W pump, working at 25% efficiency, should overcome. In system 5, the electrical consumption of the ground circuit pump was included

Table 1 Annual increase in electrical output from cooling PV-thermal arrays

System	3	4	5
Temperature difference between PV surface and glycol, which triggers cooling: K	15	5	15
Electricity generated by PV-thermal array: kWh	3001	3015	3055
Electricity generated by cooling photovoltaic cells: kWh	65	79	119
Increase in electrical generation: %	2·2	2·7	4·1
Cooling pump's electrical consumption: kWh	−15	−31	−63
Overall increase in electrical generation	50	48	56
Percentage increase in electrical generation: %	1·7	1·6	0·5
PV cooling pump with GSHP operating: h	589	1222	1186
Ground circuit pump with GSHP not operating: h	0	0	558

GSHP, ground source heat pump; PV, photovoltaic

whenever it was switched on to assist the PV cooling pump when the heat pump was not in operation. The pressure loss imposed by the ground circuit was estimated at 48 kPa, therefore a 60 W pump was assumed. The eight air panels of system 2 would impose a pressure loss of 56 kPa, requiring a 125 W pump.

Table 2 lists the heat demands for the five systems. Space heating demand varied from approximately 120 kWh in August to over 2000 kWh in December and January, while the heat demand for domestic hot water remained constant at approximately 210 kWh per month. System 2 required the least ohmic heating over the year. Systems 2 and 5 required minimal amounts of ohmic heating during July and August owing to the glycol having exceeded the maximum heat pump inlet temperature.

Table 3 lists the heat pump's thermal and electrical inputs and outputs in the five systems. The heat pump assisted by air panels consumed the most electricity but emitted the most heat. The heat pump in the control system consumed the least electricity but emitted the least heat. Configuring the PV-thermal array of system 4 to collect more heat enabled the heat pump to output an additional 956 kWh of heat per year at a cost of an additional 253 kWh of electricity. Adjusting the PV-thermal controller to recharge ground heat resulted in the heat pump outputting 56 kWh less heat in the summer but 473 kWh more heat throughout the year.

Table 4 lists where each of the five systems collected source heat for the heat pump. The control system was the least able to extract source heat for the heat pump. A large proportion of source heat was extracted from the 25 m of underground pipe leading from the heat pump to the underground thermal collectors. System 2 extracted over twice as much source heat from its air panels than from the ground. The heat pump became increasingly reliant on the air panels for source heat as the heating season progressed. In system 3 the PV-thermal array was much less effective at sourcing heat during the winter than the air panels. The PV-thermal array

Table 2 Total heat load

Heat load: kWh

System			Month												Year
No.	Description	Heater	Apr	May	Jun	Jul	Aug	Sep	Oct	Nov	Dec	Jan	Feb	Mar	Total
1	Control	GSHP	862	956	636	369	327	768	1302	1676	1182	760	523	609	9969
		Ohmic	817	165	0	0	0	0	10	209	1062	1485	1534	1428	6711
2	Air panels	GSHP	1575	1115	614	374	325	772	1305	1801	1761	1676	1332	1617	14267
		Ohmic	98	14	0	2	4	0	6	94	477	559	734	438	2427
3	PV-thermal array (biased for electricity)	GSHP	1434	1106	620	376	328	771	1302	1746	1285	810	659	885	11324
		Ohmic	238	13	0	0	0	0	8	142	968	1414	1410	1161	5354
4	PV-thermal array (biased for heat)	GSHP	1560	1102	619	377	328	770	1302	1780	1524	960	841	1118	12283
		Ohmic	115	18	0	0	0	0	10	123	704	1286	1219	929	4404
5	PV-thermal array (ground recharge)	GSHP	1529	1115	604	356	309	762	1304	1756	1301	844	824	1037	11742
		Ohmic	146	16	6	23	20	7	9	133	935	1392	1251	1011	4951

GSHP, ground source heat pump; PV, photovoltaic

Table 3 GSHP energy inputs and outputs

GSHP: kWh

System			Month												Year
No.	Description		Apr	May	Jun	Jul	Aug	Sep	Oct	Nov	Dec	Jan	Feb	Mar	Total
1	Control	Heat out	860	956	636	370	328	766	1301	1673	1179	757	521	608	9955
		Heat in	579	652	454	274	246	565	923	1131	793	508	349	407	6880
		Compressor	281	304	182	96	82	201	378	541	387	249	172	201	3075
		Ground pump	16	17	10	5	4	11	21	30	22	14	10	11	172
		Other pumps	39	30	16	8	7	18	33	50	51	48	42	43	384
2	Air panels	Heat out	1571	1114	614	375	325	771	1303	1798	1759	1672	1329	1614	14245
		Heat in	1102	803	458	284	247	580	951	1261	1219	1152	914	1117	10088
		Compressor	469	311	156	91	78	191	351	537	540	520	415	497	4157
		Ground pump	33	27	16	11	8	12	22	32	32	32	27	33	284
		Other pumps	42	27	14	8	7	17	31	48	53	54	48	51	399
3	PV-thermal array (biased for electricity)	Heat out	1432	1105	621	377	330	770	1300	1744	1282	809	656	885	11312
		Heat in	993	788	455	282	249	578	940	1199	863	545	442	601	7934
		Compressor	439	317	166	95	81	192	360	545	419	264	214	284	3377
		Ground pump	25	18	9	5	4	11	20	31	24	15	12	16	189
		Other pumps	42	28	15	8	7	17	32	49	52	48	43	45	385
4	PV-thermal array (biased for heat)	Heat out	1557	1102	620	378	329	769	1301	1776	1523	958	839	1116	12268
		Heat in	1092	794	458	285	249	581	950	1233	1034	647	567	761	8651
		Compressor	464	308	162	94	80	188	351	543	489	311	272	355	3617
		Ground pump	26	17	9	5	4	10	20	31	28	18	15	20	203
		Other pumps	42	27	14	8	7	16	31	49	53	49	45	47	388
5	PV-thermal array (ground recharge)	Heat out	1526	1115	604	356	310	762	1302	1754	1298	843	821	1035	11728
		Heat in	1065	805	450	271	237	577	947	1205	874	570	556	707	8264
		Compressor	461	310	154	86	73	185	355	548	424	274	265	328	3463
		Ground pump	24	20	12	8	7	12	19	27	21	14	14	18	195
		Other pumps	42	27	14	8	7	16	31	49	52	48	45	47	385

GSHP, ground source heat pump; PV, photovoltaic

Table 4 Sources of heat

Source heat: kWh

System			Month												Year
No.	Description		Apr	May	Jun	Jul	Aug	Sep	Oct	Nov	Dec	Jan	Feb	Mar	Total
1	Control	Ground panels	436	461	318	185	167	409	674	772	586	414	310	345	5077
		Ground pipe	183	238	172	111	98	204	340	404	247	134	85	110	2326
		Air panels	0	0	0	0	0	0	0	0	0	0	0	0	0
		PV-thermal array	0	0	0	0	0	0	0	0	0	0	0	0	0
2	Air panels	Ground panels	151	−7	−38	−46	10	216	386	498	399	242	191	147	2147
		Ground pipe	41	22	10	14	33	118	198	253	155	49	15	6	913
		Air panels	908	787	485	315	204	246	366	509	664	860	707	963	7014
		PV-thermal array	0	0	0	0	0	0	0	0	0	0	0	0	0
3	PV-thermal array (biased for electricity)	Ground panels	248	169	160	106	113	253	483	651	559	378	267	266	3653
		Ground pipe	105	97	94	73	72	134	240	337	235	116	64	76	1644
		Air panels	0	0	0	0	0	0	0	0	0	0	0	0	0
		PV-thermal array	638	521	200	102	64	190	216	209	68	50	110	258	2628
4	PV-thermal array (biased for heat)	Ground panels	166	119	109	78	98	204	409	592	532	378	248	250	3182
		Ground pipe	62	70	69	59	64	111	203	301	227	115	55	66	1404
		Air panels	0	0	0	0	0	0	0	0	0	0	0	0	0
		PV-thermal array	863	604	279	148	88	265	336	338	273	153	264	444	4054
5	PV-thermal array (ground recharge)	Ground panels	176	−52	−106	−113	−50	112	425	647	553	347	245	208	2392
		Ground pipe	66	−2	−16	−14	10	78	212	336	234	104	49	48	1106
		Air panels	0	0	0	0	0	0	0	0	0	0	0	0	0
		PV-thermal array	822	859	571	398	276	386	309	221	86	117	261	450	4758

PV, photovoltaic

in system 4 was 54% more effective than system 3 at collecting heat, while the PV-thermal array in system 5 was 81% more effective. Ground thermal recharge became possible in two of the systems: 91 kWh in system 2 and 353 kWh in system 5.

4. Discussion
4.1 Seasonal performance factor
The performance factors for the TRNSYS heating systems are presented in Table 5. Four values for seasonal performance factor (SPF) were calculated following the definitions derived in the IEE Sepemo project, described by Malenković (2012). In each calculation the numerator is the heat output, and the denominator is the electric consumption.

$$SPF_{H1} = \frac{Condenser}{Compressor}$$

$$SPF_{H2} = \frac{Condenser}{Compressor + Ground\ pump}$$

$$SPF_{H3} = \frac{Condenser + Ohmic}{Compressor + Ground\ pump + Ohmic}$$

$$SPF_{H4} = \frac{Condenser + Ohmic}{Compressor + Ground\ pump + Ohmic + Other\ pumps}$$

Subtracting the additional electricity generated by cooling the PV-thermal arrays, after taking into account the cooling pump losses, only improved SPF_{H2} by between +0·04 and 0·06, and the SPF_{H4} values by only +0·01.

4.2 Electricity savings
The TRNSYS simulations indicate that the transfer of heat from the PV-thermal modules to the heat pump would be more beneficial for improving the performance of the heat pump than for increasing the electrical output of the PV cells. For example, transferring heat from the PV-thermal modules of system 4 enabled only an extra 48 kWh of electricity to be generated, but enabled the heat pump to output an extra 2314 kWh of heating, saving 1730 kWh electricity overall when compared to the control system. Despite not generating any electricity, the air panels in system 2 saved the heating system 3074 kWh of electricity.

4.3 Accuracy of simulations
The heating system at the test house, represented in Figure 1, was instrumented. Two flow meters measured the flow of warm water over the heat pump's condenser and around the space heating circuit, while another two flow meters measured glycol flow around the ground circuit and through the PV-thermal unit. Temperature sensors were placed on the inputs and output of the heat pump, the PV-thermal unit and the air panels. Electricity meters measured the electrical consumption of the heat pump's compressor and of the heating system as a whole. The heating system was logged for over a year. A TRNSYS model was based on the test house,

Table 5 Seasonal performance factors

System		Electricity: kWh					Heat: kWh			SPF			
No.	Description	GSHP compressor	Ohmic DHW	Ohmic space	Ground pump	Other pumps	GSHP condenser	Total DHW	Total space	SPF_{H1}	SPF_{H2}	SPF_{H3}	SPF_{H4}
1	Control	3075	786	5924	172	384	9969	2501	14 179	3·24	3·07	1·68	1·61
2	Air panels	4157	165	2262	284	399	14 267	2511	14 183	3·43	3·21	2·43	2·30
3	PV-thermal array (biased for electricity)	3377	591	4763	189	385	11 324	2500	14 178	3·35	3·17	1·87	1·79
4	PV-thermal array (biased for heat)	3617	439	3965	203	388	12 283	2504	14 183	3·39	3·21	2·03	1·94
5	PV-thermal array (ground recharge)	3463	606	4345	195	385	11 742	2512	14 180	3·39	3·21	1·94	1·86

DHW, domestic hot water; GSHP, ground source heat pump; PV, photovoltaic; SPF, seasonal performance factor

which included two air panels and a single PV-thermal unit, as in Figure 1. On comparing the results from the data logs with this TRNSYS model, three factors that would impact the accuracy of the models became conspicuous.

- The ground temperature was cooler than it should have been for the location.
- The electrical consumption of the heat pump compressor was approximately 30% greater than in the models for the same thermal output.
- In the models, the glycol temperature had been allowed to fall to −5°C, but was rarely measured below 0°C at the test house.

Combined, these factors would lead to gross overestimates of the SPFs for the heating systems. According to Cantor and Harper (2011), ground source heat pump systems are usually designed so as not to allow glycol temperatures to fall below 0°C for long to avoid ground freezing. All the TRNSYS models were, therefore, altered and re-run. The results shown in Tables 1–6 were derived from the models after modification. The results of the TRNSYS model based on the test house are compared with the results derived from the data logs in Table 6. There remained differences between the performance of the model and the real system. The COP of the modelled heat pump was still higher than that at the test house, but it was thought that the electrical consumption of the test house heat pump was high owing to a leak at a diverter valve, which resulted in the heat pump reverting to domestic hot water heating mode for much of the time. Another noticeable difference was that the temperature of the glycol exiting the ground circuit was higher than predicted by the models. The data logs showed the glycol temperature very rarely dropped below 0°C in winter, while in summer the glycol

Table 6 Comparison of test house data logs with TRNSYS model

Energy	Sub-system	Data logs: kWh	TRNSYS: kWh
Electricity	Compressor	5565	3086
	Circulation pumps	1177	1025
	Ohmic heating	2974	3070
	Total	9716	7181
Source heat	Ground	9737	5378
	Air panels	2082	4537
	PV-thermal unit	387	598
	Total	12 206	10 513
Load heat	Space heat	16 281	14 190
	DHW	2457	2522
	Total	18 738	16 712

DHW, domestic hot water; PV, photovoltaic

temperature could rise to 4 or 5°C warmer than predicted by the models. This would partly explain why the real PV-thermal unit and air panels did not collect as much heat at the test house as predicted in the simulation. However, with a better control system, the logs indicated that the two air panels would have collected close to 5233 kWh, while the PV-thermal unit would have collected close to 509 kWh.

The TRNSYS PV-thermal type was modelled on a different design to the prototype at the test house. The parameters of the PV-thermal modules in the models were set to those of the prototype unit where known, but the default parameters were retained for the resistances of the substrate and back materials. The results indicate that the TRNSYS PV-thermal module dissipated more heat to the atmosphere than the test house module.

The simulation of system 2 showed the air panels to be very effective. However, the TRNSYS type used to model the air panels was limited in sophistication. The air panels were attributed a constant heat transfer coefficient of 14 W/m²K, and did not take into account such factors as direct sunlight, wind speed, heat transfer by condensation or ice build up. Heat transfer coefficients estimated from the logs ranged from 8 W/m²K in still conditions to 22 W/m²K in windier conditions.

The IVT assembly and commissioning guide indicates that a ground circuit consisting of up to nine of their compact collectors would be appropriate for a heat pump with a 4 kW thermal output. Doherty *et al.* (2004) listed some 'rules of thumb' values for sizing ground heat exchangers. For a horizontal heat exchanger, pipe lengths of 5–35 m/kW were quoted. For a slinky heat exchanger, lengths of 40–90 m/kW were quoted. The six ground panels comprised 225 m of pipe, but the spacing between the pipe-work was very tight. Assuming that the 25 m underground pipe was sufficient in length for 1 kW of thermal output, it is probable the ground circuit was undersized by 25%. Nevertheless, the higher glycol temperatures recorded in the logs indicate that the real ground circuit could collect more heat than predicted by the TRNSYS simulations.

4.4 PV-thermal against ground heat exchanger costs

Helpin *et al.* (2011) studied a solar-assisted ground source heat pump system designed for a French office building. They commented that drilling or trenching of the ground circuit was a major part of the cost of installing a ground source heat pump, approaching 50% where boreholes were used. While borehole costs amounted to approximately €100/m, the cost of solar collectors ranged from €60/m² to €100/m². They used unglazed solar collectors in their study, basically lengths of black plastic pipe laid out in parallel on a flat roof. They estimated 23% capital cost savings and that a reduction in lifecycle costs of 13% for the solar-assisted ground source heat pump when compared with one that relied on deeper boreholes for source heat.

The National Renewable Energy Centre (2007) website asserts horizontal ground circuit costs to be £250–350/kW, while a vertical ground circuit would cost £450–600/kW. Based on 2006/2007 prices, for an 8 kW ground source heat pump costing £3579, an accompanying horizontal ground circuit would cost £1470 in trenching and pipes, while a borehole ground circuit would cost £2270.

Commercially available PV-thermal systems intended to provide solar hot water are more expensive than PV arrays. For example, the Stafford Save Your Energy organisation (Save Your Energy Stafford Area, 2014) estimated costs of £1500 to £2000/kWp for a PV array, but £7000 for a PV-thermal system. The Carbon Free Group (2014) also quoted £7000 for a PV-thermal system with a 1 kWp electric output and a 4·5 kWp thermal output.

An American company, Optical Energy Technologies Inc (2014), quoted a material cost of US$2/ft^2 for a type of low-cost PV-thermal heat exchanger, designed for heating swimming pools or for pre-heating domestic hot water. The material cost per 250 W PV panel was approximately US$30. Installation costs for a 3·3 kWp array of ten PV-thermal modules were quoted at US$10 000 before rebates. These costs included the installation of the manifolds between the rubber tubing mats and a stainless steel heat exchanger.

In the UK, installers would require approval by the micro-generation certification scheme for both solar thermal and solar PV technologies. PV-thermal modules will be heavier and more difficult to install. However, in the UK there is potential for customers to benefit from both feed-in-tariff and renewable-heat-incentive payments.

PV-thermal modules have few moving parts, so should be low maintenance. However, extra manifolds would probably be required to divide and collect the heat transfer fluid between the PV-thermal modules, which might entail a leak hazard.

5. Conclusion

One conclusion drawn from the TRNSYS simulations is that extracting heat from PV-thermal modules can be a useful way of improving the performance of a ground source heat pump with a short ground circuit. Despite the extra work performed by the circulation pumps in transferring heat from the PV-thermal array to the ground when the heat pump was not in operation, ground thermal recharge did help improve the performance of the heat pump during the heating season. The PV-thermal array was more effective at improving the performance of the heat pump when it performed both as an ambient heat collector and as a solar collector. Nevertheless, according to these simulations, ambient heat collectors might be a more useful means of improving the performance of a ground source heat pump for the following reasons.

- For a similar area of wall or roof space, the air panels proved more effective than PV-thermal arrays at providing additional source heat for the heat pump during the heating season.
- During the summer months, an array of PV-thermal modules can heat the glycol to temperatures higher than a heat pump is usually designed to accommodate. Some ohmic heating may be required even in summer, particularly for domestic hot water, if the heat pump is forced to shut down.
- During the summer months the heat pump does not take much heat out of the glycol solution as during the heating season, which means the glycol solution cannot cool the PV-thermal panels as well.
- The cooling effect of pumping the glycol through the PV-thermal array increases the electrical output by approximately 4%. This amounts to about 120 kWh annually for a 3·8 kWp system, but this should be offset against the electrical consumption of the additional circulation pumps.

The benefit in transferring heat from a PV-thermal array to a ground source heat pump lies almost entirely in reducing the proportion of the annual heat demand that must be satisfied by ohmic heating. While the additional source heat improves the COP of the heat pump's thermodynamic cycle by 5%, it enables the heat pump to operate for longer.

Acknowledgements

The authors are grateful to the Engineering and Physical Sciences Research Council of the UK for the funding of this research. The authors are also grateful to Mr D. Atkins of ICE Energy Ltd for all his assistance and suggestions in improving the TRNSYS models.

REFERENCES

Bakker M, Zondag HA, Elswijk MJ, Strootman KJ and Jong MJM (2005) Performance and costs of a roof-sized PV/thermal array combined with a ground coupled heat pump. *Solar Energy* **78(2)**: 331–339.

Cantor J and Harper GDJ (2011) *Heat Pumps for the Home*. Crowood Press, Marlborough, UK, p. 34.

Carbon Free Group (2014) *PV-T collectors + geothermal heat storage*. See: http://www.carbonfreegroup.com/downloads/CFG-PV-T-info.pdf (accessed 05/06/2014).

Doherty PS, Al-Huthaili S, Riffat SB and Abodahab N (2004) Ground source heat pump – description and preliminary results of the Eco House system. *Applied Thermal Engineering* **24(17–18)**: 2627–2641.

Helpin V, Kummert M and Cauret O (2011) Experimental and simulation study of hybrid ground source heat pump systems with unglazed solar collectors for French buildings. *Proceedings of Building Simulation 2011: 12th Conference of International Building Performance Simulation Association. Sydney, Australia*.

IVT Industrier AB (2005) Compact Collector Assembly and Commissioning Guide. IVT, Sweden, 2005-02-01, Article no. 290418-9, Version 1.1.

Kjellsson E, Hellström G and Perers B (2010) Optimization of systems with the combination of ground-source heat pump and solar collectors in dwellings. *Energy* **35(6)**: 2667–2673.

Malenković I (2012) *Definition of Performance Figures for Solar and Heat Pump Systems*. Technical Report 5.1.3. Austrian Institute of Technology, Intelligent Energy Europe, Project IEE/08/593/S12.529236.

National Renewable Energy Centre (2007) *Heat Pumps*. See: http://www.narec.co.uk/cmsfiles/narec/Carbon_Mixer/62_15.10HeatPumps.htm (accessed 05/06/2014).

Optical Energy Technologies Inc (2014) *Hybrid-Photovoltaic/Thermal Solaroof*. See: http://www.opticalenergy.com/hybrid-photovoltaic-thermal-solaroof/ (accessed 05/06/2014).

Pahud D and Lachal B (2005) *Mesure des performances thermiques d'une pompe à chaleur couplée sur des sondes géothermique à Lugano*. l'Office fédéral de l'énergie, Switzerland (in French).

Save Your Energy Stafford Area (2014) *Solar PVT: hybrid panels with dual function*. See: http://www.staffordarea.saveyourenergy.org.uk/what/solar/Solar%20PVT%20hybrid%20panels (accessed 05/06/2014).

Sharp Energy Solutions Europe (2011) ND Series A2 (60 cells), June, SolarND_60A2_E0611. Sharp, Hamburg, Germany.

Varney K and Vahdati M (2013) Photovoltaic and solar-assisted ground source heat pump systems. *Engineering Sustainability* **166(1)**: 32–45.

ICE Themes Geothermal Energy, Heat Exchange Systems and Energy Piles

Craig and Gavin
ISBN 978-0-7277-6398-3
https://doi.org/10.1680/gehesep.63983.131
ICE Publishing: All rights reserved

Chapter 8
The design of thermal tunnel energy segments for Crossrail, UK

Duncan P. Nicholson MSc, DIC, CEng, MICE
Director, Ove Arup and Partners Ltd, London, UK

Qing Chen MSc, FGS
Senior hydrogeologist, Ove Arup and Partners Ltd, London, UK

Mike de Silva BSc, PhD, FCIWEM, MIEEM, CSci
Sustainability Manager, Crossrail Ltd, London, UK

Alan Winter BSc, CEng, MICE
Bored Tunnel Engineer, Crossrail Ltd, London, UK

Ralf Winterling MSc, Dipl.-Ing.
Technical Manager, Rehau A G+Co, Germany

Significant heat is generated by underground trains, particularly when braking, stopping at platforms and accelerating away from stations. A complex ventilation system including shaft, fans and under-platform extraction thus has to be designed to manage the rising temperature in tunnels and stations. This conventional approach results in high energy consumption for running the fans and neglects the possibility to use the extracted heat above ground in buildings. Lining underground rail tunnels with heat-exchange segments can provide an alternative solution to cool the tunnels and surrounding ground, and transfer the harvested heat to adjacent buildings for heating. It will also bring benefits in terms of reduction of energy consumption for tunnel ventilation operations. This chapter reports on the work carried out in designing thermal energy segments for use on the tunnelled sections of the Crossrail project in London, UK.

1. Introduction

Shallow ground has a large potential of low enthalpy energy that can be used to meet buildings' heating and cooling demands. With the advances in ground source heat pump technology, it is becoming increasingly feasible to capitalise on this geothermal energy present in the ground. The biggest advantage of the geothermal energy is that the soil temperature reaches a constant at a depth of approximately 10–15 m below surface. For example, in central London, UK, the soil temperature at 14 m below the surface remains undisturbed at around 14°C. Outside London, the soil temperature is slightly lower.

Figure 1 Schematic diagram of the tunnel energy segment system

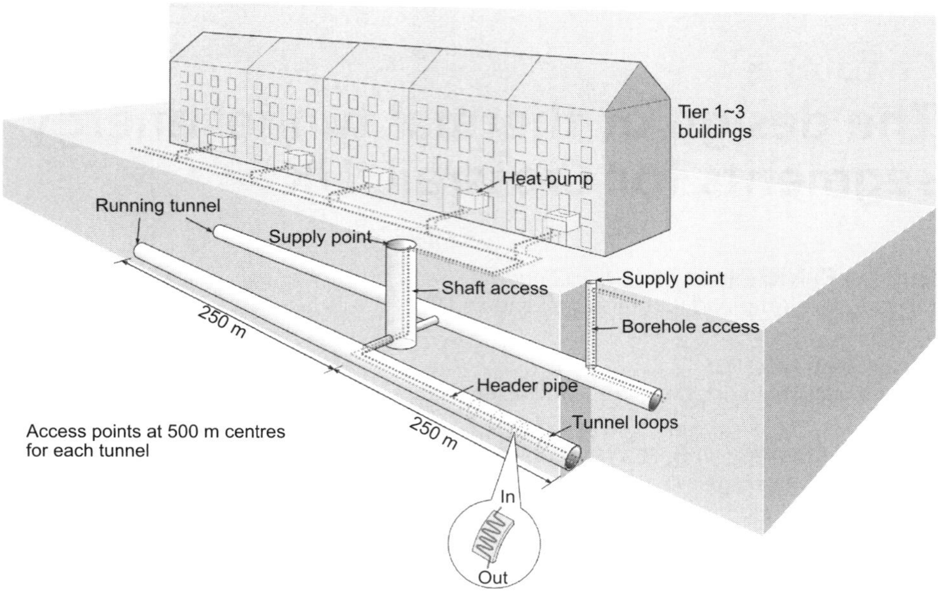

Recent development has seen the use of foundation piles, diaphragm walls and base slabs as ground heat exchangers (Adam and Markiewicz, 2009; Brandl, 2006; Fry, 2009). For example, thermal piles and walls involve attaching polymer absorber pipes to the reinforcement cages and this approach has been applied to Crossrail stations. Other application examples involving sprayed concrete linings include the tunnels at Stuttgart Metro U6 (Schneider and Moormann, 2010), Lainzer Tunnel (Adam and Markiewicz, 2009) and the metro stations (Brandl et al., 2010). Heat-exchange pipes have been placed along the lengths of the Channel Tunnel to extract heat from inside the tunnel (personal communications).

The paper by Franzius and Pralle (2011) discusses embedding plastic absorber pipes in the precast concrete tunnel segments of a conventional metro to collect heat from the adjacent ground. However, at Crossrail the primary heat source is the waste heat rejected from trains, and the surrounding ground is the secondary source. The harvested heat is transferred by way of header pipes to the ground surface, and connected to heat pumps to supply the heating energy to buildings. Figure 1 shows the conceptual design of the tunnel energy segment (TES) system.

2. Assessment of the heat inside the tunnel

Operation of the London Underground network over the past 100 years has gradually warmed the surrounding geology of London Clay. In the early 1990s, the tunnels' temperature was constant at around 15°C. Over the century, the ground around the tunnel has become unable to absorb the waste heat from trains effectively. The consequence is that today's temperature in many underground lines remains high throughout the year, with temperatures of more than 30°C in some parts of the network (Botelle et al., 2010). High air temperature presents a big challenge to the underground train operators. In future, climate change will likely exacerbate the current situation, with the mean temperatures projected to increase by 2–6°C higher than normal within the next 40 years (Defra, 2011, 2012).

To start with, an assessment of the available heat energy inside tunnels was carried out. Train heat is emitted from the motors, air conditioning systems and brakes. This waste heat is injected inside the tunnel, which warms the air and surrounding soil mass.

It is the intention that Crossrail will be a 118 railway line serving London, the central part of which is a twin tunnel underneath London from Paddington to Stratford, and a branch line will run from Whitechapel to Abbey Wood. The length of tunnel is about 21 km long, for a twin tunnel, the total length is 42 km (2×21 km). As seen in Table 1, at peak time there are 24 trains each hour in each direction, meaning one train per 2·5 min. Assuming the average train speed is 60 km/h, a train will take 21 min to complete the journey. Hence, the number of trains inside the tunnel at any one time during peak hours is $21/2\cdot5 = 8\cdot4$ trains.

Assuming one train motor generates on average 1 MW net heat, and its air-conditioning unit generates about 0·1 MW, the total heat output rate from 8·4 trains is $1\cdot1 \times 8\cdot4 = 9\cdot2$ MW. The tunnel internal diameter is 6·2 m, the total surface area of the tunnel wall will be about 410 000 m². Therefore, the peak heat output rate in terms of tunnel lining area is $9\cdot2 \times 10^6/410\,000 = 22$ W/m² of tunnel surface area. On a weekly basis, the average number of trains is 14 trains per hour. The average heat output is thus $22 \times 14/24 = 13$ W/m². In addition, the surrounding strata behind the tunnel linings would also serve as a heat source similar to the ground mass of a closed-loop ground-source heat pump design. The train's heat emission also increases dramatically during braking when the train approaches the platform and accelerates away from the station. Therefore these sections have higher heat output and are the locations where TES should be focused. For comparison, the typical heat exchange rate for closed-loop boreholes is 35 W per metre of borehole depth and about 60 W per metre run for small diameter geothermal piles (Brandl, 2006).

Based on the above simplified calculation, the base heat output from the train operation is about 7 W/m² base and the peak output is about 30 W/m² peak. On this basis a 500 m length of tunnels would be able to support about 100 family apartments with a total heat demand of 1200 MWh/annum, which would be met by a combination of a 400 kW gas boiler and a 400 kW heat pump. The full network of Crossrail tunnels comprise 84 such lengths and would consequently have the capacity to provide heat for 8400 families (30 000 people) or equivalent building heating.

3. Design of the TES system
3.1 Previous experience
There are a number of existing case studies on retrieving heat energy from tunnels. The Vienna LT22 testing plant was the first geothermally activated sprayed concrete lining tunnel in Austria. The absorber pipes were attached to a geosynthetic and placed between the primary and secondary tunnel lining (Adam and Markiewicz, 2009). A similar installation was built at Stuttgart Metro U6 in Germany (Schneider and Moormann, 2010). More recently, thermal structures have been built in TBM Tunnels. Züblin AG and Rehau AG & Co have championed this development with a field test in the Katzenberg high-speed rail tunnel in Germany followed

Table 1 Crossrail designed train frequency during normal operation (SP2)

From	To	Hours	Train frequency	No. of trains
Monday–Friday				
00:00	00:30	0·50	12	6
00:30	05:45	5·25	0	0
05:45	06:15	0·50	12	6
06:15	07:00	0·75	16	12
07:00	07:45	0·75	20	15
07:45	09:15	1·50	24	36
09:15	10:00	0·75	20	15
10:00	16:00	6·00	18	108
16:00	16:45	0·75	20	15
16:45	18:15	1·50	24	36
18:15	19:00	0·75	20	15
19:00	22:00	3·00	16	48
22:00	00:00	2·00	12	24
			Total no.	336
Saturday				
00:00	00:30	0·50	12	6
00:30	05:45	5·25	0	0
05:45	09:00	3·25	12	39
09:00	21:00	12·00	18	216
21:00	00:00	3·00	12	36
			Total no.	297
Sunday				
00:00	07:00	7·00	0	0
07:00	12:00	5·00	12	60
12:00	21:00	9·00	18	162
21:00	23:45	2·75	12	33
23:45	00:00	0·25	0	0
			Total no.	255

by the Jenbach twin track high-speed rail tunnel in Austria, which incorporated a 54 m long demonstration section with thermally activated segmental tunnel lining equipped with heat exchange pipes (Franzius and Pralle, 2011). These cross-linked polyethylene (PE-Xa) pipes were connected to a heat pump in the utility centre building to provide hot water and space heating. It should be noted that all the above schemes are taking heat from the ground mass behind the tunnel linings rather than the tunnel air.

3.2 Crossrail experience

From the start, the Crossrail TES design team have been facing demanding technical and management challenges.

- First, the TES design only started after the tunnel design had been completed and tendered. Therefore the TES design needed to cater for the heat exchange pipe requirements, while still meeting the existing design requirements for structural, fire, heat and ventilation (H&V) and so on. For example, a design method had to be developed to enable the existing tunnel H&V design to be revised to incorporate the TES. Another example, the introduction of box-out pockets at the ends of the segments to form pipe connection needed to take account of the caulking groove featured in the segment structural design.
- Second, incorporating the heat exchange pipes in the segment casting and ways of forming tunnel ring connections requires numerous discussions with tunnel contractors and design changes. The aim is to develop methods that the TES would not delay normal tunnel construction or introduce high costs.
- Third, TES needed to identify and manage a network of buildings that could use the heat.
- Fourth, the Crossrail board needed to have design risk assessments and budget costs so that they could make informative decisions.

All these challenges need to be resolved to a demanding time frame. This complex and innovative design would require effective coordination of inputs from a multidisciplinary team, including building services, costing, energy, fire, hydrogeology, geotechnics, material, mechanic and electrical engineering, structures, risk management, tunnelling and ventilation engineering, and so on.

3.3 Thermal energy segments

In order to extract heat from the tunnel, the concept design is to embed 20 mm internal diameter heat-exchange pipes (absorber) in the conventional concrete segments before casting. The segments are fabricated using a standard segmental lining manufacturing process. The absorber pipes are placed into the segment in a meandering fashion to maximise heat abstraction potential. For steel reinforced segments used on the floating track sections, the heat-exchange pipes are either tied to the steel cage or on a separate support mesh. For fibre reinforced segments, the pipes are softened in a hot water bath and then attached to a light cage before being positioned inside the segment moulds. To minimise fire risk and prevent accidental damage in the future, the pipes are placed at least 200 mm from the intrados. To meet minimum pipe bending radii, 300 mm pipe spacing has been adopted based on the PE-Xa polymer pipe specification. Figure 2 shows a typical pipe arrangement for a segment tied to a reinforcement cage before concrete is poured.

In order to form the longitudinal connections between segments, box-outs are specially designed at the ends of each segment. These pockets are 200 mm long, 100 mm wide and 70 mm deep to allow a permanent mechanical coupling of the adjacent pipe-ends to be performed. As shown in Figure 3, the Crossrail box-outs have reverse tapers in order to retain the mortar filler.

Figure 2 Absorber pipes assembled to a reinforcement segment before casting (photograph taken from Jenbach tunnel by Rehau Ltd) PE-Xa Absorber Pipe attached to reinforcement cage (photo: H8 Tunnel Austria, JV Ed. Züblin AG)

Figure 3 Reversely tapered box-out at the end of segments

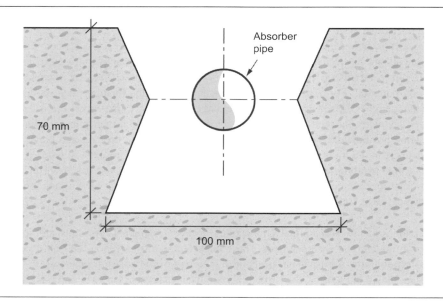

The impacts on structural integrity and fire safety introduced by these changes to standard segment design are discussed later.

During installation, the pipe connections are formed immediately after the segment ring is erected at the tunnel face. The majority of work can be carried out from the erection train towed behind the tunnel boring machine (TBM). The connection system was initially developed for thermally activated building structures in high-rise buildings for pipe connections buried in the concrete structure. The system allows permanent mechanical connections (Everloc) to be rapidly made in a small space to reduce the installation time. Once formed, these connections cannot be undone. The Everloc permanent couplings have been subjected to pressure tests and the results show that the polymer pipe fails before the joint; this provides confidence in the integrity of the couplings.

To allow a continuous flow within a lining ring, the absorber pipes in the six segments and one keystone of the ring are joined up. Five neighbouring rings are connected longitudinally to form a complete circuitry, see Figure 4. Each circuit is connected to the flow and return header pipes. Control valves are installed to isolate individual circuits in the event of a local failure and to assist with the de-airing process. The ring-to-ring connection positions cannot be predetermined because each ring is slightly tapered; the orientation of the taper is selected during tunnel construction to steer the tunnel as the work progresses. Hence the ring-to-ring pipe connections are formed by variable lengths of plastic polymer pipe mounted on the tunnel surface, see Figure 5.

There are about 11 m of absorber pipe inside one standard segment. There are seven segments plus one keystone in one complete ring, resulting in approximately 77 m of pipe per ring. There are five rings in one circuit and 31 circuits would cover about 250 m tunnel length. Tunnel header pipes would be connected to the surface through shafts and stations. Alternatively, dedicated access boreholes could be provided at preferred surface locations for the header pipes as shown in Figure 1.

It has been planned that heat from the tunnels would be connected to district heating networks supplying heat energy to blocks of buildings along the tunnel alignment. The income from the

Figure 4 Schematic diagram of pipework connections with three header pipes configuration

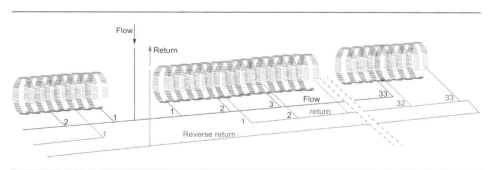

Figure 5 Schematic diagram of TES design

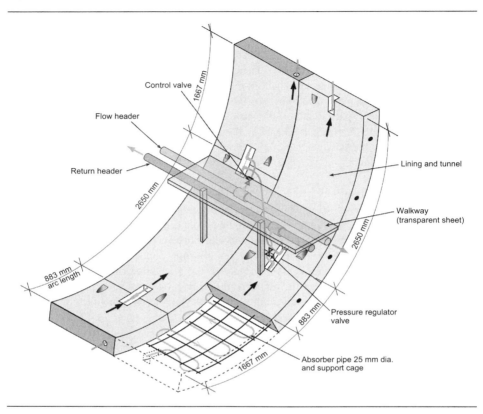

sale of heat to the district heating energy supply companies will provide a revenue stream to the tunnel operator that finances the installation and operation of the TES.

3.4 TES system design: pipework hydraulics and plant room

The three header pipe 'reverse return' arrangement was selected in preference to the two-pipe arrangement because of the advantage of self-balancing, see Figure 4. The return header is connected to the most hydraulically remote point, therefore distributing the flows and pressures more evenly across the system. This makes the reverse return arrangement more inherently balanced compared to the direct return one. A procedure for removing trapped air was developed. Pressure losses were calculated for routine operation, and flushing operations for individual circuits and the header pipes to remove air bubbles.

The flow rate through the circuit varied depending on the heat extraction rate. A flow rate of 0·06 l/s would achieve 10 W/m^2 heat extraction rate and 0·12 l/s would achieve 30 W/m^2. The pressure losses for all fittings are calculated in accordance with Chartered Institution of Building Services Engineers (CIBSE) Guide C (CIBSE, 2007) except for the long radius

U-bends that form the meander of the pipes within the segments. The calculations for normal operations included a 50 kPa pressure drop for fittings on the circulation pump, and an additional 100 kPa pressure drop for the heat exchanger. During the air bubble flushing operation, the heat exchanger would be by-passed, so only 50 kPa pressure drop has been included in the calculations. The system is balanced using a reverse return arrangement with valves on each circuit and an additional valve at the connections between the flow and return pipes. A schematic diagram of the assumed layout is shown in Figure 5. A design margin of 15% is applied on the calculated pressure drop across the pipe network. The flow in the circuits needs to maintain a turbulent state to ensure efficient heat exchange.

For normal operation header pipe lengths of 250 m were planned. Increasing the length to 500 m would result in the system's maximum pressure loss rising from 7·5 bar to 9·1 bar, assuming the header pipes are of 90 mm internal diameter.

To ensure the smooth running of the TES system, it needs to be flushed and purged regularly to remove any trapped air from the circuits. This could be done in two steps: first, the ring circuit will be flushed individually. This procedure involves closing all the valves except the one leading to the target circuit. Based on the recommendations of MIS 3005 standard (MIS, 2013), the minimum flow rate of 1·0 m/s in the loops is assumed; second, air is to be removed from the system flushed by the header pipe by closing all the circuit valves. The maximum pressure head loss is expected to reach 8·3 bar for a 250 m long section, and 12·6 bar for a 500 m section.

To ensure the hydraulic performance could be maintained, it is recommended that the header pipe length is between 250 m and 400 m on each side of an access point (e.g. ventilation shafts or boreholes). Increasing the header pipe length leads to a significant increase in pressure loss during operations, and greater risks of pipe and joint integrity.

4. Potential market for the tunnel heat

The London heat map and its building inventory along the Crossrail alignment were used to identify buildings adjacent to the tunnel alignment that could use the heat from the TES. These buildings were mapped on geographic information system (GIS). Two selection criteria were considered. First, the buildings should have a sufficiently large heat demand to justify connection to the system. The minimum threshold of heat demand was based on the available heat from the tunnel, which is classified by annual heat demand of 600 MWh, 1200 MWh, 1800 MWh and 2400 MWh. Second, by building type: the buildings should have a heat load profile that complements the efficient heat pump operation; that is, relatively constant heating base load. These two criteria were used to divide the buildings into three tiers: tier 1 – most suitable buildings, tier 2 – moderate suitability and tier 3 – low suitability buildings.

The types of buildings are

- tier 1: hotels, large residential buildings, hospitals
- tier 2: schools, higher education, public libraries, museums
- tier 3: offices, leisure centres, retails.

As a result of this investigation, 365 buildings with a heat load greater than 600 MWh/annum were identified within 100 m of the Crossrail tunnel alignment. These buildings comprised 34 tier 1, four tier 2 and 327 tier 3 buildings. Some of these are shown in Figure 6.

The technically feasible surface connection points based on ventilation shafts and tunnel access routes are shown in Figure 7. Circles of 250 m diameter centred at the access points are shown for the standard connection header pipe length. At some locations it may be necessary to increase the capacity of the header pipes where the circles did not intersect. This would enable the TES connections to the majority of the suitable buildings along the tunnel route.

5. Modelling of TES heat transfer

A two-dimensional finite-element (FE) thermal model was developed using numerical code LS-Dyna to simulate the thermal transfer behaviour of the TES, see Figure 8. LS-Dyna is a multi-physics simulation software package; the model developed here is a coupled thermal-mechanical model. As the tunnel is radially symmetrical, only a wedge through the tunnel lining and the soil behind is modelled. Within the concrete liner, the absorber pipes are buried at 200 mm depth and spaced at 300 mm. Heat transfer by conduction is modelled between the fluid inside the pipes and the concrete, the concrete and the soil, and the concrete and the tunnel air. In practice, the fluid inside the pipe will be cooler near the inlet and warmer near the outlet. For simplicity, the fluid temperature in the model is represented by an average temperature along the length of the pipe.

The thermal properties of the soil and concrete used in the model are summarised in Table 2. The heat transfer coefficient from tunnel air to the concrete wall was assumed to be 5 W/m^2 K, which was considered a conservative assumption. The tunnel is normally 15–20 m below ground; the ambient soil temperature at this depth is without seasonal variation. The radial distance of thermal influence by the tunnel on soil mass is limited. In the model, a constant thermal boundary condition of 14·8°C was set at a distance of 100 m from the tunnel. This boundary was based on monitoring in central London. The tunnel air temperature varies seasonally, from 36°C in summer to 17°C in winter, based on the tunnel ventilation simulations.

During simulation, there was no heat extraction for the first 3 years of simulation in order to establish thermal steady state within the tunnel linings and the soil mass behind it. This was followed by a continuous heat extraction for a further 7 years.

Sensitivity analyses were carried out for heat extraction rates ranging from 5 to 50 W/m^2, where the area refers to tunnel lining surface area. The resulting fluid temperature is presented in Figure 9. This figure shows that the TES system would reach quasi steady state within the first year of operation. The model predicts that during winter the fluid temperature will be 16°C for 5 W/m^2 heat extraction rate, and will reduce linearly with the heat extraction rate to −6°C for 50 W/m^2. The soil temperature does not drop below +3°C, even at the soil–tunnel interface. In order to avoid freezing, the continuous heat extraction rate of the absorber pipes must be no more than 30 W/m^2 in winter.

The design of thermal tunnel energy segments for Crossrail, UK

Figure 6 GIS study of existing energy users along the route. Tier 1 buildings at three locations – Canary Wharf, Bond Street and Liverpool Street – are shown

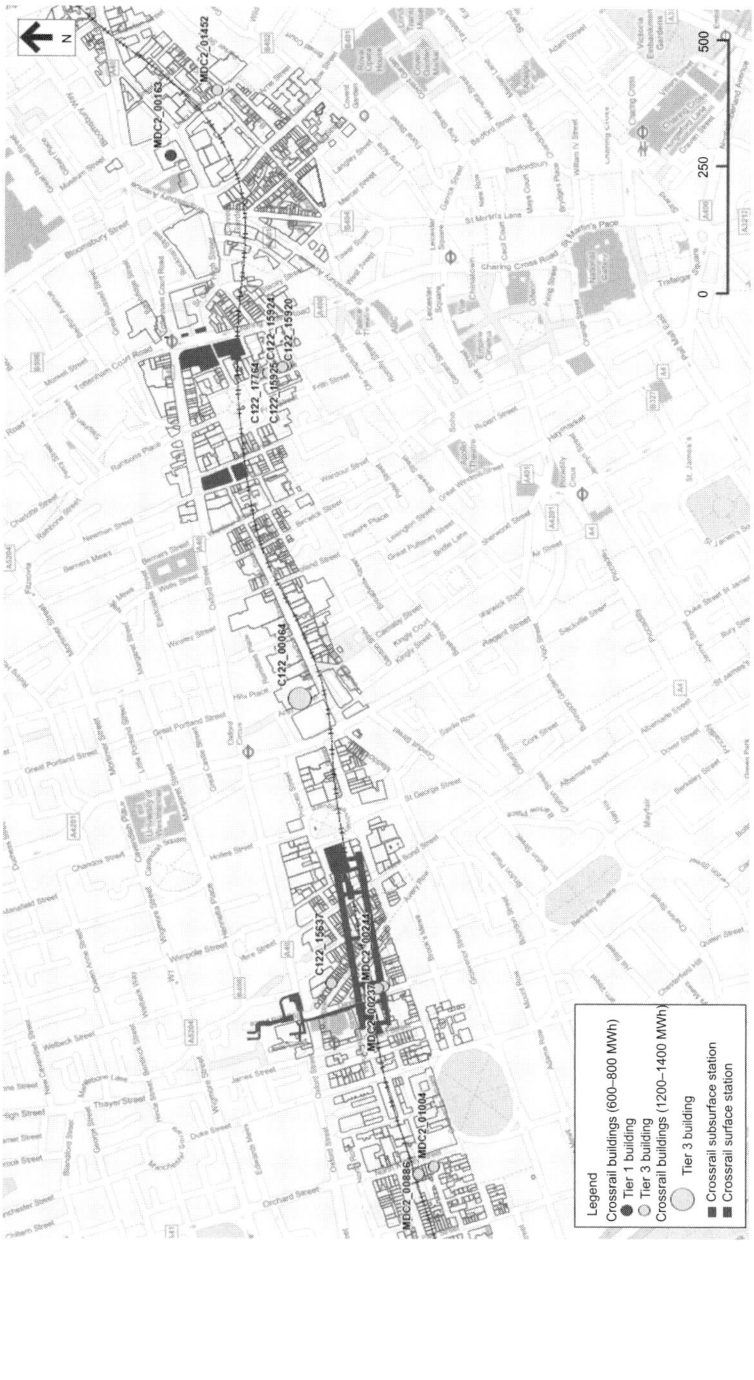

141

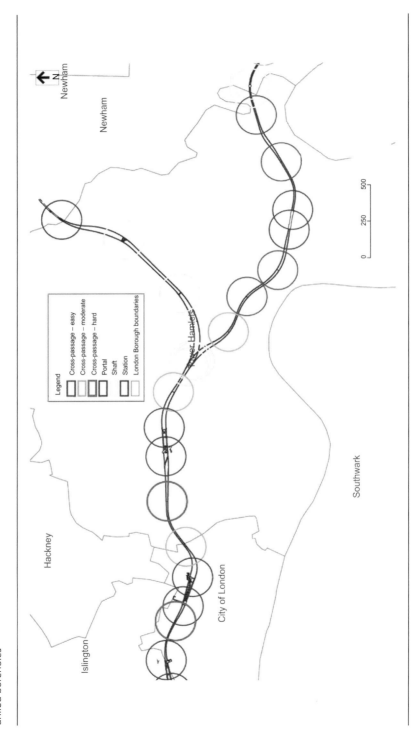

Figure 7 Analysis of surface connections. Each circle is of 250 m radius. Surface connection options are considered for shafts, cross-passages and directional drilled boreholes

Figure 8 Set-up of the numerical thermal model in LS-Dyna to simulate heat transfer to absorber pipes

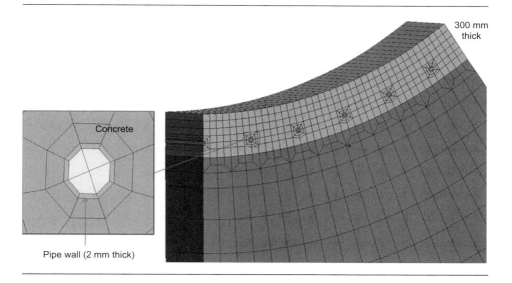

Table 2 Thermal parameters used in LS-Dyna thermal analysis

Material	Density: kg/m^3	Thermal conductivity: W/m K	Heat capacity: J/kg K	Young's modulus: kPa
Soil	2000	1·8	1000	150 000
Concrete	2500	1·33	750	31 × 10^6
Absorber pipe wall	1000	0·087[a]	1000	100
Heat transfer coefficient from tunnel air to tunnel surface		5 W/m^2 K		

[a] Pipe wall conductivity was determined in a validation study for cooled floors, it includes thermal resistance for convective transfer into the fluid inside the pipe as well as conduction through the pipe wall

Figure 9 Predicted fluid temperature inside pipes over time for various extraction rates. No heat extraction within the first 3 years

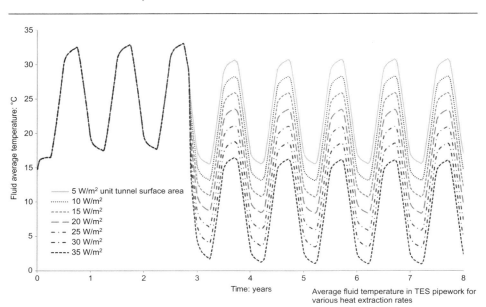

Average fluid temperature in TES pipework for various heat extraction rates

At heat extraction rates between 5 and 20 W/m^2, the outlet fluid temperature from the TES is relatively high (above 8°C in winter). This will enable the TES system to operate at high coefficient of performance (COP).

6. Tunnel cooling study

Heat produced inside the tunnel is traditionally either dissipated through the tunnel walls into the ground or discharged to the atmosphere through air movement. The latter can be achieved by means of draft relief, ventilation shafts, under-platform extract and over-track extract. An aerodynamic model is developed to take into account the geometry of the underground system of tunnels, the train movements and the main mechanism of air movement. The aerodynamic predictions are then used as a periodic forcing function in the thermodynamic model to calculate air, tunnel lining and ground temperatures, air humidity, and heat loads from trains over years. The coupled model has been used for the design of ventilation systems for Crossrail.

To understand the cooling effect of the TES inside the tunnel, the thermodynamic model described above was modified to include the TES pipework, see Figure 10. The heat transfer between the air, the tunnel wall and the ground is simulated using a one-dimensional heat transfer model, which divides the tunnel linings and the ground into concentric rings in each element. The model predicts the air temperature, the tunnel wall temperature and the inlet/outlet temperature of the TES circuit for various heat extraction rates. Heat transfer between air and the flow/return headers is considered insignificant and can be ignored. For details of the

Figure 10 The ventilation model set-up, showing heat transfer from air to tunnel and surrounding soil

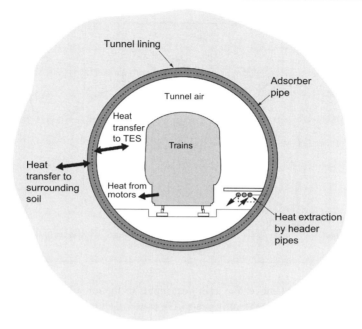

numerical study on TES cooling efficiency, see Biotto *et al.* (2013). The new model was validated by comparing the results with the LS-Dyna numerical model.

Thermal simulations were carried out for the design full-load train service pattern SP2 (240 m long trains, 24 trains per hour service frequency during peak hours), with TES operating at a continuous predefined heat extraction rate.

The ventilation model predicts a reduction in peak tunnel air temperatures of 2–4°C (dry bulb (DB)) with the pipework installed along the entire length of the running tunnels and a continuous heat extraction rate of 10 W/m^2, see Figure 11. The air temperature reduces by approximately 2°C DB at stations, and 2–4°C DB in the running tunnels. Greater temperature reductions were achieved in areas such as tunnels east of Stepney Green junction; this was mainly attributable to the lower tunnel temperature within this section.

Most of the greater heat emission activities by the trains – braking, acceleration and long residence times – happen close to the stations. One alternative design to improve the efficiency of the system is to install the TES pipework only in the close proximity of the stations. The peak tunnel air temperature in these areas is predicted to drop by up to 4°C DB at 20 W/m^2 heat extraction rate.

Figure 11 Predicted tunnel air temperature along eastbound route from Royal Oak portal to Canary Wharf Victoria Dock portal during summer peak hours, with full installation along the running tunnel and a fixed heat extraction rate of 10 W/m². The inlet and outlet fluid temperatures are also shown

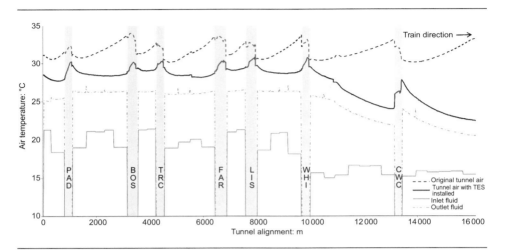

It should be noted that whenever the train heat inputs to the tunnels are lower than this full service pattern (for example in the early years of operation), a lower heat extraction rate would be required to keep the circulation fluid temperatures above freezing. Lower inlet temperature would have an impact on output from the heat pumps.

When the Crossrail is in operation, the bored tunnels are expected to be warmer than the surrounding ground due to heat emitted by the trains. Under the current design some of the heat is removed by air exchange through the draught relief ducts, and some dissipates through the tunnel walls to surrounding strata by thermal conduction. TES will offer significant benefits as it will prevent heat build-up in the ground, and reduce the tunnel air temperature. This will lead to savings in the tunnel ventilation system operational cost and future proof the tunnel design for an increase in temperature due to climate change or higher train frequencies.

7. Durability and operational considerations

7.1 Pipe durability

The PE-Xa grade pipe, comprising high-pressure cross-linked polyethylene, or equivalent, has been specified because

- tight pipe bending radii can be achieved
- it has robustness against notches, scratches and puncture loads during segment production and installation
- it allows for segment joint extension at the box-outs
- permanent mechanical connections can be formed
- its durability requirements are in line with the tunnel design life (e.g. 120 years) at the pipe operating pressure and temperatures in accordance with DIN 16892 (DIN, 2000).

7.2 Thermal effects on tunnel structure

In order to understand the thermal effect on the tunnel structure, a three-dimensional FE model was developed using LS-Dyna to represent a tunnel with a typical soil profile. The tunnel element in the model comprises a 6·2 m dia. 300 mm thick concrete ring made of seven segments and a keystone; see Figure 12. The model is a coupled thermal–mechanical model.

The assumed stratigraphy and the material properties used in the model are summarised in Table 3. Within the model, the made ground and the river terrace deposits were modelled using the elastic Mohr–Coulomb model, these formations were considered drained at all stages during the analysis. The London Clay and the Lambeth Group were modelled using the constitutive model Brick available in LS-Dyna. These clay-type materials were assumed to behave in an undrained manner. The inputs for Brick model are listed in Table 4. The concrete is modelled as a linear elastic material; the joint between segments is of a 3·5 m radius convex. Special interface elements are used to model the segment joint to allow rotational movement between the segments. The thermal properties used in the model are given in Table 2. The thermal expansion coefficient of the soil skeleton and the pore water was assumed to be $3·0 \times 10^{-5}$/K.

The coupled thermal–mechanical analyses were carried out for heat extraction rates from 0 (no heat extraction) to 30 W/m². Under natural conditions (i.e. no heat extraction), the warmed air in the tunnel causes it to expand, and the maximum computed tunnel movement is 1·1 mm expansion outward in the summer and 1·6 mm in the winter. There is negligible deformation at the crown and invert of the tunnel. Heat extraction reduces these displacements. At an extraction rate of 30 W/m² the tunnel shrinks at the axis level and expands at the crown during summer, and shrinks towards the centre during winter. The changes in tunnel diameter are summarised in Table 5. It is concluded that the 30 W/m² heat extraction would result in a further 1 mm deformation compared to the case of no heat extraction.

Figure 12 Set-up of the thermal-mechanical FE model for the Crossrail TES in LS-Dyna

Table 3 Geotechnical parameters used in the LS-Dyna analysis

Stratigraphy	Top elevation	Initial stress		Stiffness parameters			Strength parameters			
Stratum	mATD	ρ: kN/m³	Ko	E'_v: MPa (top)	E'_v: MPa (base)	Poisson ratio	c': kPa	φ'_p: degrees	c_u: kPa (top)	c_u: kPa (base)
Made ground	125	18	0·5	10	10	0·2	0	25	—	—
River terrace deposits	122	18	0·5	25	25	0·25	0	40	—	—
London Clay	118·5	20	1·2	24	29	0·2	5	25	104	137
Lambeth Group	94	20	1	70	70	0·2	10	28	350	350
Thanet Sand	76	21	1	250	250	0·2	0	36	—	—

Table 4 Brick parameters for London Clay

Strain	G/Gmax
3.04×10^{-5}	0.92
6.09×10^{-5}	0.75
0.000101	0.53
0.000121	0.29
0.000820	0.13
0.00171	0.075
0.00352	0.044
0.00969	0.017
0.02223	0.0035
0.0646	0

$\lambda = 0.1, \kappa = 0.02, \iota = 0.0019, \beta G = 4, \beta_\varnothing = 3, G_{vh}/G_{hh} = 0.5, \mu = 1.3$

Table 5 Thermal effect on tunnel dimensions

	Tunnel diameter change: mm	
Simulation stage	Crown-invert	Axis level
Summer (no extraction)	+0.1	−1.1
Winter (no extraction)	−0.5	−1.6
Summer (30 W/m² heat extraction)	+0.3	−1.9
Winter (30 W/m² heat extraction)	−0.6	−2.6

Hoop stress in the tunnel lining without thermal effects is predicted to be 4 MPa at the crown depth of 16 m. Hand-calculated hoop stress in the tunnel lining, assuming the soil density of 20 kN/m^3, is approximately the same. As expected, the maximum hoop stress occurs at the convex to convex segmental joints. When no heat is extracted, the hoop stress (average through the segment thickness) shows a seasonal variation, being maximum in summer when it is 7% greater than the as-constructed condition. This rises by a further 2% when 30 W/m^2 heat extraction takes place. In winter, the average hoop stress is negligibly different from the as-constructed condition, whether heat is extracted or not. Although the heat extraction induces a relatively large temperature gradient across the thickness of the segments, the hoop stresses at both surfaces are predicted to remain compressive.

The computed tensile stress for no heat extraction and 30 W/m^2 during summer and winter is shown in Figure 13. For the ground conditions and geometry adopted in the analysis, the

Figure 13 Comparison of maximum principal stress distribution in segments without and with heat extraction of 30 W/m² during summer

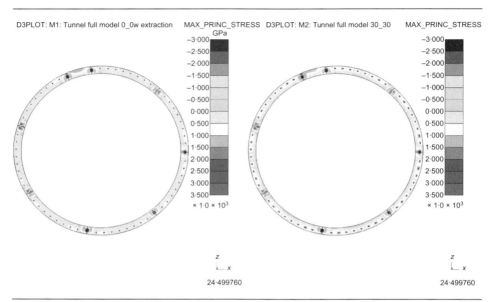

computed maximum tensile (bursting) stress occurs at the convex to convex joints. It is likely that the tensile bursting stress will be proportional to the contact force at the joints, which in turn is proportional to the average hoop stress. Therefore it is assumed that the bursting stress will rise by 7% compared to the condition of no heat extraction, and by a further 2% if 30 W/m² is extracted.

Tensile stress is also expected around the embedded pipes when heat is extracted: it reaches a maximum of 3·5 MPa when 30 W/m² is extracted. The resulting tensile stress is below the design tensile stress limit of 5·0 MPa.

7.3 Fire safety

The piping is made of PE-Xa, which is a form of polymer comprising molecular chains made up of hydrogen and carbon atoms. Through the cross-linking process, additional links are built between neighbouring carbon atoms. The Crossrail fire safety strategy for the central tunnels (Crossrail, 2009) states that materials within the tunnels should be chosen based on the London Underground standard 1–085 fire safety performance of materials (London Underground, 2011). Section 5.6.1 of the Crossrail tunnel fire Safety strategy states that

> Any materials which do not comply with this standard will be risk-assessed to determine whether they are acceptable.

The London Underground standard 1-085 has fire safety requirements for materials with regard to flammability, smoke emission and toxic fume emission. The proposed TES inclusion of exposed cross-linked polyethylene pipe with the tunnel is not fully compliant with the standard in relation to flammability and potentially smoke emission. The Crossrail fire safety strategy for the central tunnels allows a risk assessment to be carried out in lieu of strict compliance with the 1-085 standard.

The embedded absorber piping is connected to a set of three header pipes, which are made of PE-Xa and are 90 mm in diameter. To mitigate fire risk the design is based on burying the header pipes in mass concrete.

8. Operational and commercial benefits

It is important to assess the fundamental financial viability of the TES system. The elements to consider with respect to costs are initial capital tunnelling cost, surface connection cost, cost incurred to connect the system to buildings and operational cost.

The benefits that TES brings to the tunnel operators are outlined below.

8.1 Operational cost saving

Excess heat can cause operational difficulties. For example, metro tunnels may see a large heat input from trains, whereas power cables may raise the internal temperature of service tunnels. Therefore, tunnel cooling often needs to be provided. The most cost-effective solution to achieve cooling would be natural ventilation; however, this is rarely sufficient and conventionally fans are used to provide forced ventilation. Cooling the tunnel by means of the TES system will lead to savings in the operational cost of the ventilation fans. The heat capacity of the fluid in the absorber pipes embedded in the tunnel segments is much larger than air and hence allows for more efficient cooling than forced ventilation.

8.2 Capital cost saving

The physical size of the heat pumps may be significantly smaller than the air ventilation fans and associated air ducts that would normally be required for cooling the tunnel, hence allowing a more compact design and resulting in capital cost savings.

In some instances, using TES instead of forced air ventilation may even allow the tunnel diameter to be reduced, where it is governed by the required air flow area. In a similar way, intermediate ventilation shafts may be downsized or removed because of the reduced air flow. In the future if these opportunities are exploited, significant capital cost savings may be achieved. At Crossrail the TES design options could not be developed fast enough to enable reduction in air ventilation and under platform exhaust (UPE) usage to be incorporated into the design.

8.3 Carbon saving

The provision of heat energy to buildings adjacent to tunnels reduces their heat requirement – for example, the need for those buildings to heat air or water using conventional sources (gas, electricity, heating oil). On new construction work, the TES heat could be used at the planning stage as part of the renewable heat incentive or the part L renewable heating component. Accordingly, the use of TES provides a reduction in carbon output for the building being supplied with this heat energy. For example, 10 W/m^2 tunnel heat extraction provides enough heat to supply a continuous summer low temperature demand of 250 kW, and approximately

170 m^3/d domestic hot water load preheat (approximately equivalent to a 500-bed luxury hotel), achieving carbon savings of about 42%.

8.4 Heat revenue

Selling the heat harvested from the tunnels to nearby properties will generate some direct income. In addition, income may be possible through the renewable heat incentive and/or the selling of carbon offsets.

With energy prices rising and supply uncertainty, the TES may offer an alternative solution. Compared with geothermal piles and vertical borehole loops, the unit energy price is lower for TES systems.

9. Case study: Fisher Street to Tottenham Court Road

A case study design of a 860 m long demonstration TES tunnel between Tottenham Court Road station and Fisher Street shaft was developed for construction to show that all the technical issues could be addressed and costs could be quantified. In order to make full use of the tunnel length available the chosen tunnel was divided into two sections of about 430 m, with two sets of header pipes coming in and out of the Fisher Street shaft, as shown in Figure 14.

The pressure drop at the maximum extraction rate of 30 W/m^2 is calculated as 7·5 bar during normal operation, and 10·3 bar during flushing. The effect of the locally installed TES on the tunnel temperatures is illustrated in Figure 15. Owing to the space limitations the plant room layout was modified to fit to an area of 4 m by 10 m. The plant room is to accommodate the primary circuit – that is, circulation and standby pumps and the secondary and tertiary circuits are located at the end-user's premises.

In order to reduce the fire load, the headers are re-designed to be buried in mass concrete beneath the evacuation walkway. As a result, the fire load introduced by the TES over a 20 m length of tunnel duration 15 min is reduced to 375 kW, and this is 4·2% of the designed train fire load of 8·8 MW over a 20 m long carriage. Pure polyethylene complies with the London Underground material requirements in relation to toxic fume emissions.

The instrumentation and monitoring specifications were also developed both inside the tunnel and the plant room. The cost of the TES is relatively low compared to the tunnel construction budget. Excluding the one-off equipment modification cost, the manufacture and installation TES would cost between £400 and £530 per metre run of tunnel. Compared to conventional closed-loop ground source energy boreholes, the TES is more cost effective.

The following technical issues have been addressed for the development of the demonstrator TES section

- an assessment of the local heat available from the trains and the soil
- the development of the demonstrator section ventilation model that included TES
- the TES and header pipe layouts to extract heat at Fisher Street shaft
- the pipe layouts for the steel cage reinforced segments for the floating track slabs

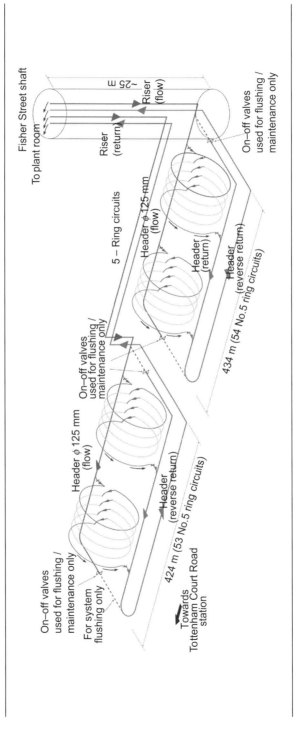

Figure 14 Sketch of the demonstration section design

Figure 15 Predicted tunnel air temperature along westbound route from Pudding Mill Lane portal to Royal Oak portal during summer peak hours, with TES only installed between Fisher Street shaft and Tottenham Court Road and a fixed heat extraction rate of 10 W/m^2

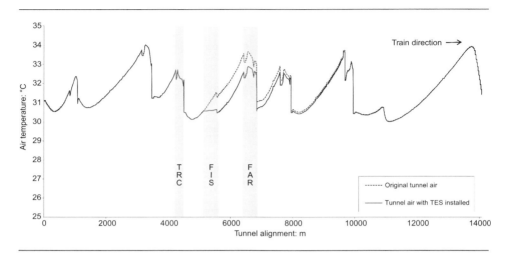

- the connection details between segments and rings and header pipes
- the space proofing and plant room arrangements
- the stress analysis for the thermal effects on tunnel linings
- the assessment of fire risks
- the construction processes for the segment moulds and their tunnel erection
- the assessment of surface buildings within 100 m from the Crossrail route for heat use
- the development of drawings, specifications, quantities and budget costs.

10. Summary and conclusion

This chapter provides an overview of a TES system for Crossrail where heat is generated by the trains in the tunnels all year round. The system uses an embedded closed-loop water-filled pipework in tunnel segments to extract this heat. In this way it both cools the tunnels and provides heat for adjacent buildings. The water temperature supplied to the heat pumps is relatively high, enabling the system to operate at a higher COP compared with conventional ground source heating systems.

A major challenge for the TES project was the coordination of the multidisciplinary team to deliver the design and costing studies for the TES project.

The TES is a viable technology for the future 'smart' cooling of tunnels; 'smart' because it extracts heat in a form that can be utilised and with a potential revenue stream and because it can be used to reduce the reliance on UPE and tunnel ventilation systems, thus saving on electrical energy. These two benefits provide significant carbon and financial savings at a time when there is a focus on carbon reduction and when fuel costs are anticipated to increase between 50~100% decade on decade in real terms. The project did not go ahead because of time constraints and high mould modification costs. Nevertheless, the development of this technology on Crossrail has provided a vital building block in understanding the benefits of the system such that it can be implemented at scale on future projects.

It is considered that the technical and commercial problems associated with the TES have been sufficiently addressed to warrant a demonstration tunnel to prove the system for use in future tunnels. Thinking one step ahead to future proofing during new metro design and construction is important.

The project received the 2011 'Technical Innovation of the Year' by International Tunnelling Awards and the 2011 'Innovative use of Equipment' by Tunnels & Tunnelling International Awards.

Acknowledgements

The authors would like to thank all who were involved in this project, in particular Anton Pillai, David Eckford, David Whitaker, Eren Musluoglu, Mohammad Tabarra, Richard Strut and Steve Doran. The authors are also grateful to Duncan Wilikson and Rob McCrae for their insightful guidance and support. This work is funded by C Ltd.

REFERENCES

Adam D and Markiewicz R (2009) Energy from earth-coupled structures, foundations, tunnels and sewers. *Géotechnique* **59(3)**: 229–236.

Biotto C, Eckford D and Chen Q (2013) Efficient tunnel cooling using tunnel wall heat extraction. *Proceedings of the 15th International Symposium on Aerodynamics, Ventilation & Fire in Tunnels*. Barcelona, Spain.

Botelle M, Payne K and Redhead B (2010) Squeezing the heat out of London's Tube. *Proceedings of the Institution of Civil Engineers – Civil Engineering* **163(3)**: 114–122.

Brandl H (2006) Energy foundation and other thermo-active ground structures. *Géotechnique* **56(2)**: 81–122.

Brandl H, Adam D, Markiewicz R, Utenberger W and Hofinger H (2010) Massivabsorbertechnologie zur Erdwarmenutzung bei der Wiener U-Bahnlinie U2, Osterr. *Ingenieur-und Architekten-Zeitschrift* **155(7–9 and 10–12)**: 1–7.

CIBSE (Chartered Institution of Building Services Engineers) (2007) *Guide C: Reference Data*. CIBSE, London, UK.

Crossrail (2009) *Fire Safety Strategy for the Central Tunnels, C124-MMD-08-TPL-CR001-0006, rev1*. Crossrail, London, UK.

Defra (Department for Environment, Food and Rural Affairs) (2011) *Climate Resilient Infrastructure: Preparing for a Changing Climate*. Defra, London, UK. Presented to parliament, Cm8065, May 2011.

Defra (2012) *The UK Climate Change Risk Assessment 2012 Evidence Report*. Defra, London, UK, Presented to parliament pursuant to Section 56 of the Climate Change Act 2008.

DIN (Deutsches Institut für Normung) (2000) DIN 16892: Crosslinked polyethylene (PE-X) pipes – general requirements, testing. DIN, Berlin, Germany.

Franzius JN and Pralle N (2011) Turning segmental tunnels into sources of renewable energy. *Proceedings of the Institution of Civil Engineers – Civil Engineering* **164**: 35–40.

Fry VA (2009) Lessons from London: regulation of open-loop ground source heat pumps in central London. *Quarterly Journal of Engineering Geology and Hydrogeology* **42(3)**: 325–334.

London Underground (2011) *Fire Safety Performance of Materials, 1-085, Issue A3*. London Underground, London, UK.

MIS (2013) MIS 3005: Heat pump standard. See http://www.microgenerationcertification.org/mcs-standards/installer-standards (accessed 26/04/2014).

Schneider M and Moormann C (2010) GeoTU6 – a geothermal research project for tunnels. *Tunnel* **02/2010**: 14–21.

ICE Themes Geothermal Energy, Heat Exchange
Systems and Energy Piles

Craig and Gavin
ISBN 978-0-7277-6398-3
https://doi.org/10.1680/gehesep.63983.157
© ICE Publishing: All rights reserved

Chapter 9
Thermal response testing through the Chalk aquifer in London, UK

Fleur Loveridge MSc, PhD, FGS, CGeol, CEng, MICE
Research Fellow, University of Southampton, Southampton, UK

Gary Holmes BEng, PhD
Engineering Manager, WJ Groundwater Ltd, Bushey, UK

William Powrie MA, MSc, PhD, CEng, FICE, FREng
Dean of the Faculty of Engineering and the Environment, University of Southampton, Southampton, UK

Toby Roberts PhD, CEng, FICE, CGeol, FGS, MCIWEM
Managing Director, WJ Groundwater Ltd, Bushey, UK

Thermal conductivity of the ground is an important parameter in the design of ground energy systems, which have an increasing role to play in providing renewable heat to the built environment. For larger schemes, the bulk thermal conductivity of the ground surrounding the system is often determined in situ using a thermal response test. Although this test method is commonly used, its limitations are often not fully understood, leading to an over-simplistic interpretation that may fail to identify key facets of the ground thermal behaviour. These limitations are highlighted using data from an instrumented thermal response test carried out in a 150 m deep borehole in east London. It is shown that a single, unique value of bulk thermal conductivity may not be appropriate, as stratification of the ground can lead to differences in thermal performance, depending on the direction of heat flow. Groundwater flow within the Chalk aquifer is also shown to have an important effect on the long-term heat transfer characteristics.

Notation

f	friction factor
h_i	heat transfer coefficient (between fluid and pipe)
Nu	Nusselt number
Pr	Prandtl number
p	power used to calculate p-linear average
q	heat flux per unit depth
R_b	borehole thermal resistance (mK/w)
R_{grout}	grout thermal resistance (mK/w)
R_p	pipe thermal resistance (mK/w)
R_{pcond}	pipe conductive resistance (mK/w)

R_{pconv}	pipe convective resistance (mK/w)
Re	Reynolds number
r	radial coordinate (m)
r_b	borehole radius (m)
r_i	pipe internal radius (m)
r_o	pipe outside radius (m)
S_{cv}	volumetric heat capacity (mJ/m³K)
t	time since start of test (s)
t'	time since start of recovery phase (s)
t_{min}	minimum time after which line source approximation is valid (s)
u	integration parameter
α	thermal diffusivity (m²/s)
γ	Euler's constant
ΔT_f	change in fluid temperature (°C)
ΔT_g	change in ground temperature (°C)
ΔT_{in}	change in loop inlet temperature (°C)
ΔT_{out}	change in loop outlet temperature (°C)
ΔT_p	change in pipe temperature (at r_o) (°C)
$\Delta T_{p\text{-linear}}$	change in fluid temperature (p-linear average) (°C)
ΔT_{rb}	change in temperature at $r = r_b$ (°C)
λ	thermal conductivity (W/mK)
λ_{fluid}	thermal conductivity of fluid material (W/mK)
λ_{pipe}	thermal conductivity of pipe (W/mK)

1. Introduction

The use of ground energy systems to provide renewable heat energy to buildings is increasing, with the UK government's renewable heat incentive (DECC, 2011) set to accelerate installation of systems for new building developments. Ground energy systems work by seasonal storage of heat in the ground. In closed-loop systems plastic pipes are cast into the ground (the ground loop), often in deep boreholes, and fluid is circulated through the pipes in order to transfer heat to or from the ground. The pipes are connected to the building heating and cooling system via a heat pump. In winter, a small input of electrical energy to the heat pump increases the temperature of the fluid to a level suitable for the delivery of usable heat to the building. In summer the heat pump can be used to reduce the temperature of the fluid returning from the air-conditioning system before recirculation through the ground loops.

1.1 Thermal response tests

Thermal conductivity is a key parameter for the design of closed-loop ground energy systems. It is often determined in situ by carrying out a thermal response test (ASHRAE, 2002; Sanner et al., 2005). The test involves circulating a heated fluid around the ground loop in a single borehole heat exchanger for a period of 2–3 days. Changes in the fluid inlet and outlet temperatures are recorded over time, together with the heating power input. By assuming that the borehole heat exchanger is acting as an infinite line heat source, the thermal conductivity can be assessed. Although this is not a perfect representation of the real conditions in the ground, it has been shown that this assumption is appropriate in many cases. However, the limitations and any uncertainty resulting from the rest should also be assessed (Banks, 2008), and reported alongside the derived thermal conductivity.

For an infinite line heat source with a constant heat injection rate per unit depth of the borehole, q (W/m), the temperature change in the ground, ΔT_g (°C), with time, t (s), is given by (Carslaw and Jaeger, 1959)

$$\Delta T_g = \frac{q}{4\pi\lambda} \int_{r^2/4\alpha t}^{\infty} \frac{e^{-u}}{u} du \cong \frac{q}{4\pi\lambda}\left[\ln\left(\frac{4\alpha t}{r^2}\right) - \gamma\right] \quad (1)$$

where λ and α are the ground thermal conductivity (W/mk) and diffusivity (m²/s) respectively, r is the radial coordinate, and γ is Euler's constant (= 0·5772). To determine the average temperature change of the fluid (ΔT_f), the heat transfer within the borehole must be accounted for and therefore an extra term must be added

$$\Delta T_f = qR_b + \Delta T_g$$
$$\Delta T_f = qR_b + \frac{q}{4\pi\lambda}\left[\ln\left(\frac{4\alpha t}{r_b^2}\right) - \gamma\right] \quad (2)$$

The first term gives the temperature change between the fluid and the edge of the borehole, and is calculated based on the thermal resistance of the borehole, R_b. The second term in Equation 2 is the temperature change at the borehole edge ($r = r_b$), calculated according to Equation 1.

In accordance with Equation 2, the gradient of a graph of fluid temperature change against the natural logarithm of time can be used to determine the thermal conductivity, λ. It is also possible to determine the borehole thermal resistance R_b from the y-axis intercept, provided an assumption is made regarding the value of volumetric heat capacity (S_{cv} in J/m³K) used to derive the thermal diffusivity

$$\alpha = \frac{\lambda}{S_{cv}} \quad (3)$$

1.2   Limitations

The line source theory underlying Equation 2 is based on some key assumptions. The first is that the borehole is infinitely long and thin. Although this is not the case in reality, analysis shows that for typical borehole geometries, where the length-to-diameter ratio is greater than 500, the finite length of the borehole does not become important until heat injection has continued for some decades (Loveridge and Powrie, 2013). Similarly, for a small-diameter borehole (<150 mm), the effect of the finite size of the cross-section results in less than 5% error in predictions of temperature change, provided the time period is greater than half a day (Philippe *et al.*, 2009).

The line source approach also assumes that the rate of heat transfer, q, is invariant along the length of the borehole, that the ground is homogeneous and isotropic, and that there are no external influences such as advection due to groundwater flow. How closely these conditions are approached for any individual test will affect the reliability of the test result. Typically, thermal response tests are considered to be accurate to within 10% (Pahud, 2000; Signorelli *et al.*, 2007; Spitler *et al.*, 2000) when analysed in this way. However, there is a tendency in practice to make a rapid, single-value determination of the ground bulk thermal conductivity, with little consideration given as to whether the boundary conditions of the interpretation method are met.

This chapter examines in detail the results from a 150 m deep thermal response test carried out in east London. It is shown that although reasonable results can be obtained from using a line source method to interpret a thermal response test, these must be tempered by an understanding of the limitations of the test method and the interpretation techniques. In particular, a single-value approach to thermal conductivity may not always be appropriate.

2. Test details

2.1 Site description

As part of a new development in east London, a field of borehole heat exchangers was installed to 150 m depth. The boreholes are 127·5 mm in diameter, and were constructed through the full sequence of London Basin deposits (Table 1). Each borehole contains a single U-loop of plastic pipe, of internal diameter 33 mm. The boreholes are spaced at approximately 5·5 m, and were backfilled with fine to medium chert gravel below the base of the Lambeth Group. This

Table 1 Ground conditions and thermistor levels

Top of stratum: mbgl	Main stratum	Description	Thermistor levels: mbgl
0	Made ground (MG)	Fine to coarse brick and concrete gravel	0·5
		Soft to firm black sandy gravelly clay	2·0
3·3	Alluvium (Al)	Very soft clayey silt, sandy clay and peat	5·5
6·2	River Terrace Deposits (RTD)	Medium dense silty fine to coarse sand and fine to coarse gravel (mainly flint)	8·5
11·2	London Clay (LC)	Stiff thinly laminated fissured silty clay with silt partings	17·5
23·5	Lambeth Group (LG)	Laminated Beds: silty fine sand	25·5
		Lower Shelly Beds: fissured silty clay	38·5
		Lower Mottled Beds: clayey sandy silts and silty fine sands	
		Upnor Formation: very dense green sand	
43·3	Thanet Sands (TS)	Very dense, slightly silty fine sand	48·5
56·1	Chalk (Ch)	Medium density (grade B3 chalk)	63·5
		Subhorizontal and subvertical medium-spaced clean fractures	78·5
			98·5
			118·5
			133·5
			146·5

high-permeability backfill was used to take advantage of the potential for flowing groundwater to enhance the heat transfer characteristics of the borehole. Above the base of the Lambeth Group the boreholes were backfilled with thermally enhanced grout, comprising a bentonite and silica sand mix. Fourteen thermistors were installed over the full depth (Table 1) of one of the boreholes, being attached to the U-loop during installation. The borehole was then subjected to a thermal response test to determine the ground thermal conductivity and borehole thermal resistance.

2.2 Thermal response test rig

A schematic of the thermal response test rig is shown in Figure 1. The rig contains a pump for recirculating the fluid in the ground loop, a flowmeter to measure the recirculation flow rate, and a header tank to maintain the fluid level in the ground loop. Three electrical heaters, in this case of 6 kW, 3 kW and 2 kW, can be used in any combination to apply a fixed heat input to the fluid. A power meter is used to measure the electrical power input to the heaters and the pump.

Figure 1 Thermal response test rig

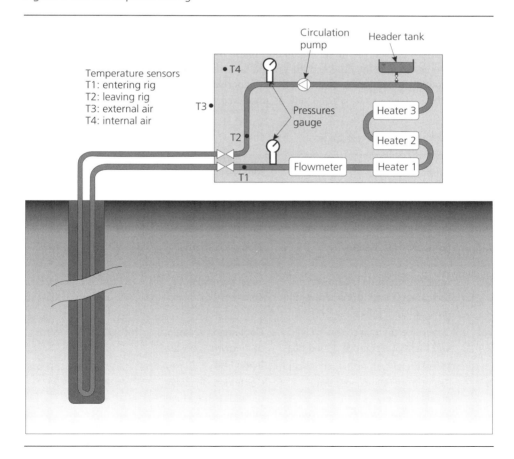

Temperature sensors measure the fluid temperature entering and leaving the ground loop, the external atmospheric temperature and the rig internal temperature. The temperature sensors, along with the flowmeter and power meter, are connected to a data logger for automated monitoring. Both the pipework within the rig and the rig enclosure are insulated to minimise external temperature effects.

2.3 Thermal response test procedure

A thermal response test typically consists of five stages, as shown in Table 2. To establish the ground thermal profile, a mobile thermistor string with measuring points at 5 m intervals is slowly lowered into one side of the ground loop, so as to cause as little disturbance as possible to the water column. The string is left in place until the readings have stabilised, and is then lowered progressively further until the whole depth of the borehole has been covered (stage 1). In this case, as the borehole is also equipped with permanent, cast-in-place thermistors within the borehole backfill (Table 1), these can be used to provide a further check on the ground temperature profile, albeit with less resolution. Once the mobile thermistor string has been removed, the thermal response test rig is connected to the ground loop. It is important that there is no trapped air in the loop, therefore the fluid is recirculated at a high flow rate to purge any air from the system (stage 2). To minimise external environmental effects, the pipes between the rig and borehole are kept as short as possible, and are well insulated.

After purging of the air, the initial circulation phase of the test starts (stage 3). This has two purposes: to ensure that the fluid and the ground have reached equilibrium, and to confirm the average ground temperature over the depth of the borehole. Initial circulation should continue until a thermal steady state is achieved, as measured by equal inlet and outlet temperature

Table 2 Thermal response test stages

Stage	Purpose	Method	Typical duration: h	Duration this test: h
1	Set up; establish initial thermal (temperature) profile in ground	Lower thermistor string down borehole	–	–
2	Purge air from system	Fluid circulation at high flow rate	–	–
3	Establish equilibrium between fluid and ground. Confirm average ground temperature	Initial circulation of fluid	2–12	15
4	Determine thermal conductivity and thermal resistance during heat injection	Continued circulation of fluid with constant heat input	50–60	53
5	Confirm thermal conductivity during recovery	Continued circulation of fluid with no heat input	12–24	21

(T1 and T2 in Figure 1). These temperatures should also correspond to the mean value of the results from the thermistor string used in stage 1. Stage 3 typically requires 2–12 h. In this test, circulation was allowed to run overnight for approximately 15 h.

With circulation continuing, a fixed heat input is applied (stage 4) – in this case 8 kW. This phase should continue for a minimum of 50 h (Sanner *et al.*, 2005); in the current test, heat was applied for approximately 53 h. During the recovery phase (stage 5), circulation of the fluid continues with no heat input. The duration of this phase of the test is typically 12–24 h, and in this case was approximately 21 h.

3. Test results

Figure 2 shows the test data; the heated fluid was subjected to a change in temperature of approximately 16°C before the heaters were switched off. While the nominal applied heating power was 8 kW, the actual applied power can be calculated from the temperature difference between the inlet and outlet pipes, the flow rate measured on the rig, and the specific heat capacity of the fluid. An hourly moving average of the actual applied power is also shown in Figure 2, with values in the range 8 ± 0.025 kW for the main period of heat injection (the early part of the test data is not analysed, as described in Section 3.1 below).

Figure 3(a) shows the undisturbed ground temperature profiles measured by the mobile thermistor string during stage 1 of the thermal response test, along with the permanent thermistor readings prior to the thermal response test. The permanent thermistors show greater scatter than the mobile thermistor string, but the trend is the same. There is an elevated temperature near the surface, reflecting the late summer period during which the test was

Figure 2 Thermal response test data

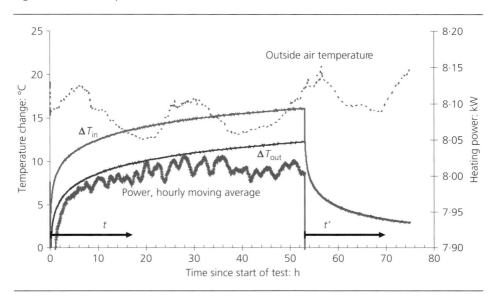

Figure 3 Temperature profiles: (a) undisturbed ground temperature; (b) theoretical fluid temperature profile at end of heat injection

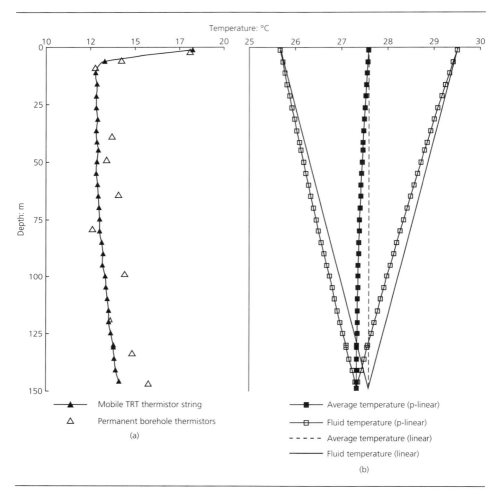

carried out. Below this, there is a gradual increase in temperature with depth. The natural temperature gradient in the ground is greatest at depth, where it is about 1·5–2°C per 100 m. This is slightly less than the median UK geothermal gradient for 0–100 m of 2·2°C per 100 m (Busby *et al.*, 2011). Higher up, between approximately 20 m and 60 m depth, there is little temperature change. This may reflect movement of groundwater, or be part of a trend towards a reversal of the geothermal gradient due to heat losses from the urban environment. Similar and more extreme effects have been observed at other urban sites (e.g. Banks *et al.*, 2009). The mean temperature measured by the thermistor string over the test depth was 13·4°C. This is consistent with the fluid temperatures measured during the stage 3 initial circulation, which were between 13·3°C and 13·5°C.

3.1 Single-value interpretation

The standard technique for interpreting of thermal response data is to take the straight-line portion of the plot of ΔT_f against $\ln(t)$ and use the gradient and intercept to calculate the thermal conductivity of the ground and the thermal resistance of the borehole respectively. The early portion of the dataset is neglected, as at small values of time the mathematical simplification in Equation 1 is not valid. In addition, the assumption of a constant borehole thermal resistance is dependent on a thermal steady state within the hole, and this may take several hours to develop. Therefore interpretation commences after a minimum time, t_{min}, given by

$$t_{min} = \frac{5r_b^2}{\alpha} \quad (4)$$

To estimate t_{min}, an assumption must be made regarding the thermal diffusivity, α. Taking a nominal value of $\alpha = 1 \times 10^{-6}$ m²/s for soils and rocks, t_{min} becomes 5·6 h.

3.1.1 Average fluid temperature

The average temperature of the thermal fluid is often taken as the mean of the loop inlet and outlet temperatures.

$$\Delta T_f = \frac{1}{2}(\Delta T_{in} + \Delta T_{out}) \quad (5)$$

This is on the basis that the rate of change of temperature of the fluid around the loop is uniform, and consequently the mean of the temperatures in the up and down sides of the U-loop is equal along the length of the borehole (Figure 3(b)). Unless the loop flow velocities are high, this is unlikely to be the case in reality. Rather, heat transfer becomes less efficient around the pipe loop, and thus the rate of change of the fluid temperature reduces. Consequently the true average fluid temperature decreases with depth. If this is not taken into account, then the borehole thermal resistance may be overestimated. To allow for this, Marcotte and Pasquier (2008) proposed using a power-linear (or p-linear) relationship to describe the fluid temperature changes with length around the pipe loop (Figure 3(b)). They found the best fit to numerical model data when the power p tended towards a value of -1. On this basis, the average fluid temperature for use with Equation 2 becomes (Marcotte and Pasquier, 2008)

$$\Delta T_f = |\Delta T_{p\text{-linear}}| = \frac{p(|\Delta T_{in}|^{p+1} - |\Delta T_{out}|^{p+1})}{(1+p)(|\Delta T_{in}|^p - |\Delta T_{out}|^p)} \quad (6)$$

Another consequence of the mean fluid temperature decreasing with depth is that the rate of heat transfer with depth is no longer constant. This means that stratification of the ground can influence the test results if a simple line source interpretation is used. This is discussed further in Section 3.2.

3.1.2 Test recovery data

Heat flow around a borehole heat exchanger is strongly analogous to groundwater flow to a well (e.g. Loveridge and Powrie, 2013). To interpret the recovery portion of the test data, the results have been analysed using the same techniques as applied to pumping tests. Taking the time from the start of the recovery part of the test, termed t' (Figure 2), it can be shown by superposition that the temperature of the fluid, ΔT_f (measured as a change from the undisturbed ground temperature), is given by

$$\Delta T_f = \frac{q}{4\pi\lambda} \ln\left(\frac{t}{t'}\right) \tag{7}$$

Thus the recovery portion of the test data can also be used to determine the thermal conductivity by taking the gradient of a graph of the fluid temperature against $\ln(t/t')$. The borehole thermal resistance cannot be obtained from this part of the test.

3.1.3 Results

The changes in average fluid temperature with $\ln(t)$ or $\ln(t/t')$ are shown in Figure 4. Using Equations 2 and 7 applied to the relevant portions of the heat injection and recovery data (the straight-line section from t_{min} to the end of the test phase under interrogation), single lumped values of thermal conductivity and thermal resistance have been calculated using both a mean temperature (Equation 5) and a p-linear fluid temperature (Equation 6). For the heat injection phase of the test the derived value of thermal conductivity is approximately 1·95 W/mK, with a small difference depending on the average fluid temperature used (Table 3). For the recovery phase of the test this increases to approximately 2·07 W/mK. During recovery, the temperature difference between the inlet and outlet is sufficiently small that the two measures of average fluid temperature are the same. It should be noted that the recovery graphs produce a small intercept value, when in theory this should be zero (Equation 7). This is a reflection of the small heat input that will be generated by the circulation pump as well as imperfect boundary conditions. Forcing the best-fit line through the axis origin fails to recognise these factors, and is therefore not appropriate.

The derived thermal conductivities are in the upper part of the range typically reported for the Chalk (between 1·8 W/mK and 2 W/mK; Headon *et al.*, 2009). The difference between the values derived from heat injection and recovery is about 6%. While this is within the generally

Figure 4 Changes in fluid temperature during heat injection and recovery

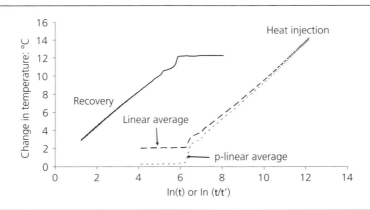

Table 3 Results of single-value interpretation

Test phase	ΔT_f	Graph gradient	Graph intercept	R^2	Thermal conductivity: W/mK	Borehole thermal resistance: mK/W
Heat injection	½($\Delta T_{in} - \Delta T_{out}$)	2·1731	−12·257	0·9997	1·97	0·087
	p-linear	2·2147	−12·912	0·9998	1·94	0·082
Recovery	½($\Delta T_{in} - \Delta T_{out}$)	2·0669	0·307	0·9999	2·07	NA
	p-linear	2·0670	0·323	0·9999	2·07	NA

Note: thermal resistance values assume a ground volumetric heat capacity of 2·69 MJ/m^{3}K

reported accuracy of the test, it is nonetheless significant, and will be explored further later in the chapter.

The borehole thermal resistance is approximately 0·085 mK/W, which is at the lower end of the range of typical values for UK construction (Banks, 2009). This reflects the high thermal conductivity of the saturated siliceous gravel with which most of the borehole is backfilled, and may also include a contribution from both flowing groundwater (advection) and free convection cells developing within the gravel pore spaces. Research suggests that the latter mechanism, where temperature changes induce density-driven groundwater movements, can become important in coarse saturated soils. Pore spaces larger than a few millimetres and high temperature gradients, such as those next to heat exchanger pipes, are required for the effect to become significant (Farouki, 1986).

3.2 Dynamic interpretation

It is also possible to interpret the test data over a range of different timescales, starting from t_{min} and gradually increasing the analysis end time. This allows assessment of whether conditions are changing during the test, and also the uncertainties associated with a simple, single-value interpretation. Given that this analysis can be carried out without recourse to additional fieldwork, and is relatively rapid, it is recommended that the approach be adopted more routinely.

Figure 5 shows the derived thermal conductivity and borehole thermal resistance with time from the start of each phase of the test. In each case the start time of the analysis is t_{min}, and the end time varies. During heat injection the derived thermal conductivity is at first fairly stable, with only a small variation. These variations occur on a 24 hour cycle, and are likely to be related to small heat losses to the air at night when the surrounding temperature is less. However, at the end of the test period there is a marked fall in the thermal conductivity. It is not clear what is causing this change, but it must reflect either a true change in thermal properties (e.g. reduced heat flow in the borehole due to grout cracking) or a change in boundary conditions. An example of the latter would be greater sensitivity to power fluctuations later in the test (when the rate of temperature change reduces), although in this case the power fluctuations are too small (±0·3%) to make this sort of impact (Figure 2). Alternatively, the fall in λ may

Figure 5 Changes in derived thermal conductivity and thermal resistance with time

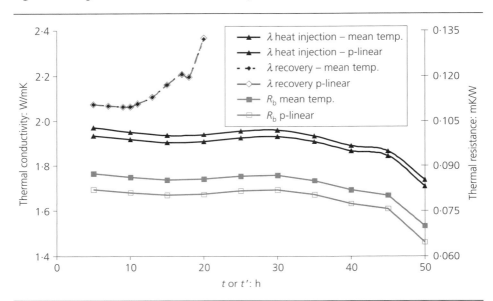

occur because the soil around the upper part of the borehole (above the water table) has dried out during heating. A similar pattern is seen in the changing values of thermal resistance (Figure 5). While this behaviour cannot be satisfactorily explained, it serves to illustrate the uncertainties that can be associated with the test, and the dangers of restricting interpretation to a single value of thermal conductivity.

During the recovery phase of the test, the difference in behaviour compared with a conventional analysis is more significant. First, the derived thermal conductivity values are much higher, and, second, the values increase markedly with time. On the first point, theoretical differences between thermal conductivity derived from heat injection and heat extraction tests have been shown by Signorelli *et al.* (2007). They used a three-dimensional numerical model to demonstrate that the heat flow around borehole heat exchangers deviates slightly from the simple one-dimensional radial flow assumed by the line source model. Instead, the presence of a natural temperature gradient within the ground (Figure 3(a)) leads to a vertical component of flow.

During heat injection the average fluid temperature will decrease with depth (Figure 3(b)). The result is a greater temperature difference between the fluid and the ground at the top of the borehole than at the base (Figure 6(a)). Consequently, the thermal conductivity calculated from the test results will be biased towards the strata surrounding the top of the borehole. During a heat extraction test the situation is reversed: the average fluid temperature increases with depth, but at a lesser rate than the geothermal gradient. Therefore the greatest temperature difference between the fluid and the ground is at the base of the borehole, and the results will be biased to the strata at this location (Figure 6(b)). This means that in stratified ground, where a borehole

Figure 6 Schematic temperature profiles during: (a) heat injection; (b) extraction; (c) recovery (from heat injection)

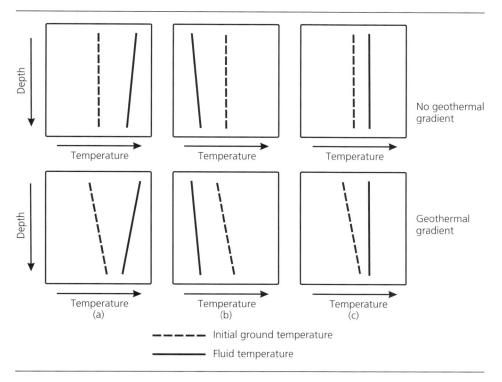

heat exchanger passes through materials of different thermal conductivities, performance will be different in heat injection and in heat rejection. During recovery tests, when there is no applied heat flux, the fluid temperature is constant with depth. In these cases, the bias is considerably reduced, explaining the difference between the results from the two stages of the test (Figure 6(c)).

Figure 5 therefore implies that the top of the borehole is surrounded by soil or rock of lower thermal conductivity than the lower parts. The lower two-thirds of the borehole passes through saturated Chalk, while the top passes through a number of strata, including a significant thickness of London Clay. This would be expected to have a lower conductivity than the Chalk, with typical values for saturated clay being around 1·6 W/mK (Banks, 2008).

As well as being higher, the thermal conductivities derived from the recovery curve increase markedly with time (Figure 5). This is usually an indication of the presence of groundwater flow around the heat exchanger (Sanner et al., 2008). However, it is unusual that this is noticeable only in the recovery phase of the test. This may reflect the fact that the main flow would be expected to be in the Chalk, in the lower part of the borehole, whose contribution would be less significant during the heat injection phase of the test.

3.3 Borehole instrumentation

Differences in behaviour due to vertical variation of ground properties can be investigated by considering the temperature changes with depth in the borehole, as measured by the permanent thermistors installed within the backfill (Table 1). Figure 7 shows the temperature changes in the borehole during heat injection compared with the p-linear fluid temperature (ΔT_f) and the theoretical temperature at the edge of the pipes (ΔT_p). The latter is calculated on the basis of the pipe resistance (R_p)

$$\Delta T_p = \Delta T_f - qR_p \qquad (8)$$

Further details are given in the appendix.

In Figure 7 all the curves for the thermistors parallel the fluid and pipe temperatures, but at different offsets, depending on the distance of the thermistor from the pipes. Ideally, the thermistors would match the pipe temperature, but because of the high temperature gradients close to the pipes even a small variation in position will cause a noticeable difference in the value of the temperature readings. However, despite this, the thermistors will still record an accurate change in temperature resulting from the heat injection, and it is this value (i.e. the gradient of the lines in Figure 7) that is important for interpretation. The straight, parallel nature of the curves in Figure 7 also demonstrates this, and shows that the line source approach is still approximately valid, even for small sections of the borehole. This is because, overall, the principal flow direction is still radial. There will be small variations in this flow path, but any vertical components of the flow are secondary, and it is still possible to use each individual thermistor to calculate the thermal conductivity (e.g. Fujii *et al.*, 2006). This is done by taking the gradient of the lines and using Equation 2 to determine λ in exactly the same way as for the

Figure 7 p-linear temperature change in borehole during heat injection

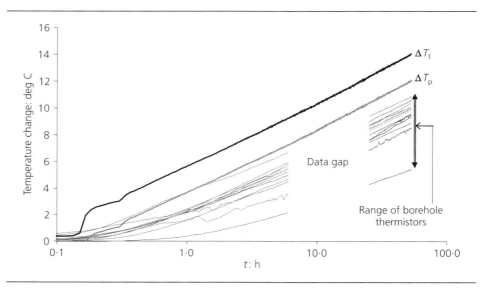

fluid temperatures. The main disadvantage of this approach is that, without the use of temperature sensors within the pipes (which in practice is very challenging, and was not feasible on this site), it is not possible to calculate the small changes in applied heat flux that occur with depth. This will add a small additional error to the results, especially near the borehole ends.

Figure 8 compares the resulting thermal conductivities calculated from the individual thermistors with the values presented in Table 3. Owing to a datalogging problem (see Figure 7) during the heat injection phase, the test has been interpreted from $t = 25$ h to $t = 53$ h during this phase. Large values of thermal conductivity are determined at shallow depth. However, these will not be true reflections of the thermal conductivity, as temperature gradients and heat flow paths in this area are influenced by the ground surface temperatures, and hence the infinite line source analysis is not valid. This will also be true for the base of the hole, but to a lesser extent.

The thermal conductivities determined from recovery are again greater than from heat injection. For both phases of the test the average of the results (neglecting the uppermost and lowermost values, owing to potential end effects) is greater than the value determined from the fluid temperatures, being 2·0 W/mK and 2·2 W/mK for injection and recovery respectively. There is also significant variability. With the exception of peaks in the Thanet Sands, near the top of the Chalk and also near the base of the Chalk, there is little consistency between the results from heat injection and recovery.

Figure 8 also shows the change in temperature in the borehole 30 h after the test has finished and the circulation pump has been switched off. Larger changes in temperature reflect a higher thermal conductivity, or perhaps the influence of groundwater flow. Distinct peaks are visible

Figure 8 Thermal conductivity and post-circulation temperature change with depth

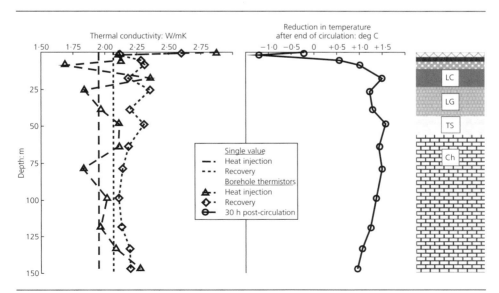

at 17·5 m depth, and between 48·5 m and 78·5 m depth. Surprisingly, the first peak corresponds to the London Clay, which would be expected to have a fairly low thermal conductivity and insignificant groundwater flow. The second, broader peak relates to the top of the lower aquifer, that is, the Thanet Sands and upper section of the Chalk. There is then a reduction in temperature change with depth throughout the Chalk, which is broadly consistent with the thermal conductivity values calculated from the thermistors during recovery, although these also show an increase near the base.

Chalk has high matrix porosity but low matrix permeability, and groundwater movement is dominated by flow in fractures. Studies in the London area (e.g. O'Shea et al., 1995) have shown that the most productive zone (and by inference the most fractured) is generally in the upper 30–40 m of the Chalk. This is consistent with the zone of higher thermal conductivity shown in Figure 8. A decrease in Chalk permeability with depth has also been reported (Williams et al., 2006), which would explain the decline in temperature change for the lower part of the borehole after the end of the test (Figure 8).

To investigate the potential influence of groundwater further, the thermal conductivities derived from the borehole thermistors have been calculated as a function of time. In most cases the results show significant scatter (which is likely to be a reflection of the additional uncertainties introduced by assessing sections of the borehole individually), and definitive conclusions cannot be drawn. Figure 9 shows these results filtered for only those thermistors within the Chalk and at 25·5 m depth within the Laminated Beds of the Lambeth Group. The latter data are included as they show some increase in thermal conductivity with time during recovery. Overall, there is still significant scatter for the heat injection phase, but there is a small increase in thermal conductivity in the recovery phase. As well as at 25·5 m depth, this is especially clear for the thermistor at 63·5 m depth. The latter is not surprising, as it is near the top of the Chalk, where greatest groundwater flow would be expected. The increase in thermal

Figure 9 Change in thermal conductivity with time at different depths: (a) heat injection; (b) recovery

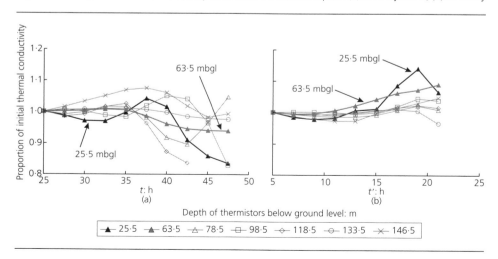

conductivity with time for the Laminated Beds is less expected, although these beds were consistently associated with water strikes during the ground investigation at the site.

For the thermistors the apparent increase in thermal conductivity with time is less than shown for the fluid temperature in Figure 5. This confirms that the groundwater flow must be predominantly along major fractures that have not necessarily been intercepted by the discrete temperature monitoring points within the borehole.

4. Discussion

When carrying out simple single-value interpretation, the differences between performance of the borehole heat exchanger during heat injection and recovery could be considered to be within the limits of accuracy of the test. However, when the test data are considered in more detail, especially with respect to variations in time, the difference in performance is more striking. While the apparent thermal conductivity during heat injection is relatively stable, it increases markedly with time during recovery. This is a strong indicator of groundwater flow, which is important, owing to its significant impact on the heat transfer behaviour of the borehole over its lifetime.

Groundwater movement within the Chalk aquifer is known to be controlled mainly by fractures with the matrix having a much lower permeability. This is consistent with the data from the borehole thermistors. Thermal conductivities derived from these data, which are specific to precise horizons, show an increase during recovery of up to approximately 10%, compared with approximately 15% for the thermal conductivities derived from the fluid temperatures. This would be consistent with groundwater movement being dominated by one or two major fractures that have not necessarily been intercepted by the thermistors.

It is surprising, however, that the influence of groundwater flow is not seen during the heat injection phase of the test. In fact, the opposite appears to be true, with some reduction in thermal conductivity values with time during this part of the test. Thermal response tests are known to be biased towards the strata at the top of a borehole during heat injection, owing to a combination of the geothermal gradient and the variation in fluid temperature with depth. For the heat injection phase to be insensitive to groundwater flow would suggest that the main flow is happening near the base, or at least in the lower half, of the borehole. This is not consistent with the upper layers of the Chalk, less than halfway down the borehole, being the most productive in terms of groundwater extraction. Nonetheless, the thermal conductivities derived at different depths and shown in Figure 8 do show some increase near the base of the borehole, even discarding the lowest value because of end effects.

It can also be inferred from Figure 10, which plots the specific capacity (well yield divided by drawdown, l/s/m) of 353 wells in the London Basin, that although the upper 60 m of the Chalk aquifer is clearly the most productive, there is a subsidiary peak beyond 100 m depth where the specific capacity increases again. This suggests a fractured horizon at greater depth in at least some locations. The presence of such a feature, possibly a zone of fracturing related to hardgrounds in the Chalk, could explain the observed test results. It is unlikely that fractures at this depth would be as wide as the potentially solution-enlarged fissures higher up the

Figure 10 Specific capacity of wells installed in the Chalk in the London Basin (after Water Resources Board, 1972)

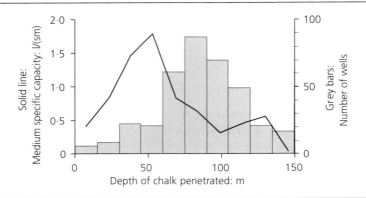

sequence. However, it is possible that such fracturing could extend for a greater thickness, and thus make a significant contribution to flow.

5. Conclusion

Thermal response testing is an important technique for determining the thermal characteristics of both the ground and the borehole heat exchanger for use in the design of ground energy systems. However, the standard test measures only a bulk value of thermal conductivity for the ground, and owing to an imperfect fit with the assumed boundary conditions of the line source model there may be uncertainties in the results. Consequently it is recommended that these uncertainties be investigated and reported in routine practice; in particular, the following

(a) Both heat injection and recovery test phases should be carried out. This enables any difference in behaviour associated with the direction of heat flow to be determined. Differences in the bulk thermal conductivity between the two test phases are a reflection of different thermal properties at the top and base of the hole. This is because the results from heat injection are biased towards the upper part of the borehole.

(b) The test results should be interpreted over a range of time periods. This will help to identify external influences on the test data, such as groundwater flow, which could have a significant impact on the long-term performance of the energy system.

In the case study presented, evidence for groundwater flow is seen only in the recovery phase of the test. This would suggest that a flow horizon is present in the lower half of the borehole, possibly towards the base. This result was unexpected, as the upper layers of the Chalk, about halfway down the borehole in this case, are more typically associated with groundwater movement. Given the presence of groundwater throughout the lower aquifer, the use of high-permeability siliceous sand and gravel as borehole backfill material has been shown to be beneficial in terms of minimising borehole thermal resistance.

Acknowledgements

This work has been carried out with support from the UK Engineering and Physical Sciences Research Council (EP/H049010/1). The authors would like to thank everyone who helped with the site works. This includes, but is not limited to, Arup Geotechnics and Tony Suckling of Balfour Beatty Ground Engineering for initial assistance, Roger Macklin and Gareth Powell at Geothermal International Ltd, Paul Turnbull and colleagues at Geo-Drill Ltd, and the staff of Foundation Developments Ltd. We are particularly grateful for the support of ISG, especially Malcolm Peart. We would like to acknowledge assistance with instrumentation installation and data logging from Joel Smethurst, James Rollinson and Harvey Skinner.

Appendix: Thermal resistance of ground loop pipes

The temperature change across a borehole heat exchanger is the product of the heat flux, q (W/m), and the thermal resistance, R_b (mK/W)

$$\Delta T_f - \Delta T_{rb} = qR_b \tag{9}$$

The thermal resistance can be represented as the sum of its component parts, the resistance of the borehole grout (or other backfill material) and the resistance of the pipes, split into a conductive resistance and a convective resistance:

$$R_b = R_p + R_{grout} = \frac{1}{q}(\Delta T_f - \Delta T) + \frac{1}{q}(\Delta T - \Delta T_{rb}) \tag{10}$$

$$R_p = R_{pconv} + R_{pcond} \tag{11}$$

The conductive resistance can be calculated using the analytical solution for the thermal resistance of a cylinder and assuming the two pipes act in parallel

$$R_{pcond} = \frac{\ln(r_o/r_i)}{4\pi\lambda_{pipe}} \tag{12}$$

where r is the pipe radius, with the subscripts i and o indicating the inner and outer dimensions. The convective resistance is calculated based on the heat transfer coefficient at the fluid pipe interface, h_i. This is dependent on the flow conditions in the pipe.

$$R_{pconv} = \frac{1}{4\pi r_i h_i} \tag{13}$$

h_i can be calculated using the Gnielinski correlation (Gnielinski, 1976) for the Nusselt number (Nu)

$$h_i = \frac{Nu \lambda_{fluid}}{2r_i} \tag{14}$$

where

$$Nu = \frac{(f/8)(Re - 1000)Pr}{1 + 12.7(f/8)^{0.5}(Pr^{2/3} - 1)} \qquad (15)$$

where Re is the Reynolds number, Pr is the Prandtl number, and f is the friction factor. For turbulent flow in smooth pipes this can be calculated using this expression from Petukhov (1970)

$$f = [0.79\ln(Re) - 1.64]^{-2} \qquad (16)$$

REFERENCES

ASHRAE (American Society of Heating, Refrigeration and Air-Conditioning Engineers) (2002) *Methods for Determining Soil and Rock Formation Thermal Properties from Field Tests*. American Society of Heating, Refrigeration and Air-Conditioning Engineers, Atlanta, GA, USA, Research Summary 1118-TRP.

Banks D (2008) *An Introduction to Thermogeology: Ground Source Heating and Cooling*. Blackwell, Oxford, UK.

Banks D (2009) Ground source heating/cooling systems: from conceptualization, through testing to sustainable design. *Proceedings of the Géotechnique Symposium in Print on Thermal Behaviour of the Ground*, London, UK.

Banks D, Gandy CJ, Younger PL, Withers J and Underwood C (2009) Anthropogenic thermogeological 'anomaly' in Gateshead, Tyne and Wear, UK. *Quarterly Journal of Engineering Geology and Hydrogeology* **42(3)**: 307–312.

Busby J, Kingdon A and Williams J (2011) The measured shallow temperature field in Britain. *Quarterly Journal of Engineering Geology and Hydrogeology* **44(3)**: 373–387.

Carslaw HS and Jaeger JC (1959) *Conduction of Heat in Solids*. Oxford University Press, Oxford, UK.

DECC (Department for Energy and Climate Change) (2011) Renewable heat incentive, Department for Energy and Climate Change, March 2011. DECC, London, UK. See https://www.gov.uk/government/uploads/system/uploads/attachment_data/file/48041/1387-renewable-heat-incentive.pdf (accessed 05/02/2013).

Farouki OT (1986) *Thermal Properties of Soils*. Series on Rock and Soil Mechanics Volume 11. Trans Tech Publications, Clausthal-Zellerfeld, Germany.

Fujii H, Okubo H and Itoi R (2006) Thermal response tests using optical fiber thermometers. *Geothermal Resource Council Transactions* **30**: 545–552.

Gnielinski V (1976) New equation for heat and mass transfer in turbulent pipe and channel flow. *International Chemical Engineering* **16(2)**: 359–368.

Headon J, Banks D, Waters A and Robinson VK (2009) Regional distribution of ground temperature in the Chalk aquifer of London, UK. *Quarterly Journal of Engineering Geology and Hydrogeology* **42(3)**: 313–323.

Loveridge FA and Powrie W (2013) Pile heat exchangers: thermal behaviour and interactions. *Proceedings of the Institution of Civil Engineers – Geotechnical Engineering* **166(2)**: 178–196.

Marcotte D and Pasquier P (2008) On the estimation of thermal resistance in borehole thermal conductivity test. *Renewable Energy* **33(11)**: 2407–2415.

O'Shea MJ, Baxter KM and Charalambous AN (1995) The hydrogeology of the Enfield–Haringey artificial recharge scheme, north London. *Quarterly Journal of Engineering Geology and Hydrogeology* **28(Supplement 2)**: S115–S129.

Pahud D (2000) Two response test of two 'identical' boreholes drilled to a depth of 160 m near Luzern. *Proceedings of the Response Test Workshop on the Framework of IEA Energy Conservation through Energy Storage Annex 12 and Annex 13*, 13 October.

Petukhov BS (1970) Heat transfer and friction in turbulent pipe flow with variable physical properties. In *Advances in Heat Transfer* (Irvine TF and Hartnett JP (eds)). Academic Press, New York, NY, USA, pp. 503–565.

Philippe M, Bernier M and Marchio D (2009) Validity ranges of three analytical solutions to heat transfer in the vicinity of single boreholes. *Geothermics* **38(4)**: 407–413.

Sanner B, Hellstrom G, Spitler J and Gehlin SEA (2005) Thermal response test: current status and world-wide application. *Proceedings of the World Geothermal Congress*, Antalya, Turkey pp. 1436–1445.

Sanner B, Mands E, Sauer K and Grundmann E (2008) Thermal response test, a routine method to determine thermal ground properties for GSHP design. *Proceedings of the 9th International IEA Heat Pump Conference, Advances and Prospects in Technology, Applications and Markets*, Zurich, Switzerland, paper 04-35.

Signorelli S, Bassetti S, Pahud D and Kohl T (2007) Numerical evaluation of thermal response tests. *Geothermics* **36(2)**: 141–166.

Spitler JD, Yavuzturk C and Rees SJ (2000) In situ measurement of ground thermal properties. *Proceedings of Terrastock 2000, Stuttgart*. vol. 1, pp. 165–170.

Water Resources Board (1972) *The Hydrogeology of the London Basin*. Water Resources Board, Reading, UK.

Williams A, Bloomfield J, Griffiths K and Butler A (2006) Characterising the vertical variations in hydraulic conductivity within the Chalk aquifer. *Journal of Hydrology* **330(1–2)**: 53–62.

Section 3
Energy piles

ICE Themes Geothermal Energy, Heat Exchange Systems and Energy Piles

Craig and Gavin
ISBN 978-0-7277-6398-3
https://doi.org/10.1680/gehesep.63983.181
ICE Publishing. All rights reserved

Chapter 10
Thermal performance of thermoactive continuous flight auger piles

Fleur Loveridge MSc, PhD, FGS, CGeol, CEng, MICE
Royal Academy of Engineering Research Fellow and Lecturer in Geomechanics, Faculty of Engineering and the Environment, University of Southampton, Southampton, UK

Francesco Cecinato MSc, PhD
Lecturer in Geomechanics, Department of Civil, Environmental and Mechanical Engineering, University of Trento, Trento, Italy

Foundation piles are being increasingly equipped with heat exchangers to efficiently harvest shallow geothermal energy. For buildings in urban areas, continuous flight auger (CFA) piles are common owing to their speed, cost-efficiency and low noise levels. To construct a thermoactive CFA pile usually requires separate central installation of the heat exchanger. However, the energy performance of this type of pile has not been investigated systematically, with most studies focused on rotary piles where the heat exchanger is attached to the reinforcing cage. In this work, insights are provided about the main influences on the energy efficiency of thermoactive CFA piles, with a focus on the implications of using CFA construction techniques rather than rotary boring. An innovative three-dimensional numerical model, able to capture the different aspects of transient heat transfer, is employed together with analytical methods to evaluate the transient and steady-state behaviour of energy piles in a number of design situations. Attention is given to understanding the role of possible pipe-to-pipe interaction, which cannot be systematically investigated with standard methods. Finally, practical guidelines on the optimal choice of design parameters to maximise the energy efficiency of CFA piles, without altering the geotechnical arrangements, are provided.

Notation

c	distance between pipe and pile edges
c_p	specific heat capacity
D	diameter
E_{tot}	total exchanged energy
L	pile length
$\dot{m}$	mass flow rate
n_p	number of pipes
Q	exchanged power
q	exchanged power per unit length

R	thermal resistance
s	centre-to-centre pipe shank spacing
T	temperature
t	time
v	fluid velocity
$\mathbf{x}$	spatial coordinates vector
ρ	density
λ	thermal conductivity

Subscripts

c	concrete
f	fluid
g	ground
in	inlet
m	model
out	outlet
s	steel, solid
2	two-dimensional
3	three-dimensional

Introduction

Every year, the floor area of the European building stock increases by approximately 1%, resulting in additional energy consumption of over 4·5 Mt of oil equivalent (BPIE, 2011). At the same time, the European Union has ambitious carbon dioxide emissions reduction targets (EC, 2009) which are in conflict with such increases in demand. One approach for reducing both energy and carbon dioxide emissions from buildings is to adopt shallow ground energy systems, where ground heat exchangers are combined with a heat pump to improve energy efficiency, potentially reducing demand by around 75%, depending on the system coefficient of performance.

Further financial and embedded carbon economies can potentially be made by using the foundations piles of a building to host the heat exchanger part of the ground energy system, so called thermoactive piles or energy piles. This innovation was pioneered in the 1980s in Austria and Germany (Brandl, 2006). However, progress towards more global adoption of thermoactive piles and other geostructures has only taken place more recently (e.g. Barla and Perino, 2014; Laloui and Di Donna, 2012). This has triggered a renewed interest in thermoactive pile research, and for the first time, there has also been a focus on maximising the energy efficiency of these systems (Bozis *et al.*, 2011; Cecinato and Loveridge, 2015; Gao *et al.*, 2008; Wood *et al.*, 2010).

For piles to become thermoactive, polyethylene pipes must be embedded within the pile concrete to allow the heat transfer from the ground to the energy system to occur by means of a circulating fluid. However, there are a number of different ways in which foundations piles can be equipped with pipes, and these depend largely on the construction method of the pile itself. In most cases, rotary bored piles are the most common type of pile used as heat exchangers. However, particularly in the UK, continuous flight auger (CFA) piles are increasingly being used given their prevalence in the building development sector, for example, see Loveridge and Powrie (2013a). This chapter will examine the advantages and disadvantages of the use of CFA piles for thermoactive foundations. While the main focus is energy efficiency implications, the following section will also discuss the construction considerations.

Construction techniques

The use of CFA piles has grown in recent years owing to their many advantages, but not least owing to their speed (and therefore reduced cost) of installation compared with rotary bored piles. In addition, CFA piles are quiet to install and produce low levels of vibrations, making them suitable for city centre sites where new building development may be concentrated. They are, however, not suitable for all ground conditions and are practically limited in size and depth owing to plant capabilities. A review of situations where CFA piles will be economic and those for which they are unsuitable is given by Brown (2005).

When it comes to thermoactive piles, one important difference in the construction method affects how the heat exchange pipes are introduced into the concrete. In the construction of traditional rotary bored piles, the steel reinforcing cage is hung from the pile casing while concrete is placed inside the pile bore, usually by means of a tremie pipe. In CFA piling, the concrete is pumped into the pile bore by means of the hollow stem of the auger, which is then withdrawn as concreting progresses. As a result, the steel reinforcing cage must be plunged separately into the wet concrete after the auger has been withdrawn.

The most practical and economic way to introduce heat transfer pipes to the foundation is by means of attachment of the pipe U-loops to the reinforcing cage. Installation of the pipe loops over the full depth of the pile is essential as the heat transfer length is a key factor in the thermal efficiency of piles used as heat exchangers (Cecinato and Loveridge, 2015). However, for most building development projects (unless the piles are required to carry tension), the steel reinforcement is not required over the full depth of the pile. Consequently, depending on the construction method of the pile, additional specific measures are required to ensure the pipes reach the full depth of the pile bore.

For rotary bored piles where the steel cage is installed ahead of the concrete, it is possible to attach full-length pipe loops to the reinforcement and allow these to hang beneath the cage during concreting. In this way, the pipes may reach the full depth of the pile even if the cage does not. Sometimes additional weight may be required to be attached to the pipes at the base to prevent buoyancy within the concrete owing to the presence of fluid within the pipes, but otherwise the combined reinforcement cage and pipe installation process is straightforward. If the steel cage is constructed in one piece, it is then possible to attach the pipes during pre-fabrication offsite. This means that there is no impact on the piling programme during construction. If the piles are of sufficient depth that the steel cages require splicing then the pipes will instead need to be attached to the cage sections during their installation. Nevertheless, a full-depth pipe installation based on attachment to the reinforcement cage is still easily accomplished.

For CFA piles, however, a different approach must be adopted. Because the steel reinforcing cage is plunged into the wet concrete, it is only possible to insert pipes with the steel cage if the pipes are limited to the length of the cage. Because the reinforcing cage is rarely full depth for a building foundation, an alternative approach to installation of the pipes is usually adopted to maximise the available heat transfer length and hence energy efficiency. To permit a full-depth installation, the pipes must be installed separately following insertion of the pile cage into the concrete. Typically, the pipes are attached to an additional steel bar for weight and rigidity and then plunged into the centre of the pile. This additional operation during construction means that

Figure 1 Typical cross-sections for (a) rotary bored and (b) CFA piles (after Loveridge and Powrie, 2013b)

Pipes installed inside prefabricated steel cage	Pipes installed outside cage during construction	Pipes and steel bar plunged into centre of pile after concrete is poured
(a)		(b)

Labels: Shear links (horizontal steel); Main reinforcing steel; Heat transfer pipes; Steel bar for stiffness

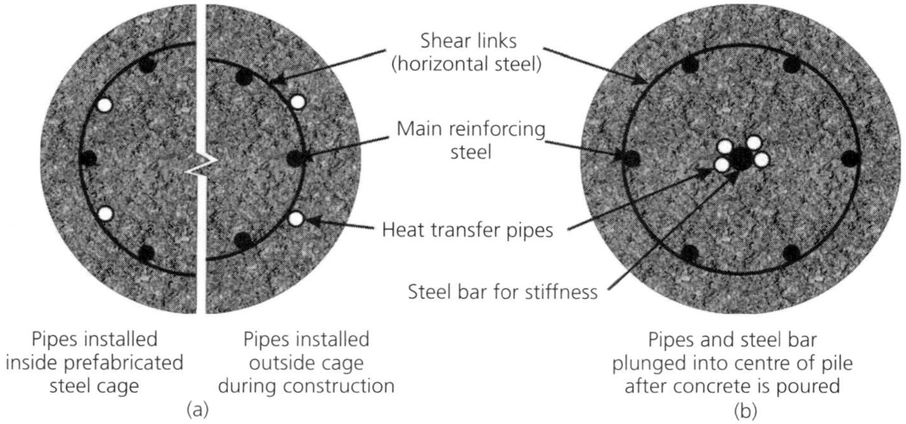

there can be some supplementary programme time required for converting CFA piles to thermoactive piles. Additional construction considerations may also arise including the need for the concrete to remain workable until the loops are installed (more problematic in certain ground conditions) and whether a handling crane will be available to lift the loops (Amis *et al.*, 2014).

Owing to these different construction approaches, rotary bored piles tend to have their pipes spaced apart around the steel cage (Figure 1(a)), which is typically only 50–75 mm from the ground. Typical pipe spacings are between 250 and 300 mm (Loveridge and Powrie, 2013b). CFA piles, on the other hand, will more typically have their pipes installed closer together in the centre of the pile (Figure 1(b)) and pipe spacings rarely exceed 60 mm. Practically no more than four pipes (two U-loops) are usually installed in the centre of CFA piles, while many more pipes may be placed in rotary bored piles depending on their diameter.

Thermal performance assessment
Background
Recently, a systematic assessment of the thermal efficiency of rotary bored piles was carried out based on numerical simulation (Cecinato and Loveridge, 2015). This assessment ranked a number of key design parameters in the order of their impact on the thermal performance of the pile. The most important parameter was found to be the number of pipes installed in the pile cross-section, followed by the length of the pile. The latter parameter highlights the importance of ensuring the pipe loops are installed over the full depth of the pile since the piled foundation will rarely be extended in depth to accommodate greater energy availability. Following the pile depth, its thermal conductivity and the diameter of the cross-section were also found to be important. Of these four parameters, the number of pipes installed is the most straightforward to engineer for thermal performance, followed by the concrete conductivity, although the latter will also be influenced by the economics of available aggregate sources, and often closer

sources may be chosen over those which are more thermally advantageous but require greater transport distances. Like the pile length, its diameter is unlikely to be adjusted only to satisfy the thermal design.

Based on the results of the study undertaken for rotary bored piles, it is likely that the number of pipes installed and the concrete conductivity would influence the efficiency of CFA piles. Additionally CFA piles would be expected to be less energy efficient than rotary piles for two reasons. First, the pipes installed will always be closer together (Figure 1), and second, practically fewer pipes may be installed within a CFA pile. Additionally, there have been concerns that close proximity of the pipes in CFA piles, as well as their installation with a high thermal conductivity material (the steel bar), would make this arrangement vulnerable to pipe-to-pipe interactions. Adverse pipe-to-pipe interactions have been studied for borehole heat exchangers (e.g. Lamarche et al., 2010; Muraya et al., 1996), resulting in the common practice of using spacers between the two shanks of borehole U-loops. However, for the case of thermoactive piles, no assessment of the potential for pipe-to-pipe interactions has been made, and the thermal performance of CFA more generally has not previously been quantified.

Numerical implementation

Model description

The model reproduces the three main heat transfer mechanisms taking place in thermoactive structures, namely thermal convection between the heat transfer fluid and the pipe wall, thermal conduction in the concrete and thermal conduction in the ground.

The convection-diffusion equation that applies to the heat exchanger fluid is

$$\dot{m}c_{pf} \nabla T = h\Delta T \tag{1}$$

where c_{pf} is the fluid specific heat capacity, $\dot{m} = \rho v A$ is the mass flow rate, A is the pipe cross-sectional area, v is the fluid velocity, ρ is the fluid density, h is the 'film' (or convective heat transfer) coefficient, and $\Delta T = (T_s - T_f)$, the temperature difference between the solid interface (pipe wall) and the fluid.

It is assumed that (*i*) convection owing to fluid flow occurs as a quasi-static phenomenon, and (*ii*) conductive heat transfer along the flow direction can be neglected compared to both the radial heat transfer at the fluid/pipe wall interface and the convective transfer. In addition, the contribution of friction heat dissipated by viscous shear is neglected (cf. Cecinato and Loveridge, 2015). These simplifications have been shown to be appropriate with the modelling approach validated against field data and analytical solutions, both for borehole heat exchangers (Choi et al., 2011) and rotary bored piles (Cecinato and Loveridge, 2015).

The heat transfer through the pipe wall, concrete, and the ground is governed by standard transient heat conduction, as

$$\rho_s c_{ps} \dot{T} = \nabla (\lambda_s \nabla T) \tag{2}$$

where ρ_s, c_{ps} and λ_s are respectively the density, specific heat capacity and thermal conductivity of the considered solid material, and $\dot{T}$ is the temperature time rate.

The transient heat convection-diffusion problem was solved employing the software ABAQUS to integrate three-dimensional (3D) transient conduction through the solids, complemented by writing bespoke user subroutines to model the convective heat transfer at the fluid/solid interface and the temperature changes in the fluid along the pipe. Each solid material (soil, concrete and steel) was defined by specifying its density, specific heat and conductivity. At each time step, alongside the standard ABAQUS calculation of heat diffusion in the concrete/ground, the necessary convection computations were performed in a semi-coupled way involving (i) the calculation by means of subroutine FILM, at each pipe segment, of the radial heat flux; (ii) the calculation by means of subroutine URDFIL, at each pipe node, of the fluid temperature change. Further details of the ABAQUS model can be found in Cecinato and Loveridge (2015).

The 3D FE mesh was created manually in an axisymmetric fashion using 6-node linear triangular prism and 8-node linear brick diffusive heat transfer elements (Figure 2). The size of the domain (5·5 m dia. and 29 m depth) was chosen by numerical experimentation to be much larger than the area actually affected by heat transfer. It should be observed that the approach outlined implies the schematisation of the pipes within the FE mesh as lines of nodes, where the heat exchange resulting from convection-diffusion in the pipes is concentrated. The 3D nature of the pipes (i.e. the relevant diameter, in addition to length) is properly accounted for by means of the user subroutines, by multiplying the heat flux corresponding to each pipe node by the corresponding lateral surface area of each pipe segment. As a consequence, each pipe node in the 3D mesh lies in the barycentre of a pipe segment. Correspondingly, if a two-dimensional (2D) cross-section of the domain is considered, each node lies in the centre of the circular pipe cross-section (Figure 2).

Simulation settings
A single CFA energy pile is represented in the mesh. Whenever two pipes (a single U-loop) are installed, significant calculation time saving is made by exploiting the symmetry of the problem, since only half of the domain is considered. However, the complete domain must be considered whenever two or more U-loops connected in series are present.

As boundary conditions, the inlet fluid temperature is prescribed with a constant inlet temperature of 20°C, after a short initial ramp lasting 5' to avoid possible numerical problems owing to the abrupt temperature change. As initial (undisturbed) ground temperature, a value of 12°C (averagely representative of central Europe) was chosen. The initial ground temperature is also taken as the farfield boundary conditions. A total simulation time of 4 days was set for all analyses, which could be typical of a thermal response test on a large diameter pile, and short enough to save computational time.

The key outputs from the simulations are the outlet temperature history $T_{out}(t)$ (i.e. the temperature of the fluid as it exits the pile, at ground level), the fluid temperature history at every node constituting the pipes $T_f(\mathbf{x},t)$ and the pipe and pile wall temperature histories $T_{pipe}(\mathbf{x},t)$ and $T_{pile}(\mathbf{x},t)$, where $\mathbf{x}$ is the spatial coordinates vector and t time.

Figure 2 FE mesh for a two pipe (single U-loop) thermoactive CFA pile, with indication of the main geometrical quantities (concrete cover c = 400 mm); only half of the domain is considered in this case, for symmetry reasons. (a) Overview of the pile area (horizontal dimension = 1·15 m), (b) enlargement of bar and pipes area (horizontal dimension = 0·5 m)

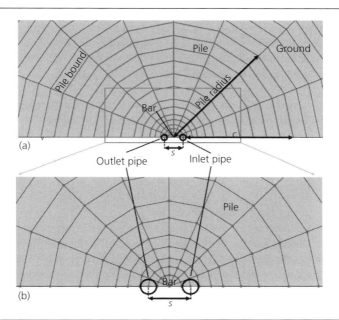

The energetic efficiency can be assessed by considering the total exchanged energy in a given time. The total exchanged power Q can be calculated from each simulation as

$$Q(t) = \dot{m}c_{\text{pf}}[T_{\text{in}}(t) - T_{\text{out}}(t)] \tag{3}$$

where $T_{\text{in}}(t)$ is the design inlet temperature history. The output variable representing energy exchanged can be then computed as

$$E_{\text{tot}} = \int_0^{t_f} Q(t)\,dt \tag{4}$$

where t_f = 4 days, the total simulation time.

Cases considered

Sensitivity analysis

A sensitivity study was performed with the above calculation hypotheses, by carrying out a number of simulations while varying the parameters that are potentially easier to engineer for CFA piles, keeping constant at typical values the model parameters that do not exhibit a high variability or cannot be easily engineered. To focus on the features that are peculiar of CFA piles, and in light of the existing knowledge on the sensitivity of rotary piles to a large parameter set (Cecinato and Loveridge, 2015), for a first CFA parametric study, the decision was made to vary the number of pipes, n_p, and the fluid velocity, v, only. We consider n_p to take two possible values, $n_p = 2$ (a single U-loop) and $n_p = 4$ (a double U-loop). Practically, owing to limited space availability around the steel bar used in CFA construction, greater numbers of pipes are not usually installed. All double U-pipe settings consider pipes connected in series, as this is the most common design option in practice. The fluid velocity is a relatively free design parameter that can be considered to vary within a wider range of values, $0.2 \leq v \leq 1.2$ m/s. To systematically and efficiently investigate the variable parameter space, a total of eight simulations were performed, as summarised in Table 1.

Among the parameters that are kept constant, the pile length L and diameter D_{pile} are set to 25 m and 900 mm, respectively; these are at the upper bound for typical CFA piles but provide convenient sizes for comparison to rotary construction. The concrete and the ground are assumed to be fully water-saturated and to take typical values (Table 2). The pipes are considered to be attached to a central steel bar with diameter $D_{bar} = 40$ mm and are assumed to take a symmetric arrangement (Figure 2). All fixed model parameters are summarised in Table 2.

Comparison to rotary bored piles

For proper comparison with rotary pile situations, three extra simulations (called run 3rot, run 5rot and run 7rot; Table 3) were carried out keeping the same settings of run 3, run 5 and run 7 (Table 1), but changing the pipe position to achieve a concrete cover of $c = 75$ mm, as is typical of rotary bored piles where pipes are attached to the reinforcement cage (Figure 3). These simulations investigate the effect of increased pipe shank spacing, s, in both single and double U-loop cases and with different fluid velocity scenarios.

The numerical model was also used to examine pipes installed with an intermediate shank spacing between CFA and rotary piles (called run 7int; Table 3). While such an arrangement is unlikely to ever be constructed in practice, these simulations were conducted to aid interpretation of the other results.

Table 1 Variable parameter sensitivity settings for investigation of the energy efficiency of thermoactive CFA pile; refer to Table 2 for constant parameter settings

	Run							
Variable parameters	1	2	3	4	5	6	7	8
Fluid velocity, v: m/s	0·2	0·4	0·8	1·2	0·2	0·4	0·8	1·2
Number of pipes, n_p	2	2	2	2	4	4	4	4

Table 2 Constant parameter values adopted in the CFA sensitivity analysis; refer to Table 1 for variable parameter sensitivity settings

Parameters	Value	Units
Pile diameter	900	mm
Pile length	25	m
Central bar diameter	40	mm
Pipe external diameter	30	mm
Pipe wall thickness	2·7	mm
Soil thermal conductivity	2	W/mK
Initial soil temperature	12	°C
Ground specific heat	1600	J/(kg K)
Concrete specific heat	1000	J/(kg K)
Steel specific heat	473	J/(kg K)
Soil density	1900	kg/m^3
Concrete density	2210	kg/m^3
Steel density	7801	kg/m^3
Soil conductivity	2	W/mK
Concrete conductivity	3	W/mK
Steel conductivity	43	W/mK

Table 3 Variable parameter settings for additional simulations; refer to Table 2 for constant parameter settings

	Run						
Variable parameters	3rot	3_Lc	3_Lh	5rot	7rot	7int	7_unsym
Fluid velocity: m/s	0·8	0·8	0·8	0·2	0·8	0·8	0·8
Number of pipes	2	2	2	4	4	4	4
Pipe positions	Rotary	CFA	CFA	Rotary	Rotary	Intermediate	CFA, bunched
Shank spacing, s: m	0·72	0·07	0·07	0·72	0·72	0·37	0·03
Steel conductivity: W/mK	43	3	73	43	43	43	43

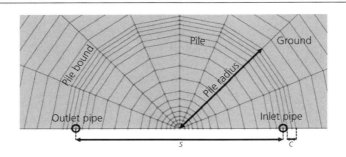

Figure 3 FE mesh for a two pipe (single U-loop) thermoactive rotary bored pile for comparison with the CFA case, with an indication of the main geometrical quantities (concrete cover c = 75 mm, pipe shank spacing s = 720 mm); only half of the domain is considered in this case, for symmetry reasons (horizontal dimension = 1·2 m)

Pipe positioning

While the analyses described above consider a centrally (and axially) symmetric arrangement for the pipes, the situation exists in practice, in the absence of spacers, when the pipes are embedded into the pile concrete in a bunched fashion, so that they are not regularly distributed around the steel bar (Figure 4(a)). Bringing the pipes closer to each other compared to a symmetric arrangement, the question may arise whether this will have an impact on the overall energy performance. Thus, a further simulation was carried out (called run 7_unsym; Table 3) with the same settings as in run 7, but applying a worst case unsymmetrical pipe arrangement (Figure 4(b)).

The impact of the steel bar

Finally, the sensitivity of results to the presence of the central steel bar or the variation in its thermal conductivity was investigated by running two simulations, called run 3_Lc and run 3_Lh (Table 3). The same settings as in run 3 were adopted, except for the conductivity of the central bar material, that was set to 3 W/mK in the former case (i.e. considering the same conductivity as concrete), and to 73 W/mK in the latter case (i.e. considering the bar to be composed of an extremely conductive steel).

Results

Energy performance

Results of sensitivity study

In Figure 5, the results of the numerical sensitivity analysis described above are reported, in terms of (*a*) outlet fluid temperature evolution and (*b*) evolution of exchanged power per metre depth of the pile. It can be observed that the largest temperature change is achieved when the fluid velocity is smallest (run5 and run1); however, by virtue of Equation 3, more thermal power is exchanged at larger mass flowrates. Figure 5(b) shows that the number of pipes also increases the exchanged power, hence in principle the efficiency of the geothermal structure. To better quantify the energy performance, in Figure 6, the total energy exchange E_{tot} (Equation 4) for each simulation is shown. It emerges that the energy efficiency of CFA piles is

Figure 4 Unsymmetrical pipes arrangements in CFA piles: (a) a practical example of a 600 mm pile (pipes protected in pile break-out zone); (b) FE mesh for simulation of this type of pipe arrangement

(a)

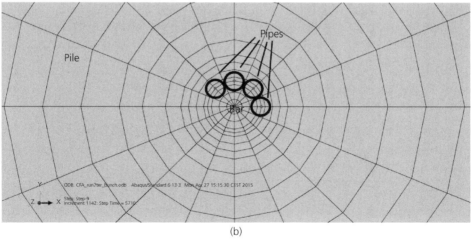

(b)

an increasing function of both fluid velocity and of the number of pipes, the latter being more influential in absolute terms. It is also clear that the energetic benefit of increasing the velocity is weaker in the upper range of typical velocities. The biggest increase in energy exchanged is from $v = 0.2$ m/s to $v = 0.4$ m/s. This corresponds to an increase in Reynolds number from 4900 to 9800, that is, the exit from the transient zone into fully turbulent flow. In other words,

Figure 5 Results of the thermoactive CFA pile sensitivity analysis: (a) fluid outlet temperature evolution; (b) evolution of exchanged power per unit length

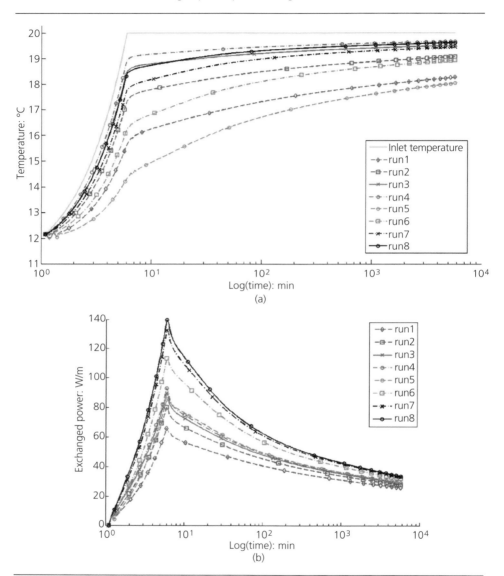

Figure 6 Results of the thermoactive CFA pile sensitivity analysis: total energy exchange for the different simulation runs

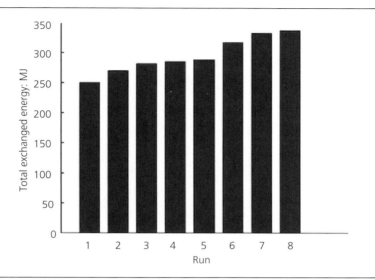

as already observed for rotary bored piles by Cecinato and Loveridge (2015), provided that the fluid velocity is large enough to ensure turbulent flow, increasing it further would only have a secondary impact in the energy efficiency. Additionally, for any operational system, it must also be considered that increasing the flow rate will require a corresponding increase in the pumping energy which will have a detrimental effect on the overall system performance. These two factors will need to be balanced during system design.

Comparison to rotary bored piles

Figure 7 shows the comparison between CFA and rotary bored piles with the same properties, but with different shank spacing. Both the exchanged power and outlet fluid temperature are presented for each simulation. In all cases, the rotary arrangements show a larger temperature change, and a corresponding larger exchanged power, implying a larger energy exchange in the considered time window. To corroborate this outcome, in Figure 8, the total energy exchanged is shown for all cases. It can be observed that the energetic improvement when switching between the CFA and the rotary configuration is substantial, increasing by up to a factor of two in the four pipe case (run 7).

Figure 7 Comparison of thermoactive CFA piles with rotary piles: (a–c) exchanged power evolution; (d–f) fluid outlet temperature evolution

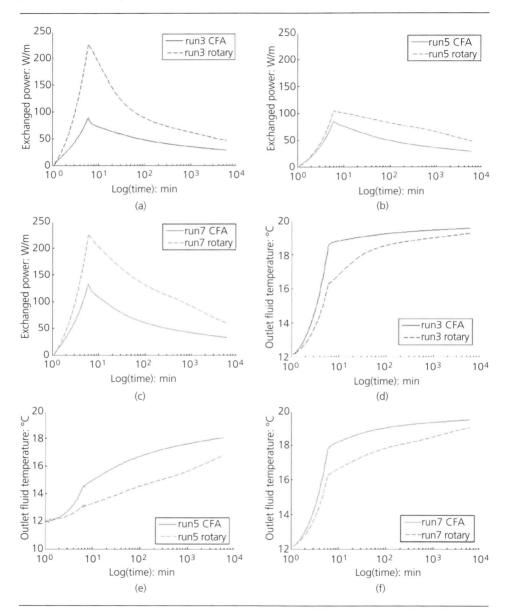

Figure 8 Comparison of thermoactive CFA pile with rotary piles: total energy exchange for the different simulation runs

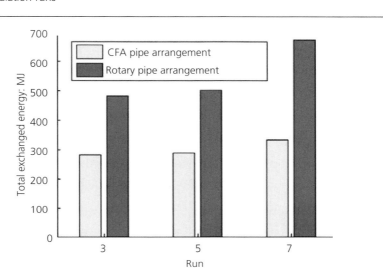

Effect of central bar and pipe arrangements

By comparing the output of simulations run3 (standard CFA settings), run3_Lc (central bar as conductive as concrete), and run3_Lh (central bar made of highly conductive steel), a negligible difference is observed, both in terms of exchanged power and of total energy. Variation between the three simulations is within 1%. Hence, it can be concluded that the presence of the steel bar does not affect the pile thermal performance.

On the other hand, different pipe arrangements appear to have a major impact on the energy performance. As can be seen in Figure 9(a), a significant decrease of exchanged power is obtained with run7 settings switching from a symmetrical CFA pipe arrangement to an unsymmetrical CFA arrangement (run7_unsym). This could be owed to pipe-to-pipe interactions, in that part of the heat may be exchanged between adjacent pipes, not contributing to outwards heat diffusion (the potential for pipe-to-pipe interactions is further discussed in the next section). Keeping the symmetrical pipe configuration and increasing the shank spacing, from a CFA situation (run7) through an intermediate case (run7_interm), to a rotary configuration (run7rot), the exchanged power substantially increases. This is also reflected by the evolution of average pile wall temperature over time, shown in Figure 9(b) for run7 at different pipe configurations. Larger pile wall temperature changes correspond to larger overall exchanged energy (Figure 9(c)).

Figure 9 Comparison of different pipe arrangements in terms of (a) exchanged power evolution, (b) average pile wall temperature evolution and (c) total exchanged energy for run7, including unsymmetrical CFA and intermediate pipe arrangements

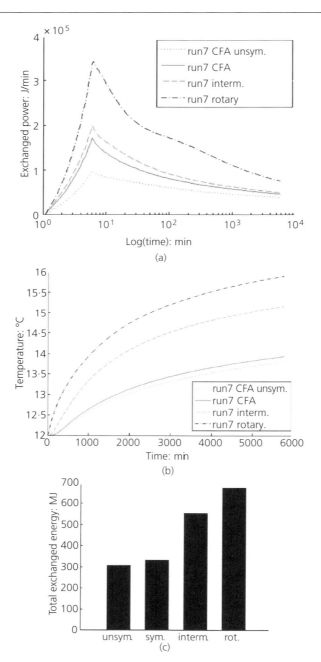

As a further example of the model simulation capabilities, the distribution of temperature around the pile cross-section at the end of the simulation time is shown in Figure 10. Typical temperature contours are shown at pile mid-depth (z = 12·5 m), for a two-pipe arrangement (run 3) in both CFA (Figure 10(a)) and rotary (Figure 10(b)) configuration, and for a four-pipe arrangement (run7) in both symmetrical CFA (Figure 10(c)) and unsymmetrical CFA (Figure 10(d)) configuration. It can be noticed that fewer symmetrical cross-sectional temperature distributions occur in the two-pipe rotary case and in the unsymmetrical CFA arrangement. It is also observed that in the rotary case, when shank spacing is significantly larger, the ground temperature increase is also larger, all other things being equal, owing to the proximity of the pipes to the ground and to the wide separation between the two pipes which keeps pipe-to-pipe interactions to a minimum.

The potential for pipe-to-pipe interactions

The potential for pipe-to-pipe interactions in CFA piles has been assessed by considering the numerically calculated concrete thermal resistance of the pile in comparison with that determined from analytical solutions. The concrete resistance (R_c) is a steady state parameter that is defined as

$$R_c = (\bar{T}_{pipe} - \bar{T}_{pile})/q \tag{5}$$

where $\bar{T}_{pipe}$ is the average temperature on the outside of the pipes, $\bar{T}_{pile}$ is the average pile wall temperature, and q is the exchanged power per metre depth of the pile. When defined from the outputs of the numerical model an additional subscript 'm' will be used to distinguish the value from analytically derived thermal resistance.

Analytically, the concrete resistance can be determined in two dimensions (R_{c2}) using the line source method (Hellstrom, 1991). This approach is easy to implement for a single U-loop and has been complemented by using the simplified method of Loveridge and Powrie (2014) for the case of two U-loops. However, to include the effects of pipe-to-pipe interactions, a pseudo-3D version (R_{c3}) must be determined (Hellstrom, 1991), which is only readily applicable to the single U-loop case. Both the line source method and the calculation of R_{c3} according to Hellstrom have been validated by Lamarche *et al.* (2010) for borehole heat exchangers using numerical simulation. Full equations for the analytical calculation of the concrete resistance are given in Appendix A.

To make appropriate comparisons between the numerically and analytically determined pile resistances, it is first important to determine whether the concrete in the piles has approached a steady state. The concrete resistance from the model (R_{cm}) was calculated dynamically over the simulation time, examples of which are plotted in Figure 11. It can be seen that while the rotary piles reach steady state more rapidly, a steady state is still approached for the case of the CFA piles by the end of the simulation. By the end of the 4-day period, the rate of change of pile thermal resistance was a maximum of 0·004 mK/W per day.

Figure 10 Temperature contours at the end of simulation ($t = 4$ days) in a pile cross-section at mid-height ($z = 12 \cdot 5$ m) for a 2-pipe arrangement (run 3) in both CFA (a) and rotary (b) configuration, and for a 4-pipe arrangement (run7) in both symmetrical CFA (c) and unsymmetrical CFA (d) configuration

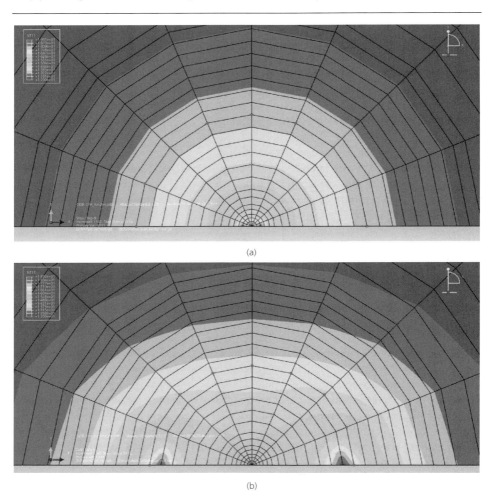

R_{cm}, determined at the end of the 4-day simulations, is compared with the analytical calculations in Figure 12. Considering first runs 1, 2, 3 and 4 with only a single U-loop installed, it can be seen that the numerically determined R_c is larger than the analytical values in all cases. The 2D and 3D resistances are virtually equal for runs 3 and 4 with the highest flow rates, while for runs 1 and 2, the 3D resistance is slightly higher. Hence, based on the analytical calculations, pipe-to-pipe interactions seem to have a small impact even at lower flow rates and to be negligible at higher flow rates. However, the significantly higher values of numerically derived resistance could also suggest pipe-to-pipe interactions are occurring to a significant degree in all four cases. A similar result is seen for the four-pipe CFA piles, although it has not been possible to determine R_{c3} in this case.

Figure 10 Continued

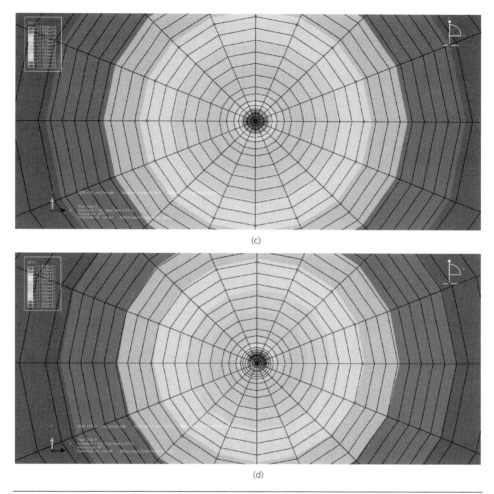

(c)

(d)

The discrepancy between the numerical concrete resistance and the analytical 3D resistance can potentially be explained by two options. First, as the analytical resistance is only pseudo-3D, it is possible that the fully 3D numerical simulation is providing a better indication of the true extent of pipe-to-pipe interactions for CFA piles. This point of view is supported by considering the resistances determined for the rotary simulations. Here, where pipe-to-pipe interactions are not expected owing to the larger shank spacing, the numerical and analytical resistances are much closer together (Figure 12). The case of the intermediate shank spacing (run 7int), halfway between a rotary and a CFA arrangement, also shows similar results between the analytical and numerical calculations.

Figure 11 Example evolution of the thermal resistance with time for CFA and rotary piles containing single and double U-loops

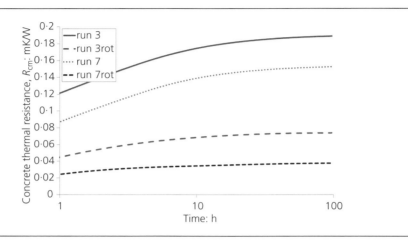

Figure 12 Comparison of numerical and analytical pile steady state concrete resistance, R_c

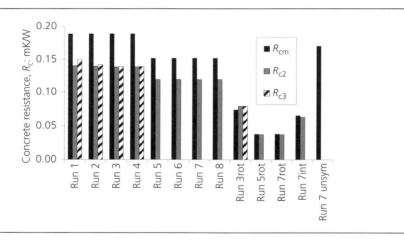

Alternatively, since the numerical model used does contain a simplification of the pipe problem (see 'Model description' section), it is possible that the analytical solutions provide a better estimate of the degree of pipe-to-pipe interactions. If this is the case, then the numerical simplification only becomes of significance when the pipes are close together as is the case for the CFA analyses. Other researchers have used similar one-dimensional representations of pipes in the simulation of ground heat exchangers (e.g. Ozudogru et al., 2014; Signorelli et al., 2007). However, none of these models have been validated in terms of either field data or resistance calculation. Rees and He (2013) modelled borehole heat exchangers using a single layer of cells to represent the fluid. With this approach, they found some short-term transient errors (up to 10%), but that the prediction of steady state resistance compared well with 2D analytical calculations, including for shank spacings closer than those considered for the CFA piles in this study. This may suggest that insignificant pipe-to-pipe interactions were occurring in their study. Therefore, the study of Rees and He (2013) lends some support to a possible overestimation of pipe-to-pipe interactions by the simulations for CFA piles presented above. On the other hand, the numerical model presented can be considered to capture reasonably the dependency of resistance on different pipe arrangements (rotary, intermediate shank spacing, CFA and unsymmetrical CFA), as can be seen from the different arrangements for run 7 shown in Figure 12.

Discussion

This chapter has presented the results from a range of simulations covering a number of different arrangements of CFA and rotary bored thermoactive piles. It is clear that the arrangement of pipes that results from the CFA construction process is less favourable in terms of energy performance than the same pile geometry constructed by rotary boring. Over 4 days, the difference in energy exchanged may approach a factor of two, being represented by a reduction in concrete thermal resistance from 0.15 to 0.19 mK/W for CFA piles to 0.04 to 0.08 mK/W for rotary piles. This seeming disadvantage of CFA piles must be balanced against their speed of installation, low cost and reduced noise and vibration characteristics. These construction advantages for CFA piles in many urban sites tend to outweigh requirements for small amounts of additional programme time and the reduced energy efficiency of thermoactive CFA piles compared with traditional rotary bored piles. Consequently, it is unlikely that the inclusion of the pipe loops would influence the choice of construction method.

Instead, given a construction decision to install CFA piles rather than rotary bored piles, it is important to consider how to maximise the potential for energy performance when converting the piles to become thermoactive. Here, in line with other studies, analysis suggests that maximising the number of pipes and maintaining turbulence in the fluid will both contribute to improving energy output. Investigation of the presence of the steel bar, used to permit installation of the pipes in the concrete, showed that this high conductivity material does not have a significant effect on the energy exchanged. However, our analysis is showing for the first time the detrimental effects of allowing the U-loops to bunch together on the steel bar. This occurrence is common in construction and was seen to increase the concrete resistance of the pile by around one third for the case considered in this study. Such unnecessary effects could be removed by the use of spacers around the steel bar. This technique has been used successfully in test piles (Brettmann *et al.*, 2010) and could be applied on a more routine basis.

CFA piles also offer another advantage compared to rotary bored piles that results directly from their increased thermal resistance. The resulting larger temperature difference between the heat transfer fluid and the ground means that the range of temperature change experienced by the ground and the ground-pile interface is reduced in CFA piles. This can be observed, for example, in Figure 9(b) (by comparing the curves representing the average pile wall temperature for the CFA and rotary cases) and has positive knock-on effects for the geotechnical design which must consider whether any detrimental change in soil or interface properties may result from these temperature changes. Additionally, a larger volume of concrete between the pipes and the ground provides a buffer to the most extreme temperatures which the fluid may experience. These extreme temperatures are often short lived and the concrete buffer may permit lower temperature limits to be set for the fluid without the risk of subzero temperatures reaching the ground. Such wider temperature limits will have a positive impact on the overall energy performance of the ground energy system.

One potential disadvantage for thermoactive CFA piles has been the concern that the potential for pipe-to-pipe interactions may lead to an increase in thermal resistance and hence a decrease in energy performance. The analysis carried out in this study tentatively suggests that there may be an increase in resistance (by up to 0·05 mK/W or +25%) for CFA piles owing to this mechanism. However, there remains a discrepancy between this result and calculations made by analytical methods which suggest only a small effect (maximum 6% increase in concrete thermal resistance) from pipe-to-pipe interactions and only at low flow velocities. Both approaches contain simplifications and it is not clear at present which approach is most appropriate. It would be beneficial to validate both approaches against thermoactive pile field data for CFA installations and also benchmark simulations containing a full representation of the heat transfer fluid. Nonetheless, the results presented in this study are conservative and the discrepancies in resistance are small compared to the bigger differences (>0·1 mk/W) between CFA piles and rotary bored piles.

Conclusions

The following conclusions are presented from this study.

1. Thermoactive CFA piles will be less energy efficient than rotary piles on a like-for-like basis with equal numbers of pipes.
2. Rotary bored piles also offer the opportunity to install more pipes in the cross-section, further increasing opportunities for maximising energy efficiency.
3. CFA piles are substantially cheaper than rotary bored piles to construct. However, some additional construction tasks are required to convert CFA piles to thermoactive structures, whereas this is not necessarily the case for rotary bored piles.
4. CFA piles are in common use for urban building developments. If converting to thermoactive piles then four pipes should be used instead of two, and care should be taken to ensure turbulence is maintained within the fluid.
5. The current practice of installing the heat transfer pipes with a steel bar for CFA piles does not appear to be detrimental to thermal performance. However, performance would be improved by adding the use of spacers to prevent bunching of the pipes on one side of the bar.
6. One additional potential advantage of CFA piles is that as the pipes are further from the pile edge, the soil-pile interface will experience a reduced temperature change compared with a rotary bored pile. This means that any influence on the geotechnical design will be reduced.

Acknowledgements

The first author is funded by the Royal Academy of Engineering under their Research Fellow scheme.

The second author acknowledges financial support from European Union FP7 project under contract number PIAPP-GA-2013-609758-HOTBRICKS.

Appendix: Calculation of the concrete resistance

The 2D pile concrete resistance is given by (Hellstrom, 1991)

$$R_{c2} = \frac{1}{4\pi\lambda_c}\left[\ln\left(\frac{D_{pile}}{D_{pipe}}\right) + \ln\left(\frac{D_{pile}}{2s}\right) + \sigma\ln\left(\frac{(D_{pile}/s)^4}{(D_{pile}/s)^4 - 1}\right)\right] \quad (A1)$$

The pseudo-3D resistance is given by (Hellstrom, 1991)

$$R_{c3} = R_{c2}\eta\coth(\eta) \quad (A2)$$

where

$$\eta = \frac{L}{\dot{m}c_p\sqrt{R_{c2}R_a}} \quad (A3)$$

and R_a is the internal resistance, with

$$R_a = \frac{1}{\pi\lambda_c}\ln\left(\frac{2A_1(A_2^2 + 1)^\sigma}{A_2(A_2^2 - 1)^\sigma}\right) \quad (A4)$$

with $A_1 = D_{pile}/D_{pipe}$, $A_2 = D_{pile}/s$ and $\sigma = (\lambda_c - \lambda_g)/(\lambda_c + \lambda_g)$.

In the above expressions, λ_g and λ_c are the thermal conductivity of the ground and the concrete respectively, D_{pile} is the outer diameter of the heat transfer pipes, D_{pile} is the diameter of the pile, L is the pile length and s is the pipe centre-to-centre shank spacing.

REFERENCES

Amis T, McCartney JS, Loveridge F et al. (2014) Identifying best practice, installation, laboratory testing, and field testing. *DFI Journal* **8(2)**: 74–83.

Barla M and Perino A (2014) Energy from geo-structures: a topic of growing interest. *Environmental Geotechnics* **2(1)**: 3–7.

Bozis D, Papakostas K and Kyriakis N (2011) On the evaluation of design parameters effects on the heat transfer efficiency of energy piles. *Energy and Buildings* **43(4)**: 1020–1029.

BPIE (Building Performance Institute Europe) (2011) *Europe's Buildings under the Microscope: A Country by Country Review of the Energy Performance of Buildings*. BPIE, Brussels, Belgium, ISBN: 9789491143014. See http://www.bpie.eu/eu_buildings_under_microscope. html.VRFQFfmsX4I (accessed 24/03/2015).

Brandl H (2006) Energy foundations and other thermo active ground structures. *Géotechnique* **56(2)**: 81–122.

Brettmann TPE, Amis T and Kapps M (2010) Thermal conductivity analysis of geothermal energy piles. *Proceedings of the Geotechnical Challenges in Urban Regeneration Conference, London, UK. 26–28 May 2010*.

Brown DA (2005) Practical considerations in the selection and use of continuous flight auger and drilled displacement piles. *ASCE Géotechnical, special publication* **129**: 251–261.

Cecinato F and Loveridge FA (2015) Influences on the thermal efficiency of energy piles. *Energy*, http://dx.doi.org/doi:10.1016/j.energy.2015.02.001.

Choi JC, Lee SR and Lee DS (2011) Numerical simulation of vertical ground heat exchangers: Intermittent operation in unsaturated soil conditions. *Computers and Geotechnics* **38(8)**: 949–958.

EC (European Community) (2009) *Council Directive 2009/28/EC of 23rd April 2009 on the promotion of the use of energy from renewable sources and amending and subsequently repealing Directives 2001/77/EC and 2003/30/EC.*

Gao J, Zhang X, Liu J, Li K and Yang J (2008) Thermal performance and ground temperature of vertical pile foundation heat exchangers: a case study. *Applied Thermal Engineering* **28(17)**: 2295–2304.

Hellstrom G (1991) *Ground Heat Storage, Thermal Analysis of Duct Storage Systems, Theory.* Department of Mathematical Physics, University of Lund, Sweden.

Laloui L and Di Donna A (2012) Understanding the behaviour of energy geo-structures. *Proceedings of the Institution of Civil Engineers Civil Engineering* **165(1)**: 14.

Lamarche L, Kajl S and Beauchamp B (2010) A review of methods to evaluate borehole thermal resistance in geothermal heat pump systems. *Geothermics* **39(2)**: 187–200.

Loveridge F and Powrie W (2013a) Performance of piled foundations used as heat exchangers. *18th International Conference for Soil Mechanics and Geotechnical Engineering, Paris, France, 2–5 September 2013.*

Loveridge F and Powrie W (2013b) Pile heat exchangers: thermal behaviour and interactions. *Proceedings of the Institution of Civil Engineers Geotechnical Engineering* **166(2)**: 178–196.

Loveridge F and Powrie W (2014) 2D thermal resistance of pile heat exchangers. *Geothermics* **50**: 122–135.

Muraya NK, O'Neal DL and Heffington WM (1996) Thermal interference of adjacent legs in a vertical U-tube heat exchanger for a ground coupled heat pump. *ASHRAE Transactions* **102(2)**: 12–21.

Ozudogru TY, Olgun CG and Senol A (2014) 3D numerical modeling of vertical geothermal heat exchangers. *Geothermics* **51**: 312–324.

Rees SJ and He M (2013) A three-dimensional numerical model of borehole heat exchanger heat transfer and fluid flow. *Geothermics* **46**: 1–13.

Signorelli S, Bassetti S, Pahud D and Kohl T (2007) Numerical evaluation of thermal response tests. *Geothermics* **36**: 141–166.

Wood CJ, Liu H and Riffat SB (2010) An investigation of the heat pump performance and ground temperature of a pile foundation heat exchanger system for a residential building. *Energy* **35(12)**: 3932–4940.

ICE Themes Geothermal Energy, Heat Exchange Systems and Energy Piles

Craig and Gavin
ISBN 978-0-7277-6398-3
https://doi.org/10.1680/gehesep.63983.205
ICE Publishing: All rights reserved

Chapter 11
City-scale perspective for thermoactive structures in Warsaw

Grzegorz Ryżyński MSc
Senior Specialist, Polish Geological Institute, National Research Institute, Warsaw, Poland
(corresponding author: grzegorz.ryzynski@pgi.gov.pl)

Witold Bogusz MSc
Research Assistant, Building Research Institute, Warsaw, Poland

This chapter presents the current conditions of and perspective on development of the use of shallow geothermal energy as a renewable energy source in highly urbanised areas in Poland, using the Warsaw municipality as an example. The type of geothermal energy source considered in the chapter is the thermoactive foundation elements of high-rise buildings and underground structures. The rising demand for new office, commercial and housing spaces in Warsaw fosters the development of new engineering projects, mostly involving deep foundations and structural elements well suited for shallow geothermal energy utilisation. In this study, the preliminary assessment has been made, considering the city-scale impact of the thermoactive foundations. New as well as planned construction areas for new engineering projects in Warsaw were evaluated on the basis of spatial management plans and analysis of local geotechnical and geothermal conditions. Furthermore, the legal issues concerning the use of thermoactive foundations are discussed, considering the current state of the national regulations and the typical construction process.

Notation

A_b	area of the horizontal projection of the building: m^2
A_{dw}	area of the outer surface of the diaphragm wall: m^2
A_T	outer surface area of metro tunnel: m^2
h_b	height of the building: m
h_f	length of the diaphragm wall: m
h_p	length of the geothermal piles: m
h_s	depth to the foundation slab: m
n_p	assumed number of energy piles for the structure
n_{sl}	number of sublevels of the structure
P_{BHE}	specific ground heat extraction rate for borehole heat exchangers: W/m
P_{dw}	specific ground heat extraction rate for diaphragm wall: W/m^2
P_s	specific ground heat extraction rate foundation slab: W/m^2
P_T	specific ground heat extraction rate metro tunnel lining: W/m^2
Q_{dw}	assumed ground heat extraction for the diaphragm wall: W
Q_p	assumed ground heat extraction for the piles: W
Q_s	assumed ground heat extraction for the foundation slab: W

Introduction

Thermoactive structures are recently becoming increasingly popular as the preferred geothermal energy systems worldwide. This technology of low-enthalpy geothermal systems is based on the integration of heat exchanger pipes with the foundation elements of buildings and infrastructure. This type of solution allows the cost of installation of the ground source exchangers for geothermal systems to be minimised, as geotechnical foundation structures are required for structural reasons; the cost of drilling boreholes for ground source heat exchangers is significantly reduced or practically eliminated. The technology of thermoactive structures is especially attractive for urbanised areas, where the space for borehole heat exchangers is highly limited. A vast majority of high-rise buildings and underground city infrastructure (such as underground car parks, metro tunnels and stations) have significantly developed foundation structures, which can be thermally activated for heating and cooling purposes. The data presented by Adam and Markiewicz (2009) show that in Austria alone, more than a thousand high-rise and other buildings were using earth-coupled geothermal systems, and in 2004, the total cumulative number of geothermal piles installed in Austria, according to Brandl (2006), stood at 22 843. Thermoactive structures are already well developed in some European countries, namely Austria, Germany, Switzerland and the UK.

The rising demand to meet the strategic targets of the EU 3×20 climate and energy package fosters the development of renewable, green energy sources, including low-enthalpy geothermal systems. The city-scale implementation of geothermal systems, especially thermoactive structures, can be a huge step towards the fulfilment of the European Commission strategic targets. In this chapter, the conditions of, and the perspective on, the development of structures equipped with thermoactive foundation elements in large metropolitan areas of Poland are discussed. As a preliminary step, a geographical information system (GIS)-based analysis has been conducted as an example for the city of Warsaw. Zhang *et al.* (2014) showed that a limited number of studies had been performed to examine the potential capacity of shallow geothermal energy at the larger scale (district/city/regional). The large-scale analysis had been focused mostly on ground source heat pumps (GSHPs), with very limited reference to thermoactive structures. In recent years, since the 1990s, Warsaw has been undergoing a very rapid development in the area of large engineering projects, including numerous high-rise buildings and underground structures, such as new metro lines and underground car parks. The rising demand for new transport infrastructure and additional office, commercial and housing spaces in Warsaw stimulates new engineering projects. These projects could offer immense potential to operate as shallow geothermal energy sources if their foundation elements (slabs, diaphragm walls and foundation piles) were to be equipped with ground source heat exchangers; however, there are still strong barriers to the development of shallow geothermal systems, including thermoactive structures. Kapuściński and Rodzoch (2010), in their report 'Low-temperature geothermal energy in Poland and worldwide', listed the main barriers for further growth of the shallow geothermal market in Poland, which include psychological, informational, educational and economic barriers. The GSHP sector developed in Poland with practically no support from the state. At the end of 2012, there were approximately 30 000 GSHPs in operation (Kępińska, 2013) with 330 MWh of installed power; however, there are no data on any thermoactive structures installed in Poland. The geothermal GSHP market in Poland is characterised by a very rapid growth in recent years, with 20–30% sales increase each year (Kępińska, 2013). The prognosis shows that the demand for thermoactive installations will increase as numerous new infrastructural and engineering projects in large Polish cities will arise.

The calculations presented in this chapter are aimed to cover the preliminary city-scale analysis. The city-scale point of view on the potential application of thermoactive structures has not been discussed in many countries up until now. This study is intended as a starting point for future research and discussions, not only in Poland, but also in other countries, where thermoactive structures have yet to be introduced, like in the majority of new EU member states.

The analysis presented in this chapter is an attempt to perform a preliminary estimation of the potential city-scale impact of the structures equipped with thermoactive foundation elements in large Polish metropolitan areas using Warsaw as an example. The analysis was performed on the basis of the spatial data on planned and existing high-rise buildings, underground car parks and new as well as existing metro lines, provided by the Office of Architecture and Spatial Planning of Warsaw Municipality. The geological model used in the analysis was prepared on the basis of GIS data from the Engineering-Geological Database of Warsaw, maintained by the Polish Geological Institute – National Research Institute, as one of the tasks of the Polish Geological Survey. The main purpose of the presented analysis has been to evaluate the amount of geothermal energy that can be extracted with the use of different types of thermoactive structures in Warsaw and to compare the results with the city heat demand, as well as Warsaw City's goals for shallow geothermal energy set in 'Warsaw City action plan goals for sustainable energy use' (Warsaw City Council, 2011).

Analysed data set

To investigate the conditions of, and the perspective on, the development of structures equipped with thermoactive foundation elements in large metropolitan areas, Warsaw was used as an example. Warsaw is now undergoing rapid growth; therefore, it is a fine example of the changes in city structure that may await other large cities in developing countries. Additionally, Warsaw is the only city in Poland with existing metro underground infrastructure, including two metro lines, which were analysed in the chapter as a potential thermoactive structure.

Warsaw is the capital city of Poland, with an area of 517 km^2 and a population of 1 707 981 inhabitants (data from 2008), according to Warsaw City Council (2011). Warsaw has 18 districts, located along the Vistula River. The western districts from the Vistula River, including the city centre, are located on the glacial plateau, while the eastern districts are located in the broad Vistula River valley. To perform the analysis presented in this study, the authors used two data sets – one with the spatial data set of high-rise buildings, planned underground car parks and metro stations and tunnel localisation, and the second one with geological and geothermal data (described later in 'Algorithm for city-scale GIS analyses of thermoactive foundation'). The general layout of Warsaw City with the localisation of the analysed buildings and infrastructure is presented in Figure 1. It can be noticed that the area of the most intensive development of high-rise buildings is limited to the city centre area, especially to the three districts: Śródmieście, Wola and Ochota, which cover 8·6% of the city area. It can also be seen that the direction of development of the new high-rise buildings may be associated with the metro line construction directions. The land around the newly built metro stations becomes more valuable; therefore, new engineering projects for high-rise buildings start where

Figure 1 General layout of Warsaw City with the existing and designed metro line location, high-rise buildings and underground car parks

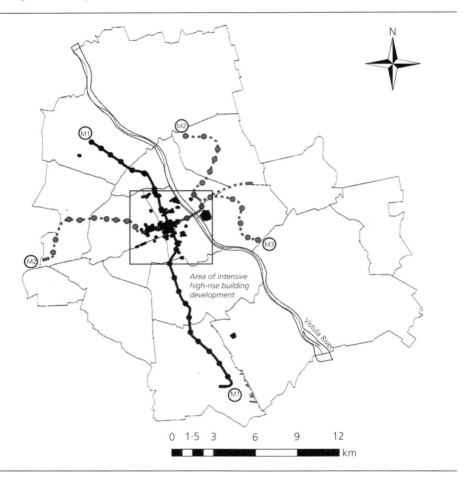

demolished old low-rise buildings existed, to maximise the usable area for office, commercial and apartment spaces.

The data provided by the Warsaw City Office of Architecture and Spatial Planning (described in Tables 1 and 2) consists of polygon and line geometry for three types of structures, with attributes including status of construction, year of construction, height, number of levels and sublevels. Moreover, some of the sites had information on usable area (about 18% of the available data set). In addition, for the purposes of the ground heat extraction calculations for all polygons of the considered type of structure, the horizontal section area and the perimeter were calculated using ArcGIS software. In total, 81 sites of high-rise buildings, 13 sites of planned underground car parks and all planned and existing metro lines and stations in Warsaw were considered in this study.

Table 1 Analysed data set for high-rise buildings

Type of structure	Status	Number	Year of construction	Height: m	Number of levels	Number of sublevels	Usable area: m²	Horizontal section area (from GIS analysis): m²	Horizontal section perimeter (from GIS analysis): m
High-rise buildings	Existing	65	1896–2014	23–237 (84)	7–60 (24)	1–5 (2·5)	5600–94 064 (44 720)	51–37 870 (2064)	28–93–1598 (475)
	Under construction	6	2016–2018	75–195 (141)	23–45 (33)	2–5 (4)	20 109–72 000 (48 665)	1132–8134 (3716)	136–563 (400)
	Planned	10	>2017	28–188 (103)	7–48 (24)	3–5 (4)	13 500–90 000 (59 640)	513–4439 (1541)	99–436 (198)

Table presents minimal–maximal and (**average**) values

Table 2 Analysed data set for underground infrastructure

Type of structure	Status	Number	Length: km	Year of construction/planned construction
Underground car parks	Planned	13		No precise data
M1 metro stations	Existing	21		1983–2008
M1 metro tunnels	Existing		23·1	1983–2008
M2 metro stations	Existing	7		2010–2015
M2 metro tunnels	Existing		6·1	2010–2015
M2 and M3 metro stations	Planned	21		2019–2022 and beyond
M2 and M3 metro tunnels	Planned		24·9	2019–2022 and beyond

To have an overview of high-rise buildings and underground infrastructure data, statistical processing was performed, and it is presented in Tables 1 and 2 and in the form of histograms in Figure 2. The average height of high-rise buildings is around 70 m, with mostly three sublevels. However, if only the planned projects are taken into consideration, it becomes apparent that there is a trend towards increasing the number of sublevels (see Figure 2, histogram for planned projects only). The histograms show that there is an increasing trend in the number of high-rise buildings since the year 1990. With the assumption that this trend for new construction projects will continue, the prognosis can be assumed that approximately 36 tall buildings will be constructed in Warsaw in the next decade (until 2025). This prediction was used in the calculations performed for the purpose of this study to simulate the potential impact of the use of thermoactive foundations at city and district scale.

For the purpose of further calculations, only the projects with their status as planned and under construction were evaluated. The data for existing structures were used for the trend analysis. Brandl (2006) stated that the installation of plastic pipes in the concrete elements is easy and economical during the construction of new buildings, while thermoactive installations in existing buildings are quite complicated. In the case of existing buildings, the installation of GSHPs located around the building outline should be considered to provide geothermal heating rather than thermal activation of existing foundation elements.

A detailed map of the Warsaw City central area, shown in Figure 3, presents how new construction projects finished in 2015 emerge in the vicinity, in the central section of the M2 metro line (6·1 km long, with seven stations, oriented east-west). The black outlined polygons are planned or under-construction engineering projects. The planned underground car parks, also presented in Figure 1, are located in close proximity to existing or planned underground infrastructure as well as to high-rise buildings.

City-scale perspective for thermoactive structures in Warsaw

Figure 2 Histograms of construction year, height and sublevel number for high-rise buildings in Warsaw

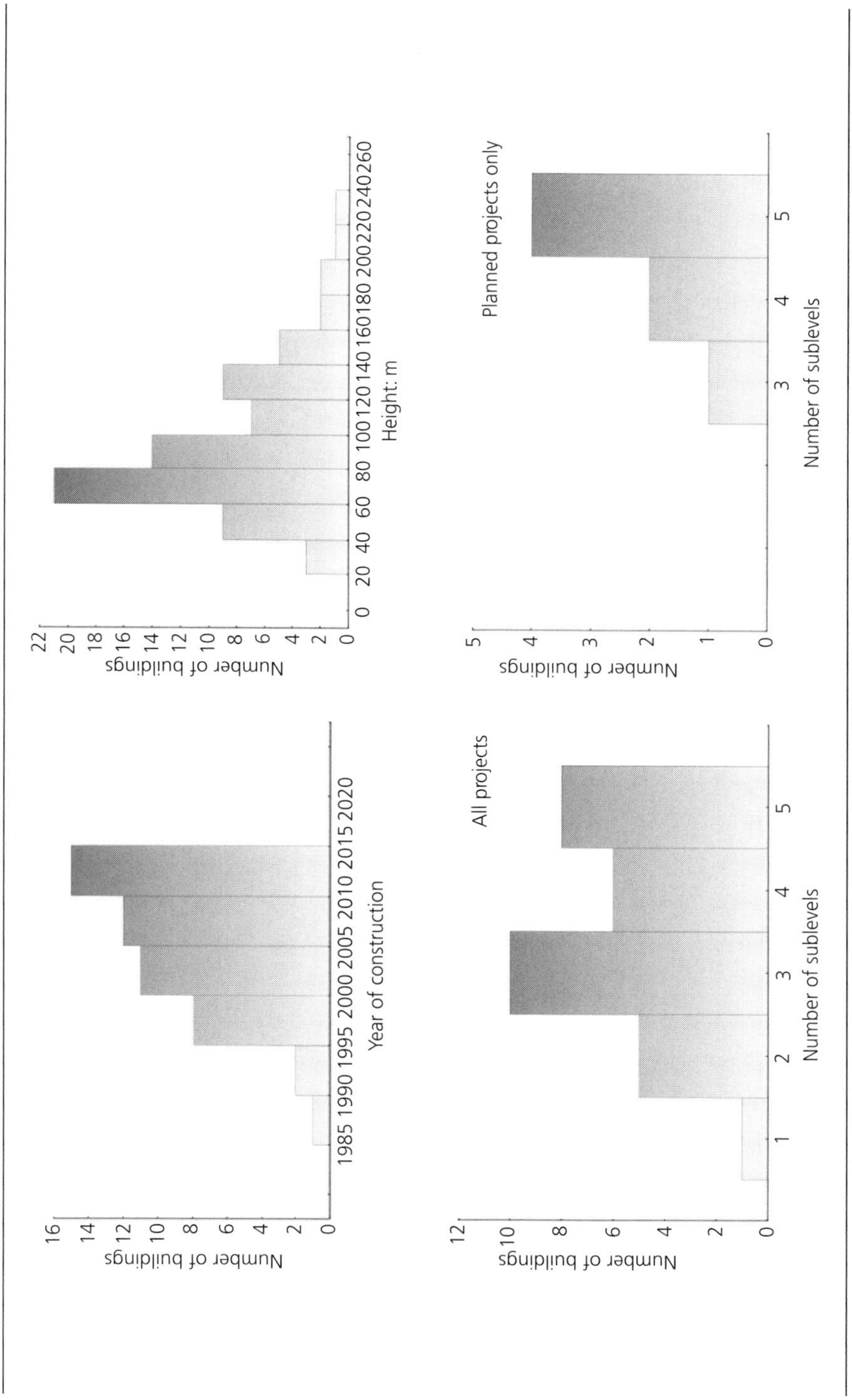

ICE Themes Geothermal Energy, Heat Exchange Systems and Energy Piles

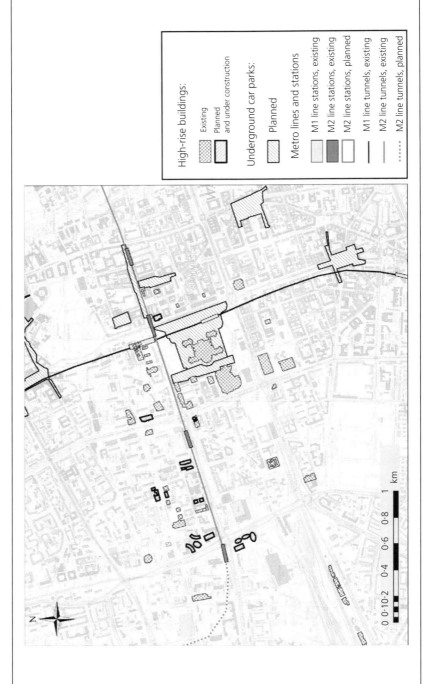

Figure 3 Detailed 1:10 000 scale map of the Warsaw City central area with existing and planned city infrastructure

City-scale analysis for structures equipped with thermoactive foundation elements

As there is no practical experience with thermoactive foundations in Poland, the city-scale analysis presented in this chapter should be considered as preliminary and valid for the feasibility study stage only. The analysis has been based on the first attribution of the simplified model of the thermoactive foundation for certain structures. The second step was the GIS processing of attributed polygons of the horizontal sections of planned high-rise buildings and underground infrastructure with available geological and geothermal data. Only planned and prognosed structures were considered.

Simplified model of thermoactive foundation for city scale analysis

Due to the scope of this study, which concentrates on the city-scale analysis, as well as due to the limited data availability, simplified models for different types of structures were assumed. The presented analysis mainly applies to the planned structures and those currently under construction, as the prospects for future development are its main concern. Three major types of structures, which may be equipped with thermoactive elements (Adam and Markiewicz, 2009; Brandl, 2006) are the focus of this study, namely high-rise buildings, underground car parks and metro lines. For these structures, foundation on thermally activated diaphragm walls and slabs was assumed, with additional thermoactive piles in the case of the high-rise buildings. The distinction between the different foundation solutions was neglected, as most of the analysed structures are still at the planning phase, with only the basic data available. The extractable power of 20 W/m^2 was conservatively assumed for the slab foundations and the diaphragm walls. Brandl (2006) reported that 1 kW of heating needs roughly between 20 m^2 (saturated soil) and 50 m^2 (dry sand) of the surface of the concrete structures in contact with soil or groundwater. Additional information about the possible heat extraction values for specific structure types have been reported by Barla and Perino (2014), Di Donna and Barla (2016) and Laloui and Di Donna (2013).

It should be noted that at the design stage, specific parameters for each structure should be derived for the specific ground conditions and for the chosen foundation type (foundation slab, classic pile foundation or pile-raft foundation). They will generally result from structural and geotechnical design calculations. In some of the cases, the compression piles might not be necessary; while in others, the tension piles might be used to counteract the uplift forces due to high groundwater pressure. For different loading and restraint conditions of the piles, the appropriate method should be used to take into account additional thermal loading. Specific aspects of the design are beyond the scope of this study and were analysed by Bourne-Webb *et al.* (2009), Dupray *et al.* (2014) and Laloui and Di Donna (2011).

Based on the available data and the following assumptions, ground heat extraction Q was calculated for different structural elements (Q is for the foundation slab, Q_{dw} is for the diaphragm wall and Q_p is for the piles).

High-rise buildings

There are 67 high-rise buildings in Warsaw, 49 of which were built before the year 1989. Currently, five more are already under construction, and 11 are planned to be built before the year 2018. The scheme of the assumed simplified foundation model used for the analysis is presented in Figure 4.

Figure 4 Scheme of the simplified foundation model used for analysis of the city-scale energy potential of thermoactive engineering structures

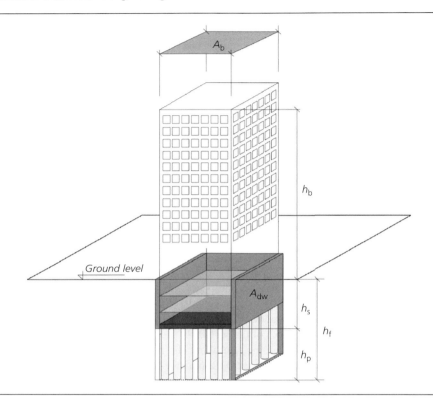

The area of the foundation slab was assumed equal to the area of the horizontal projection of the structure A_b. The area of the outer surface of the diaphragm wall A_{dw} was calculated based on the outer perimeter length obtained from the GIS data and the length of the wall h_f assumed to be up to twice the depth of the excavation h_s. The excavation depth was calculated based on the data related to the number of sublevels while assuming the average sublevel height of 3·5 m. For buildings without accurate data on the number of sublevels, the excavation depth of 10% of the height h_b was assumed.

The number of piles n_p for each building was calculated assuming a spacing of 5 m (one pile for 25 m² area of the slab foundation) as

$$n_p = A_b/25 \text{ m}^2 \tag{1}$$

The length of piles was assumed as the distance from the foundation slab to the base of the diaphragm wall ($h_p = h_s$). The energy efficiency of the pile foundation was derived depending on the general geological conditions present at the specific site, based on the Engineering-Geological Database of Warsaw (Polish Geological Institute, 2016).

Underground car parks

The average horizontal projection area of the planned car parks is approximately 37 500 m^2. The number of sublevels varies, ranging from one to five. The need for extra parking space in the city centre drove Warsaw City authorities to develop a strategic plan for underground car parks in this area. This type of underground structure was considered, for the purposes of analyses in this study, as founded on diaphragm walls and slab foundations. The simplified thermoactive foundation model was similarly assumed as for high-rise building but without the piles. Due to their high thermoactive potential and low demand for geothermal heating and cooling, these structures should be considered as an energy source for adjacent buildings.

Metro lines

The average geometrical parameters were assumed for the tunnels and the stations based on their central sections. The entire existing M1 line consists of 21 metro stations and is 23·1 km long. This line was built between the years 1983 and 2008.

The M2 line is currently under construction since the year 2010. The first central 6·1-km-long part of the line, with seven stations, was finished in 2015. The next extensions to the east and the west are to be gradually built over the coming years. The entire line will be 31 km long and will consist of 31 stations in total. The value of 22 km used in the analysis for the tunnels was obtained by subtracting the length of the existing central section and the planned stations from the planned total length of the metro line M2. The existing line consists of two tunnels made using tunnel boring machine earth pressure balance technique, 6·0 m dia. each, and a lining composed of precast concrete segments. The average tunnel axis depth near the C11 Świętokrzyska station, located near the city centre, is approximately 20 m below ground level (bgl). Similarly, the average geometrical parameters for the planned stations were based on the C11 metro station, which is approximately 135 m long and from 24·0 to 25·8 m wide. The slab foundation is located approximately 23·7 m bgl, while the diaphragm walls are approximately 28 m long.

Algorithm for city-scale GIS analyses of thermoactive foundation

To perform the city-scale analysis, a simplified geological model has been developed. The simplified geological profile of the deep foundation subsoil on the glacial plateau in Warsaw, assumed for the purpose of this study, consists of a significant cover of man-made grounds (mostly composed of World War II debris), Quaternary Pleistocene glacial tills and fluvioglacial sands with gravels of three main glaciations and Neogene Pliocene clays as a bottom layer. Pliocene clays in the area of Warsaw were tectonically deformed by the glaciers in the Pleistocene. As a result, their upper surface is heavily undulated, locally with the level difference reaching tens of metres within the scale of a construction site (see Figure 5).

The significant simplification of the geological model has been done to allow the relatively quick performance of analysis within the ArcGIS software, without the necessity for the creation of a complex new three-dimensional (3D) geological model, due to the preliminary scope of this study. Only two-dimensional data (shapefiles) from the Engineering Geological Database of Warsaw were used for the analysis, as to represent the geological model. These data consist of four spatial layers listed in Table 3. The layers from Table 3 were used to represent the main layers in the simplified geological profile, as shown in Figure 5. Layer A was obtained from the man-made ground thickness, while layers B1 and B2 were calculated

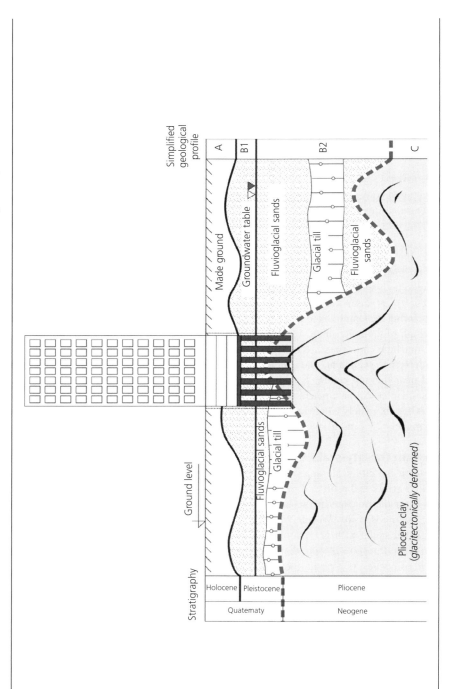

Figure 5 Simplified geological model used for the city-scale analysis of the energy potential of thermoactive engineering structures

Table 3 Spatial layers from Engineering-Geological Database of Warsaw used for GIS analyses

Spatial layers used for analysis	Interpolation method	Number of categories	Description of categories				
Pliocene clay upper surface depth	Kriging	15	Depth range: 0–140 m bgl			Each category: 10 m	
Groundwater level	Euclidean allocation	5	0·5–1·0 m bgl	1·0–2·0 m bgl	2·0–5·0 m bgl	5·0–10·0 m bgl	>10·0 m bgl
Man-made ground thickness	Euclidean allocation	5	0·5–1·0 m	1·0–2·0 m	2·0–5·0 m	5·0–10·0 m	>10·0 m
Geomorphological units	—	2	Glacial plateau			Vistula River valley	

from the difference between the man-made ground thickness and the Pliocene upper surface depth. The ground water level information was used to make a subdivision of the Pleistocene soils into layers B1 (above-ground water-level soils) and B2 (fully saturated soils). The information on the profile of layer C (Pliocene clays) was obtained from the Pliocene clay upper surface shapefile. The simplified geological profile was used as a basis for the calculations of the ground heat extraction rate (Q). The scheme of the model is presented in Figure 5.

In Figure 5, the simplified geological model was intersected in ArcGIS software with a polygon layer of the horizontal sections of the planned high-rise buildings and underground infrastructure (metro stations and tunnels and underground car parks) attributed with the simplified thermoactive foundation model as described in the section 'Simplified model of thermoactive foundation for city scale analysis'. The result of the intersection has given the set of secondary polygons within each parent polygon of the analysed structures, with unique simplified geological profile. The simplified geological profiles were then used to calculate the ground heat extraction rate for the piles (Q_p). The calculation scheme used in the analysis is presented as follows. To calculate the ground heat extraction for the considered thermoactive structures, Equations 2–4 were used. Equation 3 was also used for the calculation of heat extraction for metro stations and underground car parks

$$Q_{building} = Q_s + Q_{dw} + Q_p \qquad (2)$$

$$Q_{station} = Q_{car\ park} = Q_s + Q_{dw} \qquad (3)$$

$$Q_{tunnel} = 2 \times A_T \times P_T \qquad (4)$$

To calculate the assumed heat extractions of the foundation slab and the diaphragm walls, Equations 5 and 6 were used. The specific ground heat extraction for the tunnel lining (P_T), the foundation slab (P_s) and the diaphragm walls (P_{dw}) was assumed as the value equal to 20 W/m² to obtain conservative results.

$$Q_s = A_b \times P_s \qquad (5)$$

$$Q_{dw} = A_b \times P_{dw} \qquad (6)$$

To calculate the geothermal pile heat extraction Q_p, Equations 7–9 were used. The data for the secondary polygons, with distinct simplified geological profiles, were obtained from GIS analysis. Q_{pi} was considered as a partial heat extraction rate calculated for each secondary polygon of the horizontal sections for the analysed thermoactive structures. The specific heat extraction rate P_{BHE} in Equation 8 was taken as a weighted mean value for different geological soil layers (A, B1, B2, C) based on their lengths in profile (l_A, l_{B1}, l_{B2}, l_C) up to the thermoactive foundation depth (h_f).

$$Q_p = \sum_{i=1}^{n}(Q_{pi}) \qquad (7)$$

$$Q_{pi} = n_p(A_i/A_b)h_p P_{BHE} \qquad (8)$$

$$P_{BHE} = (P_{BHEA}l_A + P_{BHEB1}l_{B1} + P_{BHEB2}l_{B2} + P_{BHEC}l_C)/h_f \qquad (9)$$

The assumed specific ground heat extraction rates for simplified geological profiles A, B1, B2 and C, based on data provided by Rubik (2011) and PORT PC (2013) as well as the regional comparable experience of the authors, were taken as the following values

- P_{BHEA} = 20 W/m (man-made grounds, Holocene)
- P_{BHEB1} = 28 W/m (dry sands and dry glacial tills, Pleistocene)
- P_{BHEB2} = 43 W/m (saturated sands and glacial tills, Pleistocene)
- P_{BHEC} = 30 W/m (saturated Pliocene clays).

Results

The results of the analysis are presented in Table 4. The calculated heat extraction rates for all considered types of thermoactive structures has given a total value of 48·29 MW at the city scale, which is approximately 0·73% of Warsaw City's annual heat consumption, according to the data from the Warsaw City Council (2011). The results show that there is a dominant contribution of underground infrastructure (metro tunnels, stations and car parks) heat extraction (41·67 MW) in the total calculated value. These types of structures are large engineering projects, executed with major participation and supervision of city authorities, with high potential for city-scale impact as geothermal energy sources. The high-rise buildings, on the other hand, are attractive for private developers due to their high sustainability and

Table 4 Analysis results: estimated heat extraction values for considered types of structures for assumed simplified foundation models

Calculated heat extraction rates: MW					
M2 and M3 metro lines (planned)		Underground car parks $Q_{car\ park}$ (13 sites)	High-rise buildings $Q_{buildings}$		Q_{TOTAL} (with prognosis for high-rise buildings)
$Q_{stations}$ (21 sites)	$Q_{tunnels}$ (22 km)		Planned (10 projects)	Prognosis (36 projects until 2025)	
3·56	16·59	21·52	3·55	6·61	48·29

energy efficiency. The analysis indicates that the thermoactive foundations can provide, for considered buildings, from 14% to 45% of their heating demand. This confirms the observations made by Zhang et al. (2014) that tall buildings, with more than six storeys, have difficulties with meeting their own heating demands with geothermal systems only (GSHPs).

Taking into account the rising trend for new engineering projects in Warsaw (based on the histograms presented in Figure 2), the forecast value for 36 new high-rise buildings possible until 2025 was calculated to be 6·61 MW (0·34% of the annual heat consumption of three central districts of Warsaw – Śródmieście, Ochota and Wola, based on the data from 2011).

The comparison of the results of the assumed total heat ground extraction by thermoactive structures with the city heat demand from renewable energy sources (GSHPs) taken from the Warsaw City Action plan goals (2011) has been made. The total ground heat extraction from thermoactive structures (as 48·29 MW with 1800 FLEQ (full load equivalent)) would provide 42% of the Warsaw City action plan goal for shallow geothermal heating (GSHPs) at the level of 205 000 MWh/year. This comparison shows that from the point of view of the Warsaw City action plan (2011) geothermal goals, the city-scale impact of the thermoactive structures can be significant.

Conclusions

The GIS analysis allowed the preliminary evaluation of the city-scale impact of thermoactive structures using Warsaw as an example. The results showed that the thermoactive structures can utilise a significant amount of heat from the city subsoil for the purposes of fulfilling the Warsaw City goals for geothermal energy. Over 42% of its targeted value could be delivered if new engineering projects, including forecast high-rise buildings, new metro lines, tunnels and underground car parks, were designed with thermally activated foundation elements. The dominant share of underground infrastructures (metro lines and underground car parks) in the total calculated ground heat extraction value leads to a conclusion that there is a strong demand for strategic decisions of city authorities on legal background and administrative support, as well as incorporation of planned thermoactive structures into the city master plans.

Assuming that other large cities in Poland (such as Kraków, Poznań, Wrocław and Trójmiasto) will undergo a similar rapid growth to that observed in Warsaw, the future potential development for thermoactive structures in Poland is substantial.

A lack of proper regulations can be considered a major inhibiting factor in the development and promotion of the design of thermoactive structures in Poland. Currently, only the use of heat pumps is a subject of legal concern. Construction law provides only minor references to their use, namely stating that the installation of heat pumps does not require a building permit. Additionally, geological and mining law states that boreholes up to 30 m (with the exception of mining areas) for the purposes of ground heat utilisation are beyond the scope of its regulations. Theoretically, the current legal status does not prevent the use of shallow geothermal energy. However, in practice, it significantly impedes the construction process, especially when close cooperation between investor, designers of different specialities and contractors is necessary to ensure the quality and the proper functionality of the thermoactive system.

Also, legal regulations should be created to cover the environmental issues connected with the city-scale growth of thermoactive foundation systems. They should include the major environmental problems like heat pollution, excessive geothermal exploitation of city subsoil, impact on groundwater quality and temperature. The legal aspects of the GSHP use vary between different countries in that regard (Dehkordi and Schincariol, 2014).

The authors are aware that the analysis presented here has limitations due to its preliminary character. As presented, the city-scale point of view on the potential application of thermoactive structures has not yet been discussed in Poland; this chapter should be considered as a starting point for future research and discussion.

The 3D approach should be implemented in further analyses, especially for the purposes of spatial planning and preparations of the underground master plan for the city of Warsaw. Additionally, in future analyses, the dedicated geothermal maps should be developed and used.

It is also of significant importance that the nine biggest cities in Poland (and after the year 2016, 13 of them) have their own unique engineering-geological databases (with over 25 000 boreholes each, with ten and more unique geological spatial layers). Based on that, the methodology presented in this study can be applied to other cities in Poland to make the preliminary assessment of city-scale thermoactive foundation impact. The key factor for such analyses is the availability of spatial data on underground infrastructure and high-rise buildings as well as the cooperation with city authorities and their spatial planning departments. The authors will undertake such analyses in the future. With sufficient data available, such analysis can be performed practically for any city.

The presented methodology can be easily transferred to other non-Polish cities that are starting out on the path to the implementation of thermoactive structures. Complex site-specific models do not provide sufficient ease-of-use capabilities to start a discussion and activities by stakeholders (companies, investors) and city officials at the initial phases of the thermoactive-structure development process at the city scale. The target groups for the approach presented in this study are city planners and city authorities that need a simple and quick, but scientifically supported, methodology, which can be easily applied with limited available data.

Acknowledgements

The authors would like to acknowledge the city of Warsaw, the Office of Architecture and Spatial Planning for cooperating and providing the spatial planning data for the purposes of this study.

REFERENCES

Adam D and Markiewicz R (2009) Energy from earth-coupled structures, foundations, tunnels and sewers. *Géotechnique* **59(3)**: 229–236, http://dx.doi.org/10.1680/geot.2009.59.3.229.

Barla M and Perino A (2014) Energy from geo-structures: a topic of growing interest. *Environmental Geotechnics* **2(1)**: 3–7, http://dx.doi.org/10.1680/envgeo.13.00106.

Bourne-Webb PJ, Amatya B, Soga K *et al.* (2009) Energy pile test at Lambeth College, London: geotechnical and thermodynamic aspects of pile response to heat cycles. *Géotechnique* **59(3)**: 237–248, http://dx.doi.org/10.1680/geot.2009.59.3.237.

Brandl H (2006) Energy foundations and other thermo-active ground structures. *Géotechnique* **59(2)**: 81–122, http://dx.doi.org/10.1680/geot.2006.56.2.81.

Dehkordi SE and Schincariol RA (2014) Guidelines and the design approach for vertical geothermal heat pump systems: current status and perspective. *Canadian Geotechnical Journal* **51(6)**: 647–662, http://dx.doi.org/10.1139/cgj-2012-0205.

Di Donna A and Barla M (2016) The role of ground conditions on energy tunnels heat exchange. *Environmental Geotechnics*, http://dx.doi.org/10.1680/jenge.15.00030 (published online – ahead of print).

Dupray F, Laloui L and Kazangba A (2014) Numerical analysis of seasonal heat storage in an energy pile foundation. *Computers and Geotechnics* **55**: 67–77, http://dx.doi.org/10.1016/j.compgeo.2013.08.004.

Kapuściński J and Rodzoch A (2010) *Geotermia niskotemperaturowa w Polsce i na świecie. Stan aktualny i perspektywy rozwoju, Uwarunkowania techniczne, środowiskowe i ekonomiczne (Low-temperature geothermal energy in Poland and worldwide. State-of-the-art and perspectives for future development. Technical, environmental and economic conditions)*. Report for National Fund of Environmental Protection and Water Management. Ministry of Environment, Warsaw, Poland.

Kępińska B (2013) Geothermal energy use, country update for Poland. *Proceedings of European Geothermal Congress 2013, Pisa, Italy, 3–7 June 2013*.

Laloui L and Di Donna A (2011) Understanding the behavior of energy geo-structures. *Proceedings of the Institution of Civil Engineers – Civil Engineering* **164(4)**: 184–191, http://dx.doi.org/10.1680/cien.2011.164.4.184.

Laloui L and Di Donna A (2013) *Energy Geostructures: Innovation in Underground Engineering*. ISTE Ltd and John Wiley & Sons, Hoboken, NJ, USA.

Polish Geological Institute (2016) http://atlasy.pgi.gov.pl (accessed 13/02/2016).

PORT PC (Polska Organizacja Rozwoju Technologii Pomp Ciepła – Polish Organization for Development of Heat Pumps Technology) (2013) *Wytyczne projektowania, wykonania i odbioru instalacji z pompami ciepła. Część 1. Dolne źródła ciepła (Guidelines for design, construction and quality control of heat pump systems. Part 1. Ground source heat exchangers)*. PORT PC, Kraków, Poland.

Rubik M (2011) *Pompy ciepła w systemach geotermii niskotemperaturowej (Heat pumps in low-temperature geothermal systems)*. MULTICO Oficyna Wydawnicza, Warsaw, Poland.

Warsaw City Council (2011) *Plan działań na rzecz zrównoważonego zużycia energii dla Warszawy w perspektywie do 2020 roku (Action plan for sustainable energy use for the City of Warsaw in perspective to year 2020)*. Report Attachment to Warsaw City Council Resolution No. XXII/443/2011. Warsaw City Council; Warsaw, Poland.

Zhang Y, Soga K and Choudhary R (2014) Shallow geothermal energy applications with GSHPs at city scale: study on the City of Westminster. *Géotechnique Letters* **4(2)**: 125–131, http://dx.doi.org/10.1680/geolett.13.00061.

ICE Themes Geothermal Energy, Heat Exchange Systems and Energy Piles

Craig and Gavin
ISBN 978-0-7277-6398-3
https://doi.org/10.1680/gehesep.63983.223
© ICE Publishing: All rights reserved

Chapter 12
Energy piles: site investigation and analysis

Phil Hemmingway BE, PhD, MIEI
Teaching Fellow, School of Biosystems Engineering (formerly School of Civil, Structural and Environmental Engineering), University College Dublin, Ireland

Mike Long MEngSc, PhD, CEng, MIEI, MICE
Senior Lecturer, School of Civil, Structural and Environmental Engineering, University College Dublin, Ireland

Despite an increasing worldwide use of geothermal energy foundations, there is a lack of published guidelines and results from thermal response testing of such installations. In this chapter the results are presented from thermal response, thermal recovery and laboratory thermal testing performed at two sites in Ireland. Some practical issues concerned with the use of thermal response testing rigs, designed for use with deep boreholes, on relatively short piles are discussed and addressed. Given the relatively short geothermally active depth of the energy foundations tested, and the fact that the University College Dublin thermal response testing rig has been designed primarily for testing on medium and deep geothermal boreholes, thermal response tests of shorter durations than are normally used for deep boreholes were performed. The techniques used to analyse the various test results are outlined, and the resulting values of thermal conductivity obtained are within the range of those expected for the prevailing geology of the sites.

Notation

H	depth of borehole heat exchanger
h	groundwater head
k	slope of the line on a plot of average temperature against ln(time)
k_t	hydraulic conductivity
P	length of above-surface piping
Q	injected heat power
Q_r	radial heat loss
r	distance from centre of heat exchanger
r_1	outer radius of piping insulation
r_2	inner radius of the piping insulation (m)
r_b	borehole radius (m)
s	groundwater storage capacity
T_d	maximum temperature difference between circulating fluid and ambient air
t	time (s)
z	depth of heat exchanger

223

z_t	groundwater flow
α	thermal diffusivity (m^2/s)
λ	thermal conductivity
λ_p	thermal conductivity of piping insulation
∇	Laplace operator

1. Introduction

Piping can be installed into practically any subsurface engineering structure that provides a large interface area with the ground in order to harness ground thermal energy. Typical examples are bored or prefabricated piles, retaining walls, diaphragm walls, tunnel walls, basement slabs, basement walls and soil anchors (Adam and Markiewicz, 2009; Brandl, 2006; Franzius and Pralle, 2011). Although energy foundations are considered to be new, relative to other forms of renewable energy technologies such as wind, solar or biomass (Abdelaziz et al., 2011; Peron et al., 2011), they have now been installed in various engineering structures throughout the world (Boennec, 2008; Gao et al., 2008; Laloui, 2011). This is in contrast to the Republic of Ireland, where it is believed that no such systems exist (Hemmingway and Long, 2011a). In spite of growing interest in the area, there is a distinct lack of documented thermal response tests on energy foundations (Amis et al., 2010; de Moel et al., 2010). A thermal response test (TRT) is a controlled in situ test during which a known quantity of heat energy is injected into (or extracted from) a closed-loop borehole heat exchanger by way of a circulating heat carrier fluid, while the change of the fluid temperature is monitored.

Aside from the TRT operations outlined in the following paragraphs (only one of which was performed on energy piles), very little peer-reviewed literature is available in this specific area. Brettmann and Amis (2011) carried out TRTs on three continuous flight auger (CFA) piles of varying diameter (300 mm and 450 mm) and construction, and found that the measured value of thermal conductivity in each case was close to what was expected. The International Ground Source Heat Pump Association (2008) recently published design and installation standards for closed-loop/geothermal heat pump systems which suggest that the test bore should not exceed 6 in (152·4 mm) in order to carry out a TRT. However, the TRT regime performed by Brettmann and Amis (2011) indicates that the test methodology remains valid for larger-diameter heat exchangers. Franzius and Pralle (2011) carried out a TRT on the Katzenberg tunnel high-speed rail tunnel wall in Germany prior to its opening. Although neither the calculation method used to determine the thermal conductivity results nor the actual value of calculated thermal conductivity are described in detail by Franzius and Pralle (2011), the resulting calculated heat flux from the test was used in the design of a demonstrator project in Jenbach, Austria, which is described in detail in Frodl et al. (2010). Schneider et al. (2011) briefly describe the performance of thermal response and laboratory testing carried out on another tunnel in Germany equipped with heat exchanger pipes. Although in-depth detail is not provided, the testing resulted in measured values in the range 2·0–2·8 W/m K for the clay marl/limestone segment of the formation. The lower value of 2·0 W/m K observed by Schneider et al. (2011) is in close agreement with the value of 1·9 W/m K obtained by the University College Dublin (UCD) TRT rig for the testing described in Hemmingway and Long (2012c), which was performed on a borehole heat exchanger installed in a similar geological formation. Amis et al. (2010) performed a TRT on an 800 mm-wide, 36 m-deep diaphragm wall in England. The TRT was carried out in two stages, one prior to excavation and one following excavation, so that the thermal effect of removing 24 m of the 36 m of soil from the non-loop side of the diaphragm wall could be analysed. Results from the tests indicate that the thermal resistance increased by 20% and the thermal conductivity decreased by 13% (Amis, 2011) following excavation.

Following completion of a TRT, a thermal recovery test may be performed. A thermal recovery test is carried out immediately following a TRT; the heating power of the TRT rig is turned off, and the circulating fluid is allowed to continue circulation. The shape of the plot of average borehole heat exchanger (BHE) circulating fluid temperature against time for a recovery test is effectively the inverse of that plotted for the preceding TRT, and therefore it has been reported that the results can be analysed in exactly the same way (Banks, 2008). Thermal recovery tests are rarely carried out in practice, owing to time and cost constraints, and very little published information relating to the performance or evaluation of these tests is available.

Gehlin (2002) refers to thermal recovery testing being carried out in the UK, and states that the line source approach may be used in order to evaluate test results. Witte and Gelder (2006) make reference to several multi-power step TRTs on a BHE with cycles of heat addition followed by heat extraction. Raymond *et al.* (2011) carried out analytical and numerical simulations of thermal response and recovery tests. They suggest that: (*a*) the temperature of the circulating fluid inside the BHE homogenises rapidly after heat injection is stopped, and therefore performance and analysis of thermal recovery tests reduce the uncertainty of the thermal conductivity results calculated using a conventional TRT, where placement of the temperature probes may affect the results; and (*b*) the temperature evolution of a thermal recovery test agrees with the line source model. Gehlin and Spitler (2002) state that care should be taken either to make temperature measurements at the borehole inlet and outlet or, if the heat transfer rate is measured elsewhere, to minimise any unmeasured heat loss or gains. Zervantonakis and Reuss (2006) recommend placement of sensors directly in the fluid line. The temperature sensors on the UCD TRT rig are installed in the fluid line at the point of fluid exit and fluid entry from the trailer unit. Readers may refer to Hemmingway and Long (2012a) for further details of the design and construction of the UCD TRT rig. Regardless of where temperature sensors are placed within a TRT rig, it is vital that elements through which heat energy could be lost or gained (for example, all above-surface piping) are heavily insulated in order to minimise heat loss or gain, which would adversely affect test results.

2. Energy pile installations

The primary author organised and supervised the installation of several energy piles at two research sites in Co. Cork, Ireland, and later carried out thermal response, thermal recovery and laboratory thermal tests. The sites are located at the Cork Docklands (1 km east of Cork City Centre) and Carraigtwohill (approximately 15 km east of Cork City).

2.1 Cork Docklands site

Hemmingway and Long (2011b) provide a detailed description of the ground conditions at a site adjacent to the site referred to as the Cork Docklands site in this chapter. The distance between the Cork Docklands site and the adjacent site is less than 150 m, and therefore similar ground conditions are assumed. This was confirmed by observation of the spoil resulting from the drilling operations at the Cork Docklands site. The ground conditions consist of a thin layer of alluvium (approximately 1–2 m in depth) underlain by saturated sand and gravel to bedrock at approximately 42·2 m depth, with a layer of very stiff grey silty clay from 15 m to 25 m below ground level. A graphical representation of the soil strata at the Cork Docklands site is shown in Figure 1(b). The energy piles are installed to a depth of 14·5 m, and therefore the bedrock (thought to be at a depth of between 42·2 m and 60 m) (Allen *et al.*, 2003; Hemmingway and Long, 2011b; Long and Roberts, 2008; Milenic and Allen, 2005) is not reached.

Four 14·5 m-deep energy piles were installed in order to provide media for the completion of TRTs, and also in order to understand the practical issues associated with the installation of energy piles. Single-U- and double-U-shaped piping configurations were installed to a depth of 14·2 m into concrete piles of diameter 250 mm and 350 mm respectively. The dominant geological formation along the length of the installed energy piles is saturated sand with gravel, which has an expected thermal conductivity in the range 1·5–5·02 W/m K (Bristow *et al.*, 2001; Clarke *et al.*, 2008; EED, 2010; Goodrich, 1986; Midttomme and Rolandset, 1998). The database provided in the ground source energy heat exchanger design tool

Figure 1 Ground conditions at: (a) Cork Docklands site; (b) Carraigtwohill site

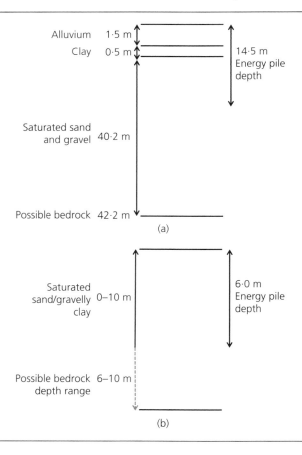

Energy Earth Designer (EED, 2010) quotes recommended values of thermal conductivity of 2·4 W/m K for saturated sand, 1·8 W/m K for alluvium (or saturated silt) and 1·6 W/m K for saturated clay. Assuming that the soil profile along the depth of the 14·2 m energy pile installation consists of 1·5 m of alluvium, 0·5 m of clay and 12·2 m of saturated sand/gravel, the depth-weighted average 'expected' thermal conductivity of the formation is 2·31 W/m K.

Polyethylene piping with an outer diameter of 40 mm and wall thickness of 3·7 mm was used for the installations. The piping was tied to either side of a central 'plunge bar' prior to insertion into the energy piles, following installation of the piles using the CFA piling process. Figure 2a shows heat exchanger piping being installed into one of the CFA piles at the Cork Docklands site. Figure 3 shows a sketch of the energy pile installation geometry for the Cork Docklands and Carraigtwohill sites. In both the single-U and double-U piping configurations, U-shaped pipes are tied onto a central steel plunge bar.

2.2 Carraigtwohill site

According to the Geological Survey of Ireland Quaternary and bedrock geology maps (GSI, 2011a, 2011b), the site is underlain by undifferentiated till, which in turn is underlain by a limestone bedrock formation. These findings are consistent with nearby borehole site investigations carried out for the Bray to Cork Gas Pipeline (GSI, 1976), which runs from west to east at a distance approximately 2 km south of the Carraigtwohill site. These site investigation records indicate that the soil in the area is made up primarily of sandy/gravelly clay (i.e. typical boulder clay), and that bedrock is located at depths ranging from 3 m to 10 m. A graphical representation of the soil strata at the Carraigtwohill site is shown in Figure 1(a). Energy piles 300 mm in diameter and 6 m deep were installed at the site (see Figure 2(b)). The underlying bedrock was not penetrated during the drilling operations, and therefore each of the piles was installed into a geological formation consisting of saturated sandy/gravelly clay. The expected thermal conductivities of moist clay and saturated sand fall in the ranges 0·9–2·22 W/m K and 1·5–5·02 W/m K respectively (EED, 2010). Therefore the expected thermal conductivity of the material present at the Carraigtwohill site may be estimated in the region of 2·2 W/m K or above.

3. Thermal test analysis

3.1 Thermal response test analysis: line source method

The most widely used theory for the evaluation of TRT results is the analytical line source method, which is developed from Kelvin's line source theory, such that Equation 1 may be written

$$\lambda = \frac{Q}{4\pi kH} \tag{1}$$

where λ is thermal conductivity, Q is injected heat power (W), H is the depth of the BHE (m) and k is the slope of the line on a plot of average temperature against ln(time). Data can be evaluated to an accuracy of within 10% if the lower time criterion shown in Equation 2 is satisfied (Florides and Kalogirou, 2008).

$$t > \frac{5r_b^2}{\alpha} \tag{2}$$

Figure 2 (a) Piping installation at the Cork Docklands site; (b) Carraigtwohill energy pile layout

(a)

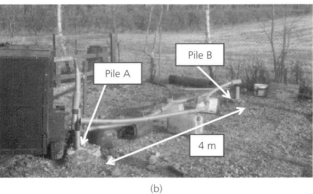

(b)

Figure 3 Energy pile installation geometry

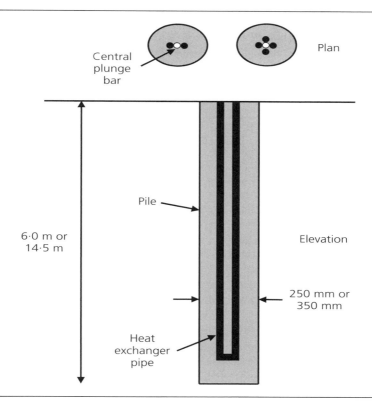

where r_b is the borehole radius (m), t is time (s) and α is thermal diffusivity (m²/s). For a detailed overview of the theory underlying the development of the analytical line source method, readers may refer to Eskilson (1987), Gehlin (2002), Hwang *et al.* (2010), Ingersoll and Plass (1948) or Signorelli *et al.* (2007).

3.2 Thermal response test analysis: geothermal properties measurement model

A second, more advanced method that may be used to evaluate the results of TRTs is the geothermal properties measurement (GPM) model developed by Shonder and Beck (2000). The GPM method was developed in order to analyse thermal properties from short-duration TRTs using a parameter estimation technique. The technique is referred to by several authors (Florides and Kalogirou, 2007; Hu *et al.*, 2012; Rey-Ronco *et al.*, 2012; Schiavi, 2009), but it does not seem to be widely applied in either academia or industry, where in both cases application of the analytical line source method appears to be the norm, owing to its simplicity of application (Saljnikov *et al.*, 2007). Shonder and Beck (2000) present an analysis of TRT data from three 50 h field tests carried out at two sites, and compare these results against those derived from year-long operating data at each respective site. The GPM calculated thermal conductivity values were found to agree within 2% and 4% respectively. Rey-Ronco *et al.*

(2012) state that the GPM method is one of the main methods for determining ground thermal conductivity from a TRT, and analyse a set of test results using the analytical line source method, the GPM method and a two-variable, parameter-fitting method. Thermal conductivity values ranging from 1·75 W/m K to 2·35 W/m K were calculated using the three methods. Previous research suggests that the GPM method is suitable for analysis of standard, best-practice, conventional TRT data. However, further investigation of the applicability of the method in the case of short-duration tests, or tests performed on energy piles, is merited. The GPM model operates by obtaining the finite-difference solution of the Fourier equation

$$\nabla^2 = \frac{1}{r}\frac{\partial}{\partial r}\left(r\frac{\partial}{\partial r}\right) + \frac{1}{r^2}\left(\frac{\partial^2}{\partial \theta^2}\right) + \frac{\partial^2}{\partial z^2} \tag{3}$$

where r is the distance from the centre of the heat exchanger, z is the depth of the heat exchanger and ∇ is the Laplace operator in the geological domain. The borehole backfill material (e.g. grout or concrete) and the geometric arrangement of the pipes are modelled as a single pipe of an effective radius. Using this approach, the model solves the Fourier equation by assuming that the temperature field along the vertical axis of the BHE is constant, and also that the heat exchanger is of a constant radius about the vertical axis (Rainieri et al., 2011).

Readers may refer to Shonder and Beck (2000) for a detailed description of the theory underlying the development and operation of the GPM model.

3.3 Thermal recovery test analysis

A direct mathematical analogy between the analysis of groundwater flow and subsurface heat flow exists (Banks, 2009). The key analogy is that between Fourier's law, which describes heat transfer by conduction, and Darcy's law, which describes groundwater flow. Both equations are shown in Table 1, which is edited after Banks (2009). By comparing the variables in each of the equations, it is easy to see that groundwater flow (z) is analogous to heat flow (Q), hydraulic conductivity or permeability (k) to thermal conductivity (λ), and groundwater head (h) to temperature (T). Similarly, the concept of a 'hydraulic skin' that accounts for head losses

Table 1 Analogy between water flow and heat flow

	Groundwater flow	Subsurface heat flow
Key physical law	Darcy's law	Fourier's law
	$z = -KAdh/dx$	$Q = -\lambda AdT/dx$
Flow	z = groundwater flow (m³/s)	Q = heat flow (J/s or W)
Property of conduction	k = hydraulic conductivity (m/s)	λ = thermal conductivity (W/m K)
Measure of potential energy	h = groundwater head (m)	T = temperature (°C or K)
Measure of storage	s = groundwater storage	S_{vc} = specific heat capacity (J/m³K or J/kg K)
Measure of bore efficiency	Well loss/hydraulic skin	Borehole thermal resistance

in the vicinity of a pumped well (Kruseman and de Ridder, 2000) may be considered analogous to the thermal phenomenon referred to as 'borehole thermal resistance', which refers to the resistance to transportation of heat from the borehole wall into the fluid inside the pipes, and is controlled *inter alia* by the type of grouting, the pipe material and the borehole and pipe geometry (Sanner *et al.*, 2011). Although not specifically analysed in the presented work, borehole thermal resistance has been shown to be an important consideration in the design of energy piles, owing to the transient nature of energy loads on energy piles, and the possible large diameter of some piles. Readers may refer to the work of researchers such as Loveridge and Powrie (2012) and Pahud (2007) for more information relating to borehole thermal resistance and energy piles.

The hydrological characteristics of an aquifer may be determined by carrying out a pumping test. In order to carry out a pumping test, water is pumped from one well, and the resulting fall in groundwater level at nearby observation well(s)/piezometer(s) is monitored (Powrie, 2004). Following completion of a pumping test, the pump is shut down, and the water levels in the observation wells gradually rise until they reach their initial levels. This is known as a 'pumping recovery test'. Data from a pumping recovery test can provide an independent check on the results of the preceding pumping test. It has been stated that the data collected from a pumping recovery test are more reliable that those collected from a pumping test, because water level recovery occurs at a constant rate, whereas a constant discharge during pumping can be difficult to achieve in the field (Kruseman and de Ridder, 2000). Figure 4 shows a plot of typical test data from a pumping test and pumping recovery test (Osborne, 1993). A pre-pumping measurement period precedes the tests, so that any regional effects on groundwater level (e.g. caused by tidal variation) can be recorded. Following the pre-pumping measurement, the pump is switched on, resulting in a period of rapid drawdown, followed by a gradual reduction in the slope of the rate of change of the drawdown until a steady state is reached where the level of drawdown is controlled by the hydraulic conductivity of the aquifer. The length of pumping test required may vary, depending on the objectives of the test, the type of aquifer, the location of suspected boundaries, and the degree of accuracy required (Osborne, 1993). In general, the length of test is determined when the test operative is satisfied that the quantity of data collected is adequate for the purposes of that particular test. At the end of the pumping test, the pump is switched off and recovery of water levels in the observation wells/piezometers commences. The equation of a trend line imposed on the latter portion of pumping test data may then be used to extend the pumping test data to show the trend that would have resulted were the pumping to continue (this is shown as the lower dashed line in Figure 4). The difference between the collected pumping recovery test data and the extended pumping test data for each respective data point is termed the 'recovery'.

The correlation between pumping/pumping recovery tests and thermal response/thermal recovery tests may immediately be seen by comparing Figure 4 (pumping test followed by pumping recovery test) with Figure 5 (TRT followed by thermal recovery test). Commencement of the heating period (or the thermal 'response' test) corresponds to a steep initial increase in the temperature of the water circulating around the BHE piping, following which a steady-state rate of temperature increase develops, where the rate of temperature increase is controlled by the thermal conductivity of the geological formation surrounding the BHE. At the end of the TRT the heaters are turned off, and the circulating fluid continues circulation.

Figure 4 Typical pumping test and pumping recovery test (Osborne, 1993): (a) map of aquifer test site; (b) change of water level in well B

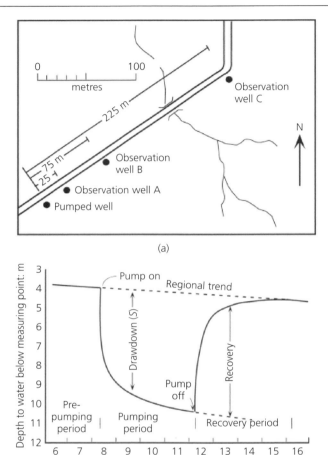

The fluid temperature then begins to return to a temperature near (but slightly above) its initial standing temperature. The reason why the temperature of the circulating fluid will not return fully to its initial standing temperature is that, after the heaters have been turned off, the circulating pump continues to impart a small amount of heat energy. This supplementary energy addition can be easily quantified and factored into the thermal response and thermal recovery test analyses, as shown in the following section.

Figure 5 Typical thermal response and thermal recovery test data

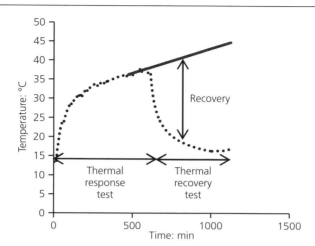

The solid line in Figure 5 shows the trend that would have resulted were the TRT (i.e. heating period) to continue, and is calculated in exactly the same way as for the pumping test that is described earlier in this section. The difference between the collected thermal recovery test data and the extended TRT data (the solid line in Figure 5) for each respective data point is termed the 'recovery', and results from the drop in power due to switching off the heaters. The calculated recovery may be plotted on a graph of temperature recovery against the natural logarithm of time, and then be evaluated using the line source method in exactly the same manner as the thermal response results.

3.4 Heat addition and loss

The heating power supplied by the resistance heaters in each of the tests presented is 3 kW. This is the minimum in heat injection rate of the UCD TRT rig: the heat injection rates (W/m) for the two tests referred to are therefore somewhat above the level of heat injection/extraction that would be realised when operating a ground source energy system (this may or may not have an impact on the thermal diffusive properties of the ground). A small amount of heat is also added to the heat transfer fluid by the circulating pump. The circulating pump on the UCD TRT rig is rated at 185 W. The heat added to the circulating fluid by the pump during the TRT may be estimated as 150 W by assuming reasonable electrical motor and mechanical efficiencies of 80%. Therefore the total heat added to the circulating fluid in all tests presented is 3150 W.

In the case of the Carraigtwohill thermal response and thermal recovery tests, there is a significant quantity of above-surface heat exchanger piping through which the heat exchanger fluid must circulate (i.e. the piping joining pile A to pile B: see Figure 2(b)). Although this piping was insulated with high-density pipe lagging to minimise heat loss, it is important to estimate the level of heat loss during the tests, so that it can be factored into subsequent calculations. The radial heat loss equation for a multi-layered cylinder given as Equation 4 by Cengel (2006) is used in order to calculate the instantaneous loss of heat due to the above-surface insulated piping on the worst-case scenario: that is, when the differential between the temperature within the piping and the ambient air is at a maximum

$$Q_r = \frac{T_d}{\ln(r_1/r_2)/2\pi P \lambda_p} \qquad (4)$$

Taking the total length of above-surface piping (P) to be 10 m, the maximum temperature difference between the circulating fluid and ambient air (T_d) to be 20°C, the outer radius of the insulation surrounding the piping (r_1) to be 31 mm, the inner radius of the insulation (r_2) to be 20 mm, and the thermal conductivity of the selected insulation (λ_p) to be 0·03 W/m K, the total heat loss is calculated as 70 W. The injected heat power used in the calculations for the Carraigtwohill TRT is therefore 3080 W.

4. Thermal test results

The thermal response and recovery tests were performed using the TRT rig and operating fundamentals (described in Hemmingway and Long, 2012a), approximately 12 months after pile construction. Temperature distributions along the length (in depth) of a test energy pile and in other elements at varying distances from a test pile presented in the research carried out by Bourne-Webb *et al.* (2009) suggest that the key assumption required for analysis using the analytical line source method (i.e. that the pile behaves as an infinitely long line source) is reasonable in the case of an energy pile. The applicability of the analytical line source method to TRT data gathered from tests carried out on energy piles is further confirmed by Brettmann and Amis (2011), who successfully evaluated results from TRTs carried out on energy piles of 300 mm and 450 mm diameters using the analytical line source method.

4.1 Cork Docklands response test

Prior to commencement of thermal response and thermal recovery tests, the undisturbed ground temperature (T_0) along the length of the borehole/pile should be established. A temperature probe was lowered down the length of the Cork Docklands energy pile, resulting in a measured average undisturbed ground temperature of 12·8°C (denoted 'Before TRT' in Figure 6(a), where the left-most vertical line represents the average temperature).

A measurement of temperature with depth was also taken 5·5 h after completion of the thermal recovery test, and is denoted '5·5 h after recovery' in Figure 6(a). This measurement confirms the findings of Hemmingway and Long (2011b), where the results of geotechnical site investigations suggest the presence of soft clay layers at approximately 6 m depth. The measurements of temperature with depth presented in Figure 6(a) show a slower rate of return to natural soil temperature conditions at depth 6 m relative to the rate of return to natural ground temperature below this point. This indicates that a thin formation of low-conductivity material (such as clay/alluvium) may be present at this point. The rate of return to natural temperature conditions below 6 m is probably higher because the material (i.e. saturated sand and gravel) is of higher relative thermal conductivity, and therefore the heat generated by the TRT is able to dissipate at a faster rate; it is

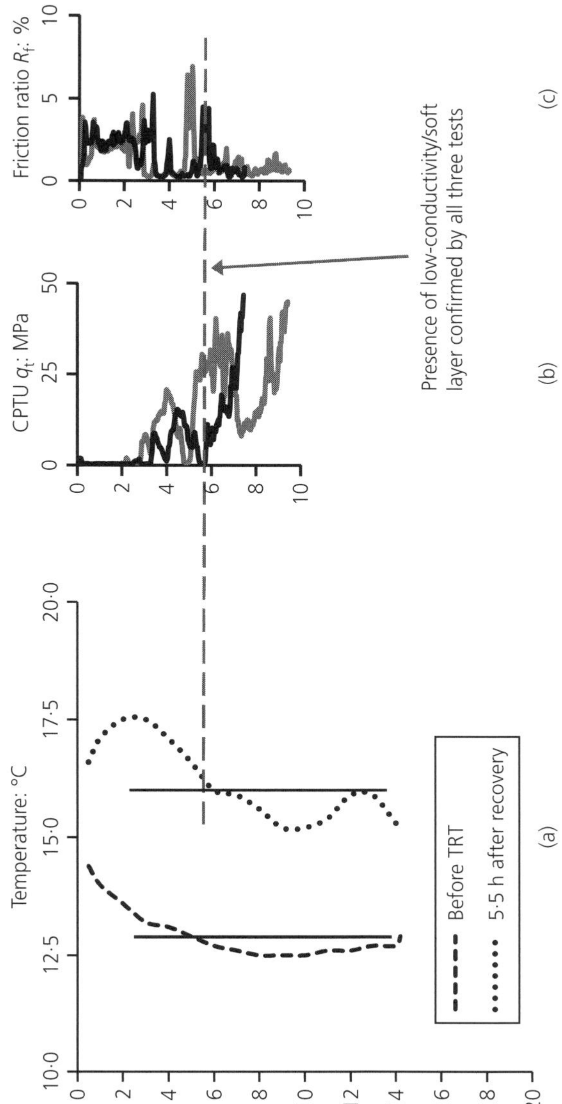

Figure 6 Cork Docklands: (a) before and after TRT temperature measurement; (b) CPTU corrected cone resistance; (c) CPTU friction ratio

also due to the presence of a complicated tidal groundwater flow regime in the area (Hemmingway (2012)). This finding is confirmed by inspection of site investigation data presented by Hemmingway (2012) that are shown in Figures 6(b) and 6(c). It is clear from the plots of CPTU cone resistance and CPTU friction ratio with depth that a soft layer (e.g. clay/alluvium) exists at approximately 6 m depth below ground level.

Figure 7 shows the results from the TRT performed on the 250 mm diameter single-U pipe energy pile installation at the Cork Docklands site. The solid black line on the graph represents the energy pile injected fluid temperature (denoted T_{Down}), the dashed line represents the return fluid temperature (denoted T_{Up}) and the grey line represents the flow rate, which remained turbulent throughout the testing period. Heat input rates of between 30 W/m (for low-conductivity formations) and 80 W/m (for high-conductivity formations) are suggested by Ashrae (2002) and Sanner et al. (2005). The net result of using a heat input rate of 3 kW (the minimum injectable heating power of the UCD TRT rig) for evaluation of the thermal properties of the ground surrounding the 14·2 m-deep energy pile is that the thermal response heating period was curtailed in order to protect against damage to the system due to overheating. The temperature profile for the injected and return fluid temperatures (T_{Down} and T_{Up} respectively) shown in Figures 7 and 10(a) appears to be quite smooth. However, there are a small number of 'dips' in the temperature, where the resistance heaters switch off for a few seconds and then switch back on again. Examination of the test data shown, and of those presented by Hemmingway and Long (2012a), indicates that this occurred only approximately once per hour during the thermal response testing, and appears to have a negligible effect on the results of the test, owing to the smooth profile of the circulating fluid temperature development over the duration of the tests. A reduction in flow during progression of the TRT is evident in Figure 7. It is believed that this reduction in flow may be due either to thermal expansion of the BHE piping as the test progressed, resulting in a gradual reduction in system pressure, or to a small leak in the closed-loop system, resulting in increasing pumping difficulty as the test progressed. No fluid leakage was observed on site. Examination of the relationship between commencement of heat energy injection and reduction in circulation fluid flow associated with the UCD TRT rig is currently

Figure 7 Cork Docklands thermal response measurements

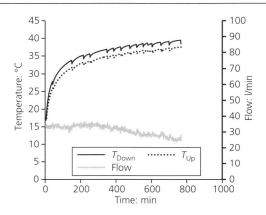

under way. Readers may refer to Hemmingway (2012) and Hemmingway and Long (2012a) for details of how these issues were resolved for subsequent tests.

Figures 8(a)–8(d) show evaluation of differing portions of the results from the TRT using the analytical line source method. Separate evaluation of data from (a) the final 8 h, (b) the final 6 h, (c) the final 4 h and (d) the final 2 h of test data was performed in order to investigate whether or not steady-state conditions had been reached. The slope (k) of the selected portion of data was determined for each of the graphs and entered into Equation 1, with the injected heat flux (Q) set to 3150 W and the depth of bore (H) set to 14·2 m. This analysis resulted in calculated thermal conductivity values of 5·33 W/m K, 5·40 W/m K, 5·03 W/m K and 3·23 W/m K for the evaluation periods included in Figures 8(a)–8(d) respectively. The continual reduction in calculated thermal conductivity in each subsequent evaluation indicates that steady-state conditions had not been fully achieved. The lower time criterion for use of the analytical line source method described by Equation 2 is 22·5 h. The ground thermal diffusivity (α) was

Figure 8 LSM (laser scanning microscopy) analysis of Cork Docklands single-U TRT: (a) last 8 h; (b) last 6 h; (c) last 4 h; (d) last 2 h

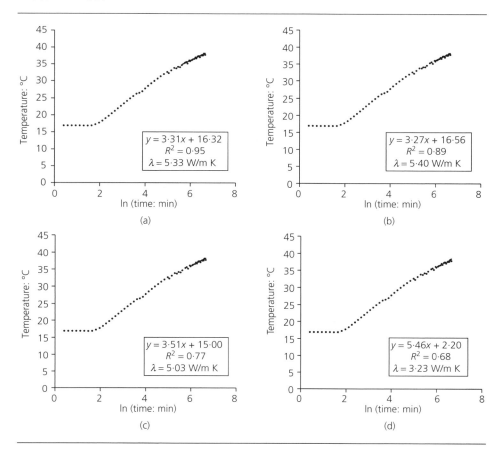

calculated by dividing the calculated thermal conductivity from the test results by the volumetric heat capacity for the soil, taken from a database such as that provided in EED (2010) (note: this value is known not to vary greatly in saturated strata; Sanner et al., 2011). This suggests that 22 h of data are required in order to evaluate test data using the analytical line source method to an accuracy of within 10%.

The best estimate of thermal conductivity available from the TRT data as calculated using the analytical line source method is 3·23 W/m K, as analysed in Figure 8(d). This is the only one of the analyses that complies fully with the Ashrae (2002) recommendation to disregard the first 5–10 h of test data prior to evaluation. The coefficient of determination (R^2 value) for the portion of data from which this slope is measured is 0·68 (as shown in Figure 8(d)), and therefore, owing to this relatively low value, further validation of the calculated thermal conductivity value would increase confidence in the evaluation accuracy. Analysis of the Cork Docklands data using the GPM method referred to in Section 3.2 results in a calculated thermal conductivity value of 5·82 W/m K. Owing to the divergence between the values obtained by the line source method and GPM method analyses, further investigation is attempted by performance of a thermal recovery test and laboratory thermal conductivity tests.

4.2 Cork Docklands thermal recovery test

Following completion of the TRT, the heater power was switched off and the heat carrier fluid was circulated for a period of 6 h. The resulting data measurements, alongside calculation of the 'extended' TRT data trend that would have resulted had the heaters been left on, are shown in Figure 9(a). The results from the recovery test are analysed using the analytical line source method, resulting in a calculated thermal conductivity value of 3·53 W/m K for the formation. As was the case for the evaluation of the data from the TRT, the calculated value is above the depth-weighted estimated value of 2·31 W/m K. The values obtained by the TRT (3·23 W/m K) and the thermal recovery test (3·53 W/m K), although not exactly the same, are in reasonably good agreement. The difference between the two may be due to the TRT not fully reaching steady-state conditions, illustrated by the evaluation of data from varying time ranges as shown in Figure 8, or to the existence of a tidal groundwater flow that is thought to be present at the site, following site investigations presented in Hemmingway (2012).

4.3 Laboratory thermal measurement of Cork Docklands soil

A laboratory thermal conductivity measurement was performed on a soil sample from the Cork Docklands site. The sample was extracted at a depth of 12 m below ground level during the site investigations described in Hemmingway and Long (2011b). A full description of the thermal probe used to perform the laboratory thermal conductivity measurement, alongside the evaluation theory, is provided in Hemmingway and Long (2012b). The excavated sample was recompacted back to conditions that are representative of those present at the site (bulk density of 2 Mg/m^3 at 15% water content). The laboratory test resulted in a measured thermal conductivity value of 3·15 W/m K. The measured thermal conductivity values from the thermal response, thermal recovery and laboratory tests are therefore 3·23 W/m K, 3·53 W/m K and 3·15 W/m K respectively. Each of these values is greater than the calculated, depth-weighted average 'expected' thermal conductivity of the tested formation (2·31 W/m K). This illustrates the importance of direct site or laboratory thermal conductivity measurements during the design of ground source energy systems, so that inherently conservative or ambitious values are not used.

Figure 9 Cork Docklands: (a) calculation of 'extended' TRT data; (b) line source analysis of thermal recovery data

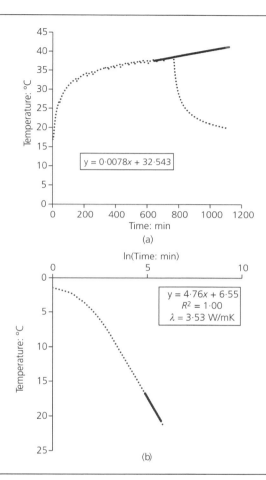

4.4 Carraigtwohill thermal response test

A group TRT was performed at the Carraigtwohill site by connecting two piles in series as shown in Figure 2(b), thereby increasing the available 'geothermally active depth' of heat exchanger to be tested. Figure 10(a) shows the results from the group TRT. Visual inspection of the figure indicates that the profile of the response test is in line with that which would be expected, with a sharp increase in the average circulating fluid temperature in the early stages of the TRT when the backfill material is heating up, followed by a stabilisation of the rate of temperature increase with time (suggesting the achievement of steady-state conditions). The TRT lasted for 10 h, at which time the electrical resistance heater was turned off.

Figure 11 shows a plot of the average circulating fluid temperature against the natural logarithm of time for the TRT data. The data are analysed in the same manner as the data collected

Figure 10 Carraigtwohill: (a) group response test data; (b) pre-TRT temperature measurement

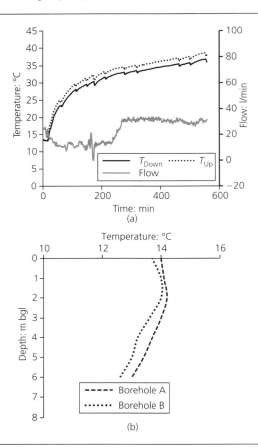

during the Cork Docklands TRT, and the resulting best estimate of thermal conductivity of 2·87 W/m K is calculated using the analytical line source analysis method. In this case the measured thermal conductivity value is higher than the calculated, depth-weighted, expected thermal conductivity of the ground formation. Analysis of the Carraigtwohill thermal response data using the GPM method referred to in Section 3.2 results in a calculated thermal conductivity value of 2·94 W/m K, which is in excellent agreement with the value calculated using the analytical line source method.

4.5 Carraigtwohill thermal recovery test

Figure 12(a) shows the data measurements from the thermal response and recovery tests at the Carraigtwohill site alongside the calculation of the 'extended' TRT data trend that would have resulted had the heaters been left on. The results of the thermal recovery test (shown in Figure 12(b)) are analysed in the same fashion as for the Cork Docklands thermal recovery data, resulting in a calculated thermal conductivity value of 2·60 W/m K. The measured values of thermal

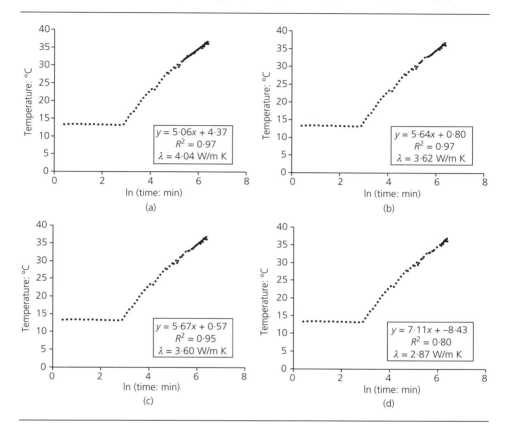

Figure 11 LSM analysis of Carraigtwohill group TRT: (a) last 8 h; (b) last 6 h; (c) last 4 h; (d) last 2 h

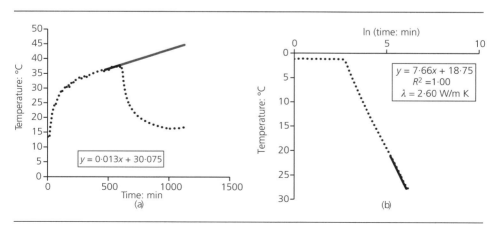

Figure 12 Carraigtwohill group test: (a) calculation of 'extended' TRT data; (b) line source analysis of thermal recovery data

conductivity are therefore 2·87 W/m K from the TRT and 2·60 W/m K from the thermal recovery test. Although the results are very similar, the small difference is thought to be due to the shortened test duration, necessitated by the combination of the shallow geothermally active depth of the energy piles (12 m) and the minimum heating power of the UCD TRT rig (3 kW).

5. Discussion

The installation of research energy piles at two sites is described in this chapter. A description of each site is provided, alongside a depth-weighted average estimated thermal conductivity value for the ground formation along the depth of the installed energy foundations. Shorter TRT durations than are normally used for deep boreholes were necessary for the tests performed at both sites, owing to the fact that the minimum heat injection rate of the UCD TRT rig is 3 kW. This necessitated completion of the TRTs when the temperature of the circulating fluid reached the upper limit allowable for protection of the rig components (about 40°C). It is important to note that the authors do not suggest that TRT durations should necessarily be shortened to those presented in this chapter. This chapter simply presents an analysis of the best available TRT results from the Cork Docklands and Carraigtwohill sites. Thermal recovery and laboratory thermal tests were performed in order to provide additional confidence in the results obtained from the preceding TRTs. The resulting values of thermal conductivity are within the range expected for the prevailing geology of the sites, but are higher than the calculated depth-weighted expected thermal conductivity values at each respective site. This illustrates the issue associated with relying on average values from generalised databases of thermal conductivity. It is recommended that thermal conductivity values are taken from databases only when: (*a*) the database has been produced to take account of regional and local geological conditions; and (*b*) the database has been developed using scientifically rigorous methods, such as proven correlations between thermal parameters and, for example, common engineering parameters. In situ measurement of thermal properties is recommended for the design of medium- to large-scale ground source energy systems. Laboratory measurements may be suitable in cases where the designer is very familiar with the geology at a particular site, and is fully satisfied that laboratory tests will appropriately represent the in situ conditions (for example, issues such as ground thermal parameter heterogeneity, groundwater flow and borehole construction require in-depth consideration).

The results of the TRTs at the Carraigtwohill and Cork Docklands sites are analysed using both the analytical line source method and the geothermal properties measurement (GPM) method. The analytical line source method is the most widely used method in industry for the analysis of TRT results. The durations of the Cork Docklands and Carraigtwohill TRTs were 13 h and 10 h respectively, necessitated by the minimum injectable heating power of the UCD TRT rig, coupled with the shallow depth of the installed energy piles. Analysis of TRT data to an accuracy of within 10% using the analytical line source method requires a TRT of duration greater than that described by Equation 2. Following this theory, in the case of the TRTs presented in this chapter, 22 h of test data were therefore required in order to evaluate them using the analytical line source method to an accuracy of 10%. The presented TRTs are of course shorter than this, and therefore additional testing and analyses are completed in order to provide further confidence in the results calculated by the analytical line source method. Additional analyses are performed by examining the TRT data from both sites using a method

Table 2. Thermal test results

Test/analysis type	Thermal conductivity, Carraigtwohill site: W/m K	Thermal conductivity, Cork Docklands site: W/m K
TRT/analytical line source analysis	2·87	3·23
TRT/GPM analysis	2·94	5·82
Thermal recovery test	2·60	3·53
Laboratory thermal conductivity test	—	3·15

known as the 'geothermal properties measurement' (GPM) method. The GPM method was developed in order to analyse thermal properties from short-duration TRTs using a parameter estimation technique.

The results obtained from the analytical line source method and GPM analyses of the TRT data obtained at the Cork Docklands and Carraigtwohill sites are shown in Table 2. For the Carraigtwohill site, the values resulting from both the analytical line source and GPM methods are in excellent agreement. However, for the Cork Docklands site, there is a divergence in results. It is believed that this analysis is complicated by the complex groundwater flow regime operating in the Cork Docklands, as described by Hemmingway (2012). However, further analysis will be required in order to confirm whether this divergence is due solely to the complex groundwater flow regime at the site, or also to the reliability of the GPM method.

In addition to analysis of the TRT data using the analytical line source and GPM methods, thermal recovery tests and laboratory thermal tests were performed, the results of which are also shown in Table 2. The thermal conductivity values calculated from the thermal recovery tests at both sites are close to those calculated from the analytical line source analyses, and the results of laboratory thermal conductivity testing on material extracted from beneath the Cork Docklands also reveals a result close to that calculated using the analytical line source method.

6. Conclusion and recommendations

The purpose of the work presented is to analyse thermal tests performed on energy pile foundations at two sites, in Carraigtwohill and the Cork Docklands, both of which are located in County Cork, Ireland. Unfortunately, the minimum injectable heating power of the UCD TRT rig, coupled with the shallow depth of the energy piles (when compared with conventional closed-loop borehole heat exchanger installations) meant that the TRTs presented had to be shorter than those typically used in industry. In order to provide further confidence in the thermal conductivity values calculated by the industry standard 'analytical line source' technique, additional interpretation methods (thermal recovery testing and analysis of the TRT data using the GPM method) were undertaken. These analysis techniques were supplemented by laboratory testing on soil samples extracted from beneath the Cork Docklands site. All results were in general agreement, except for the Cork Docklands, where some anomalies exist. The case of the Cork Docklands is complicated by the complex groundwater flow regime operating there, and further research is required in order to define the extent, characteristics and effect that this groundwater flow has on these analyses. The most widely used theory for the evaluation of TRT results in industry is the analytical line source method; it is recommended that use of the GPM method be further explored.

Acknowledgements

The authors acknowledge the help of the ESB and McDonnell Piling & Foundations Ltd for the provision of site access. The authors also acknowledge the assistance provided by Mark Adams of Oak Ridge National Laboratory (USA) for advice in relation to the operation of their GPM analysis software tool. The first author's PhD research was funded by the Irish Research Council for Science, Engineering and Technology (IRCSET) and the University College Dublin (UCD) Strategic and Major Initiatives Scheme.

REFERENCES

Abdelaziz S, Olgun C and Martin J (2011) Design and operational considerations of geothermal energy piles. *Proceedings of Geo-Frontiers 2011, Dallas, TX, USA*, pp. 450–459.

Adam D and Markiewicz R (2009) Energy from earth coupled structures, foundations, tunnels and sewers. *Géotechnique* **59(3)**: 229–236.

Allen A, Dejan M and Sikora P (2003) Shallow gravel aquifers and the urban 'heat island' effect: a source of low enthalpy geothermal energy. *Geothermics* **32(4–6)**: 569–578.

Amis T (2011) Energy foundations in the UK (Presentation). *Ground Source Live, Peterborough, UK*.

Amis T, Robinson C and Wong S (2010) Integrating geothermal loops into the diaphragm walls of the Knightsbridge Palace Hotel project. In *Proceedings of the Basements and Underground Structures Conference 2010*. Emap, London, UK.

Ashrae (2002) *Methods for Determining Soil and Rock Formation Thermal Properties from Field Tests*. American Society of Heating, Refrigerating and Air-Conditioning Engineers, Atlanta, GA, USA, Research summary ASHRAE 1118-TRP.

Banks D (2008) *An Introduction to Thermogeology: Ground Source Heating and Cooling*. Blackwell, Oxford, UK.

Banks D (2009) An introduction to 'thermogeology' and the exploitation of ground source heat. *Quarterly Journal of Engineering Geology and Hydrogeology* **42**: 283–293.

Boennec O (2008) Shallow ground energy systems. *Proceedings of the Institution of Civil Engineers – Energy* **161(2)**: 57–61.

Bourne-Webb P, Amatya B, Soga K *et al.* (2009) Energy pile test at Lambeth College, London: geotechnical and thermodynamic aspects of pile response to heat cycles. *Géotechnique* **59(3)**: 237–248.

Brandl H (2006) Energy foundations and other thermo-active ground structures. *Géotechnique* **56(2)**: 81–122.

Brettmann T and Amis T (2011) Thermal conductivity evaluation of a pile group using geothermal energy piles. *Proceedings of Geo-Frontiers 2011, Dallas, TX, USA*, pp. 499–508.

Bristow K, Kluitenberg G, Goding C and Fitzgerald T (2001) A small multi-needle probe for measuring soil thermal properties, water content and electrical conductivity. *Computers and Electronics in Agriculture* **31(3)**: 265–280.

Cengel Y (2006) *Heat and Mass Transfer: A Practical Approach*, 3rd edn. McGraw-Hill, Boston, MA, USA3rd edn..

Clarke B, Agab A and Nicholson D (2008) Model specification to determine thermal conductivity of soils. *Proceedings of the Institution of Civil Engineers – Geotechnical Engineering* **161(3)**: 161–168.

de Moel M, Bach PM, Bouazza A, Singh RM and Sun JO (2010) Technological advances and applications of geothermal energy pile foundations and their feasibility in Australia. *Renewable and Sustainable Energy Reviews* **14(9)**: 2683–2696.

EED (2010) *Energy Earth Designer Ver.3·16*. Blocon Sweden, Lund, Sweden.

Eskilson P (1987) *Thermal Analysis of Heat Extraction of Boreholes*. PhD thesis, University of Lund, Lund, Sweden.

Florides G and Kalogirou S (2007) Ground heat exchangers: a review of systems, models and applications. *Renewable Energy* **32(16)**: 2461–2478.

Florides G and Kalogirou S (2008) First in situ determination of the thermal performance of a U-pipe borehole heat exchanger in Cyprus. *Applied Thermal Engineering* **28(2–3)**: 157–163.

Franzius J and Pralle N (2011) Turning segmental tunnels into sources of renewable energy. *Proceedings of the Institution of Civil Engineers – Civil Engineering* **164(1)**: 35–40.

Frodl S, Franzius JN and Bartl T (2010) Design and construction of the tunnel geothermal system in Jenbach. *Geomechanics and Tunnelling* **3(5)**: 658–668.

Gao J, Zhang X, Liu J, Li K and Yang J (2008) Numerical and experimental assessment of thermal performance of vertical energy piles: an application. *Applied Energy* **85(10)**: 901–910.

Gehlin S (2002) *Thermal Response Test: Method Development and Evaluation*. PhD thesis, Division of Water Resources Engineering, Department of Environmental Engineering, Luleå University of Technology, Luleå, Sweden.

Gehlin S and Spitler J (2002) *Thermal Response Test: State of the Art 2001*. International Energy Agency, Paris, France, Report IEA ECES Annex 13.

GSI (Geological Survey of Ireland) (1976) *Geotechnical Site Investigation Report: Gas Pipeline from Powerhead, Bray to Cork, Aghada and Marino Point*. Geological Survey of Ireland, Dublin, Ireland, Report ID 1530.

GSI (2011a) *Geological Survey of Ireland 1: 100, 000 Bedrock Geology Map*. Geological Survey of Ireland, Dublin, Ireland. See http://www.gsi.ie (accessed 12/11/2011).

GSI (2011b) *Quaternary Geology of Cork City*. Geological Survey of Ireland, Dublin, Ireland.

Goodrich L (1986) Field measurements of soil thermal conductivity. *Canadian Geotechnical Journal* **23(1)**: 51–59.

Hemmingway P (2012) *Investigation of Ground Properties for Ground Source Energy Systems*. PhD thesis, College of Engineering & Architecture, School of Civil, Structural & Environmental Engineering, University College Dublin, Ireland.

Hemmingway P and Long M (2011a) Energy foundations: potential for Ireland. *Proceedings of Geo-Frontiers 2011, Dallas, TX, USA*, pp. 460–470.

Hemmingway P and Long M (2011b) Geothermal energy: settlement and water chemistry in Cork, Ireland. *Proceedings of the Institution of Civil Engineers – Engineering Sustainability* **164(3)**: 213–224.

Hemmingway P and Long M (2012a) Design and development of a low-cost thermal response rig. *Proceedings of the Institution of Civil Engineers – Energy* **165(3)**: 137–148.

Hemmingway P and Long M (2012b) Interpretation of in situ and laboratory thermal measurements resulting in accurate thermogeological characterisation. In *Geotechnical and Geophysical Site Characterization 4* (Coutinho RQ and Mayne PW (eds)). CRC Press, London, UK, pp. 1779–1787.

Hemmingway P and Long M (2012c) Thermal response testing of compromised borehole heat exchangers. *International Journal of Low-Carbon Technologies*, http://dx.doi.org/10.1093/ijlct/cts013.

Hu P, Meng Q, Sun Q, Zhu N and Guan C (2012) A method and case study of thermal response test with unstable heat rate. *Energy and Buildings* **48**: 199–205.

Hwang S, Ooka R and Nam Y (2010) Evaluation of estimation method of ground properties for the ground source heat pump system. *Renewable Energy* **35(9)**: 2123–2130.

Ingersoll R and Plass H (1948) Theory of the ground pipe heat source for the heat pump, heating piping and air conditioning. *Heating, Piping & Air Conditioning Journal* **20(7)**: 119–122.

International Ground Source Heat Pump Association (2008) *Closed-Loop/Geothermal Heat Pump Systems: Design and Installation Standards*. IGSHPA, Oklahoma State University, Stillwater, OK, USA.

Kruseman G and de Ridder N (2000) *Analysis and Evaluation of Pumping Test Data*, 2nd edn. International Institute for Land Reclamation and Improvement, Wageningen, the Netherlands.

Laloui L (2011) In-situ testing of a heat exchanger pile. *Proceedings of Geo-Frontiers 2011, Dallas, TX, USA*, pp. 410–419.

Long M and Roberts T (2008) Engineering characterisation of the glaciofluvial gravels of Cork City. *Transactions of Engineers Ireland* **131**: 16–28.

Loveridge F and Powrie W (2012) Pile heat exchangers: thermal behaviour and interactions. *Proceedings of the Institution of Civil Engineers – Geotechnical Engineering* **166(2)**: 178–196.

Midttomme K and Rolandset E (1998) The effect of grain size on thermal conductivity of quartz sands and silts. *Petroleum Geoscience* **4**: 165–172.

Milenic D and Allen A (2005) Buried valley ribbon aquifers: a significant groundwater resource of SW Ireland. In *Groundwater and Human Development* (Bocanegra EM, Hernández MA and Usunoff EJ (eds)). Balkema, Amsterdam, the Netherlands, pp. 171–184.

Osborne PS (1993) *Ground Water Issue: Suggested Operating Procedures for Aquifer Pumping Tests*. Environmental Protection Agency, Washington, DC, USA.

Pahud D (2007) *PILESIM2: Simulation Tool for Heating/Cooling Systems with Energy Piles or Multiple Borehole Heat Exchangers – User Manual*. Scuola Universitaria Professionale della Svizzera Italiana, Lugano, Switzerland.

Peron H, Knellwolf C and Laloui L (2011) A method for the geotechnical design of heat exchanger piles. *Proceedings of Geo-Frontiers 2011, Dallas, TX, USA*, pp. 470–479.

Powrie W (2004) *Soil Mechanics Concepts and Applications*. Spon Press, London, UK2nd edn..

Rainieri S, Bozzoli F and Pagliarini G (2011) Modeling approaches applied to the thermal response test: a critical review of the literature. *HVAC&R Research* **17(6)**: 977–990.

Raymond J, Therrien R and Gosselin L (2011) Borehole temperature evolution during thermal response tests. *Geothermics* **40(1)**: 69–78.

Rey-Ronco M, Alonso-Sánchez T, Coppen-Rodríguez J and Castro-García M (2012) A thermal model and experimental procedure for a point-source approach to determing the thermal properties of drill cuttings. *Journal of Mathematical Chemistry* **51(4)**: 1139–1152.

Saljnikov A, Goricanec D, Dobersek D, Krope J and Kozic D (2007) Thermal response test use of a borehole heat exchanger. *Proceedings of the 2nd IASME/WSEAS International Conference on Energy and Environment, Portoroz, Slovenia*, pp. 5–9.

Sanner B, Andersson O, Eugster WJ *et al.* (2011) *Geotrainet Training Manual for Designers of Shallow Geothermal Systems*. European Federation of Geologists, Brussels, Belgium.

Sanner B, Hellstrom G, Spitler JD and Gehlin S (2005) Thermal response test: current status and world-wide application. *Proceedings of the World Geothermal Congress 2005, Antalya, Turkey*.

Schiavi L (2009) 3D simulation of the thermal response test in a U-tube borehole heat exchanger. *Proceedings of the COMSOL Conference, Milan, Italy*.

Schneider M, Vermeer P and Moormann C (2011) Tunnelling as a contribution to sustainable engineering. *Proceedings of the 15th European Conference on Soil Mechanics and Geotechnical Engineering, Athens, Greece*, pp. 1815–1820.

Shonder J and Beck J (2000) *A New Method to Determine the Thermal Properties of Soil Formations from* In Situ *Field Tests*. Oak Ridge National Laboratory, Oak Ridge, TN, USA.

Signorelli S, Bassetti S, Pahud D and Kohl T (2007) Numerical evaluation of thermal response tests. *Geothermics* **36(2)**: 141–166.

Witte H and Gelder A (2006) Geothermal response tests using controlled multi-power level heating and cooling pulses (MPL-HCP): quantifying ground water effects on heat transport around a borehole heat exchanger. *Proceedings of the 10th International Conference on Thermal Storage – Ecostock 2006: Thermal Energy Storage Here and Now, Stockton, NJ, USA*, Paper 10A-3.

Zervantonakis I and Reuss M (2006) Quality requirements of a thermal response test. *Proceedings of the 10th International Conference on Thermal Storage – Ecostock 2006: Thermal Energy Storage Here and Now, Stockton, NJ, USA*, Paper 10A-2.

ICE Themes Geothermal Energy, Heat Exchange Systems and Energy Piles

Craig and Gavin
ISBN 978-0-7277-6398-3
https://doi.org/10.1680/gehesep.63983.249
ICE Publishing: All rights reserved

Chapter 13
Pile heat exchangers: thermal behaviour and interactions

Fleur Loveridge MSc, FGS, CGeol, CEng, MICE
Research Fellow, Faculty of Engineering and the Environment, University of Southampton, Southampton, UK

William Powrie MA, MSc, PhD, CEng, FREng, FICE
Professor of Geotechnical Engineering and Dean of the Faculty of Engineering and the Environment, University of Southampton, Southampton, UK

Thermal piles – that is structural foundation piles also used as heat exchangers as part of a ground energy system – are increasingly being adopted for their contribution to more sustainable energy strategies for new buildings. Despite over a quarter of a century having passed since the installation of the first thermal piles in northern Europe, uncertainties regarding their behaviour remain. This chapter identifies the key factors which influence the heat transfer and thermal–mechanical interactions of such piles. In terms of heat output, pile aspect ratio is identified as an important parameter controlling the overall thermal performance. Temperature changes in the concrete and surrounding ground during thermal pile operation will lead to additional concrete stresses and displacements within the pile–soil system. Consequently designers must ensure that temperatures remain within acceptable limits, while the pile geotechnical analysis should demonstrate that any adverse thermal stresses are within design safety factors and that any additional displacements do not affect the serviceability of the structure.

Notation

A	area (m^2)
F	dimensionless temperature response function
G	temperature response function for an infinite cylindrical heat source
H	pile or borehole length (m)
h	heat transfer coefficient (W/(m^2K))
L	thickness of material (m)
m	mass flow rate (kg/s)
Q	rate of heat transfer (W)
q	rate of heat transfer per unit length (W/m)
R	thermal resistance (K/W in conjunction with Q or mK/W in conjunction with q)
r	radial coordinate (m)
S_c	specific heat capacity (J/(kg K))
T	temperature (K)

t	time (s)
x	distance, length along pipe circuit (m)
α	thermal diffusivity (m^2/s)
γ	Euler's constant
Δ	change in value (usually temperature)
λ	thermal conductivity (W/(m K))

Subscripts

b	borehole wall pile diameter
c	concrete
cond	conductive
conv	convective
f	fluid
g	ground
i	internal diameter
in	inlet
o	outer diameter
p	pipe

1. Introduction

Rising energy prices and government policy drivers are leading to an increase in the use of ground energy systems to contribute to the heating and cooling requirements of new buildings (Preene and Powrie, 2009). Thermal piles are a specialist type of closed-loop ground energy system in which small-diameter pipes are cast into the piled foundations of a building to allow circulation of a heat transfer fluid. For rotary bored piles with a full depth cage, the pipes are usually fixed to the pile cage either during prefabrication, or on site if the cage comes in sections (Figure 1(a)). For continuous flight auger (CFA) piles, or piles where the cage is less

Figure 1 Typical thermal pile construction details: (a) pipework fixed to a rotary bored pile cage; (b) pipework installed in the centre of a pile

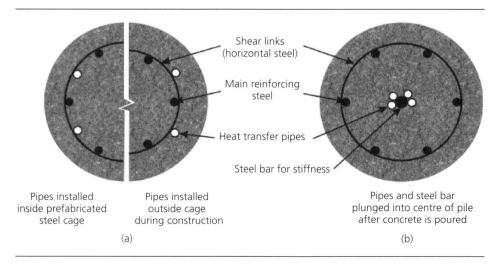

than full depth, it is common to plunge the pipe loops into the centre of the concrete, often attached to a steel bar to provide sufficient rigidity to facilitate installation of the loop within the pile (Figure 1(b)).

Below the upper few metres, the ground is essentially of constant temperature throughout the year (Figure 2). Hence in winter, circulation of cooler fluid within thermal piles allows heat extraction from the surrounding ground and in summer, circulation of warmer fluid allows injection of excess heat into the ground. A heat pump enables the temperature of the heated fluid to be increased to a more useful level by the input of a small amount of electrical energy. Similarly, in cooling mode, a heat pump allows a reduction in fluid temperature to below that

Figure 2 Typical near-surface seasonal temperature variation (calculated numerically assuming dry bulb air temperature profile for London, UK (CIBSE, 2005) and $a = 1.875 \times 10^{-6}$ m^2/s)

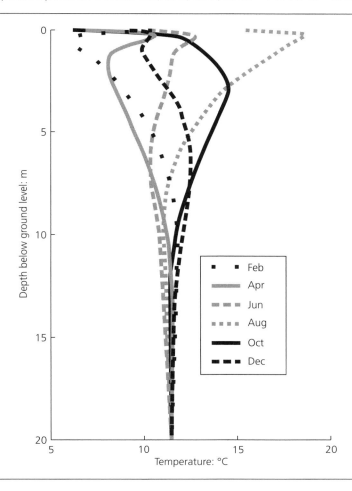

used in the air-conditioning system, increasing the effectiveness of heat transfer on reinjection into the ground. Operation philosophies may differ, as described below.

(a) For small or domestic properties there is usually only a heating demand, which is met in conjunction with a heat pump. Heat transfer is unidirectional and systems must be designed to prevent excessive temperatures developing in the ground.

(b) For larger structures, which have both heating and cooling needs, it is advantageous to balance these and make use of inter-seasonal ground energy storage. This allows greater thermal efficiency between the same ground temperature limits. In this case the heat pump must be reversible.

(c) In some circumstances it is possible to adopt so-called 'free cooling' whereby warm fluid is returned to the ground heat exchangers without passing through a heat pump. If temperatures allow, this mode of operation is highly efficient.

Ground energy systems have been in use for decades, with significant take-up (particularly in northern Europe and North America) commencing in the 1970s due to increasing oil prices. Many ground energy systems use drilled boreholes as heat exchangers and research into these systems was pioneered in the 1980s in Scandinavia (e.g. Eskilson, 1987) and North America (e.g. Bose *et al.*, 1985). The first thermal piles were installed in the 1980s (Brandl, 2006), but while design methods for borehole heat exchangers (BHEs) have matured, research into the behaviour of thermal piles has been more limited. In addition, coupling the structural and heat exchange functions of a pile means that the impact of thermal changes in the pile on its load-bearing capacity needs to be addressed. Standard design methods for either the thermal or the geotechnical aspects are not yet available and few sources of guidance are published (NHBC, 2010; SIA, 2005).

This chapter sets out the underlying thermodynamic concepts relevant to thermal pile performance. It then outlines the key thermal design aspects for BHEs. This is important as these approaches are often used as a basis for assessing the heat output of thermal piles. Lessons learnt from the study of BHEs are then used to help understand the key factors controlling pile thermal behaviour. The chapter then examines the interactions between thermal behaviour and mechanical performance of thermal piles, before introducing some more practical issues that must be considered. Finally knowledge gaps and areas where further research is required are identified.

2. Heat transfer concepts

Thermal piles, like other ground energy systems, function through the transfer of heat by way of conduction and convection. Conduction, due to the movement of atomic particles, is the primary heat transfer mechanism in solids. It is also referred to as diffusion. Convection is actually two heat transfer mechanisms: diffusion and the bulk movement of a fluid, termed advection. Convection is referred to as forced when the fluid flow is driven by external forces such as pumps. The flow may be internal (e.g. within a pipe) or external (e.g. around a fixed body).

Figure 3 illustrates a simplified heat transfer pathway for a thermal pile from the heat transfer fluid through to the ground. Forced convection occurs by way of the internal flow in the pipes; conduction occurs across the pipe walls and through the concrete to the ground. In the ground, conduction is usually the dominant process (Rees *et al.*, 2000), but if groundwater is flowing then advection can also be important (Chiasson *et al.*, 2000).

All convection is described by Newton's law of cooling, which relates the rate of heat transfer (Q, measured in W) per unit area (A in m^2) to the temperature difference (in K) across the

Figure 3 Thermal pile heat transfer concepts: (a) plan of thermal pile components; (b) temperature differences and component resistances

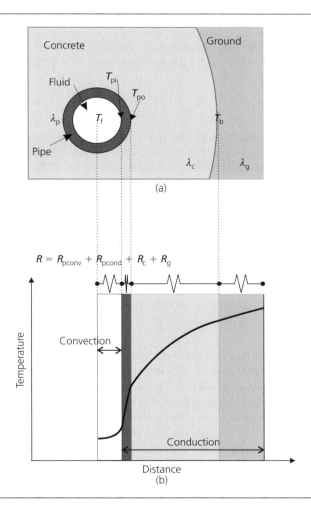

convection surface and a heat transfer coefficient, h (in W/m² K). Thus for heat transfer between the heat exchange fluid in the pipes and the pipe wall

$$\frac{Q}{A} = h(T_{pi} - T_f) \tag{1}$$

The value of h will depend on the properties of the heat transfer fluid, the nature of the flow conditions and the size of the pipe (e.g. Coulson and Richardson, 1990; Hellstrom, 1991). For water with turbulent flow the value of h is typically between 1000 and 3000 depending on the Reynolds number. There will be some degree of temperature dependency, but this is small and the impact on heat transfer is normally neglected. For laminar flow the heat transfer coefficient is an order of magnitude less than for turbulent conditions.

For steady heat conduction in one dimension, Fourier's law describes the relationship between the heat transfer rate and the temperature profile. Fourier's law is analogous to Darcy's law (Table 1) for groundwater flow, and for a temperature difference ΔT over a length L

$$\frac{Q}{A} = -\lambda \frac{\Delta T}{L} \tag{2}$$

The constant of proportionality λ is the thermal conductivity (in W/m K) and is a measure of how well a substance conducts heat. It is analogous to the Darcy hydraulic conductivity and to electrical conductance. Hence a resistance to heat transfer, R (in K/W), can also be defined

$$R = \frac{\Delta T}{|Q|} = \frac{L}{A\lambda} \tag{3}$$

Thermal resistance is a useful concept, as like electrical resistance, the component resistances of a system in series may be added to give an overall resistance (Figure 3(b)). The concept of resistance can also be used for convection, in which case

$$R = \frac{\Delta T}{|Q|} = \frac{1}{hA} \tag{4}$$

While heat transfer within a heat exchanger is often assumed to be at steady state and therefore considered in terms of its resistance, the response in the ground is usually transient. In transient conditions, heat transfer depends not only on the combination of thermal conductivity and geometry (i.e. resistance) but also on the speed at which temperatures change. This in turn is governed by the specific heat capacity of the ground, S_C (the amount of heat released per unit mass for a one degree change in temperature). Transient conduction is described by the diffusion equation, which is analogous to the groundwater diffusion equation (Table 1) and relates the change in temperature with time to the temperature gradient

$$\frac{dT}{dt} = \alpha \frac{d^2 T}{dx^2} \tag{5}$$

Table 1 Comparison between heat flow and groundwater flow

	Heat flow		Groundwater flow			
	Temperature, T		Excess (total) head, h		Excess pore water pressure, U_e	
Steady-state flow	$\frac{Q}{A} = -\lambda \frac{\Delta T}{L}$	Fourier's law	$\frac{Q}{A} = -k\frac{\Delta h}{L}$	Darcy's law	$v = -\frac{k}{\gamma_w}\frac{dU_e}{dz}$	Darcy's law (in terms of excess pressure)
	Q	Rate of heat energy transfer: W	Q	Rate of groundwater flow: m³/s	v	Darcy (superficial) flow velocity: m/s
	λ	Thermal conductivity: W/mK	k	Hydraulic conductivity: m/s	$\frac{k}{\gamma_w}$	
Transient flow	$\frac{dT}{dt} = \alpha \frac{d^2T}{dx^2}$	Thermal diffusivity equation	$\frac{dh}{dt} = \frac{T}{S}\frac{d^2h}{dx^2}$	Groundwater diffusivity equation:	$\frac{dU_e}{dt} = C_v\frac{d^2U_e}{dx^2}$	One-dimensional consolidation equation
	$\alpha = \frac{\lambda}{\rho S_c}$	Thermal diffusivity: m²/s	$\frac{T}{S} = \frac{kb}{S_s b} = \frac{k}{S_s} = D$	Hydraulic diffusivity: m²/s	$C_v = \frac{k}{m_v \gamma_w}$	Coefficient of consolidation: m²/s
	ρS_c	Volumetric specific heat capacity: J/m³K	$S_s = m_v \gamma_w = k/C_v$	Specific storage: m³/m³m	m_v	Coefficient of volume compressibility: m²/kN
Radial transient flow	$\Delta T = \frac{q}{4\pi\lambda}\int_{r^2/4\alpha t}^{\infty} \frac{e^{-u}}{u} du$	Infinite line source	$\Delta h = \frac{Q}{4\pi T}\int_{r^2 S/4Tt}^{\infty} \frac{e^{-a}}{a} da$	Theis equation		Note: q is rate of heat transfer per length of heat exchanger
	$\Delta T = \frac{q}{\lambda} \times \frac{1}{4\pi}\left[\ln\left(\frac{4\alpha t}{r^2}\right) - \gamma\right]$	Infinite line source (simplified)	$\Delta h = \frac{Q}{T} \times \frac{1}{4\pi}\left[\ln\left(\frac{4Tt}{r^2 S}\right) - \gamma\right]$	Cooper–Jacob approximation		

Note: T = transmissivity; S = storativity; b = aquifer thickness

where α is the thermal diffusivity in m²/s and is a measure of how quickly a material responds to a change in the temperature regime. α can also be expressed as $\alpha = \lambda/\rho S_c$ where ρ is the density. Extending the groundwater flow analogy, the thermal diffusivity can be considered to be equivalent to the hydraulic diffusivity in aquifer terminology or the coefficient of consolidation in consolidation theory (Table 1). Thermal conductivity and thermal diffusivity (or specific heat capacity) are the key ground parameters required for design ground energy systems, and are discussed by Busby *et al.* (2009), VDI (2009), Banks (2008) and Kavanaugh and Rafferty (1997).

In practice, the heat transfer occurring within a thermal pile is more complex than is shown in Figure 3. The heat transfer pathway is not simply linear and it is possible for the different pipes to exchange heat with each other as well as with the ground by way of the concrete. In addition, where there is a change of material type, and the interface between those materials is imperfect, additional resistance to heat flow is provided by 'contact resistance'. The major complexities are discussed further in the following sections.

3. Thermal performance of borehole heat exchangers

Borehole heat exchangers have a number of similarities to thermal piles, but also some significant differences. Consequently lessons can be learnt from the extensive research and experience on borehole design methods, as long as these are tempered with an understanding of the key differences in behaviour which will be discussed in Section 4. This section sets out some important concepts relevant to BHE behaviour. These concepts will then be extended for thermal piles in Section 4.

In the assessment of BHEs, the external response of the ground and the internal response of the heat exchanger are usually considered separately. Assuming steady state conditions in the borehole, the temperature change across the borehole and the temperature change in the ground can be summed as follows

$$T_f - T_0 = \Delta T_{\text{borehole}} + \Delta T_{\text{ground}} = qR_b + \frac{q}{\lambda}F \tag{6}$$

where T_f is the temperature of the circulating fluid and T_0 is the initial temperature in the ground. q is the rate of heat transfer per unit length and R_b is the borehole thermal resistance (in m K/W). F is a transient temperature response function, which describes the transient change in temperature in the ground in response to q. F is a function of time, distance and thermal diffusivity, but is of the same mathematical form for a given geometry. Thus the shape of the temperature response curve is independent of the actual temperatures and heat transfer rate. This type of behaviour is common to many heat transfer problems and lends itself to dimensionless analysis.

3.1 External response

The simplest method of calculating the ground thermal response is to consider the borehole to be an infinitely long line heat source (ILS) within an infinite medium. This is analogous to the radial flow of groundwater to a well (Table 1). As in the Theis equation, assuming a constant flux q, the temperature response function due to the heat source can be simplified to a log-linear relationship (Figure 4). The response function then becomes (Carslaw and Jaeger, 1959)

$$\Delta T = \frac{q}{\lambda} \times \frac{1}{4\pi} \left[\ln\left(\frac{4\alpha t}{r^2}\right) - \gamma \right] \qquad (7)$$

However, at small times the ILS approach will underestimate the temperature response. This is because it assumes that the heat source is at the centre of the borehole rather than the circumference. This shortcoming can be addressed by modelling the borehole as an infinite cylindrical heat source (ICS). The analytical solution for the temperature response function for the ICS is more complex (Ingersoll et al., 1954), but a simpler curve-fitted version can be used (Bernier, 2001). Figure 4 compares the ILS (Equation 7) and ICS (calculated numerically) temperature response functions. For typical BHE diameters (100–200 mm) the ILS will underestimate the temperature response by over 10% for approximately the first half day of heating. For the first 6 h these errors will be in excess of 25%.

For an infinite heat source the temperature change in the ground continues indefinitely. In reality, a steady state will be reached as heat extraction (or input) is matched by solar recharge (or losses) at the ground surface. Using a constant surface temperature boundary condition, Eskilson (1987) developed a finite line source (FLS) model using a combination of analytical and numerical approaches to derive a series of temperature response functions (termed g-functions) to take account of this effect. Figure 4 gives examples of FLS g-functions compared with the ICS and ILS temperature response functions. These show the ILS to overestimate temperature changes at large times; however, for typical boreholes which are longer than 100 m, it will take over 30 years for these errors to reach 10% (Philippe et al., 2009).

Eskilson (1987) also made an important step forward in BHE design by superimposing numerical solutions to account for interactions between different borehole installations. These

Figure 4 Dimensionless temperature response functions for heat exchanger design

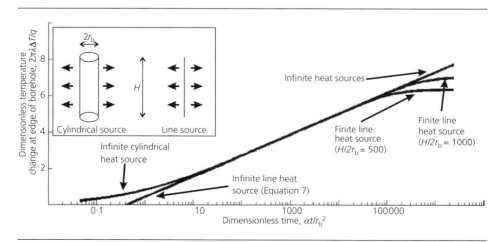

multiple borehole g-functions, which now underpin a number of commercial software packages, allow designers to take account of the reduction in available thermal capacity when multiple heat exchangers are installed close enough together so that thermal interactions will occur between the individual heat exchangers.

All the preceding discussions assume a constant and continuous heat transfer rate q. In reality q will vary with time according to the actual energy use in the building. Consequently the response will step from one temperature response curve to another depending on the actual value of q at any one time.

3.2   Internal response

The heat exchanger is usually considered to be at a steady state (Bernier, 2001; Remund, 1999; Shonder and Beck, 1999; Xu and Spitler, 2006) and the estimated resistance is used to calculate the temperature change between the fluid and the borehole edge. The standard approach is to sum the resistances of the different components (Figure 3(b)), but this is a simplification as it can neglect contact resistances and pipe-to-pipe interactions. The former are usually assumed to be negligible, although there is a lack of research to confirm this. This simple approach also neglects the heat capacity of the borehole, although this is of minor significance for BHEs which would reach a steady state within a few hours.

Standard approaches for determining the resistance associated with the fluid (R_{pconv}) and the pipe (R_{pcond}) are well known (e.g. as described by Bernier (2001) and Marcotte and Pasquier (2008)) and are equally applicable to thermal piles. The effective resistance of the grout within a BHE is more complex and depends on the geometric positioning of the pipes with respect to the hole. Consequently common empirical approaches (e.g. Remund, 1999) cannot be applied to thermal piles and new methods are required.

3.3   Fluid temperature profiles

Simple design methods assume that the rate of heat transfer between the fluid and the borehole is constant around the length of the pipe circuit and hence with depth down the heat exchanger. For this to be the case, the fluid must lose heat (and therefore change temperature) at a constant rate around the pipe circuit (Figure 5(a)). Then, for a single U-tube installed in a borehole, the mean of the up and down fluid temperatures is constant with depth. However, numerical modelling (Lee and Lam, 2008; Marcotte and Pasquier, 2008) and field measurements (Acuna et al., 2009) show that a constant-temperature boundary condition (Figure 5(b)) is more representative of reality, and this results in an exponential variation in the fluid temperature with distance x around the pipe circuit (Incropera et al., 2007)

$$\frac{T_f - T_b}{T_{fin} - T_b} = \exp\left(\frac{-x}{2R_b m S_c}\right) \quad (8)$$

where T_{fin} is the inlet fluid temperature and m is the fluid mass flow rate. As a consequence the average fluid temperature for a single U-tube, and by extension the heat transfer rate, is not constant with depth (Figure 6).

Figure 5 Boundary conditions and fluid temperatures profile for internal pipe flow: (a) constant heat transfer rate boundary condition; (b) constant temperature boundary condition

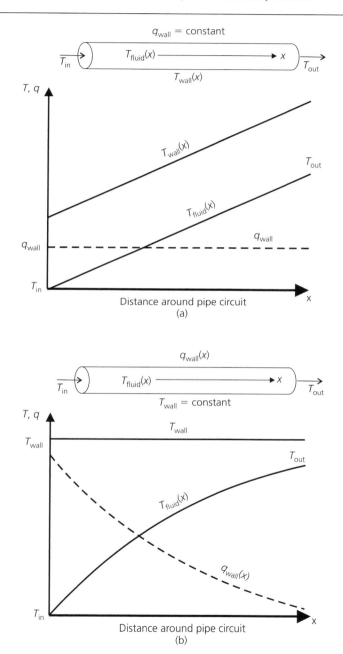

Figure 6 Fluid temperature profiles for a single U-tube in a vertical ground heat exchanger (calculated based on Equation 8 and the approach of Diao *et al.* (2004a) for interacting pipes)

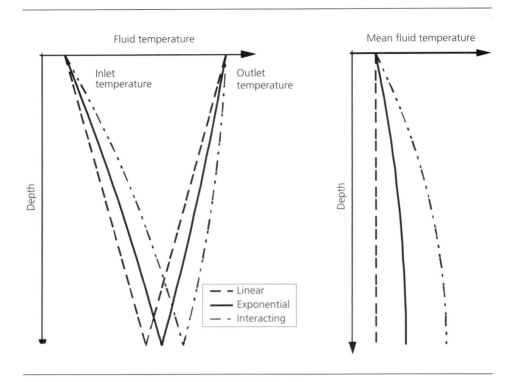

Depending on the spacing of the two shanks of a U-tube, the two pipes may also exchange heat with each other (e.g. Diao *et al.*, 2004a), thus reducing the efficiency of the system and increasing the variation of mean fluid temperature with depth (Figure 6). This is reflected in an increased borehole thermal resistance. Analytical solutions do exist for the calculation of the exact fluid temperature profile for a single U-tube (Diao *et al.*, 2004a; Hellstrom, 1991); however, to implement these solutions allowing for interference between pipes is complex and requires knowledge of the precise internal geometry of the pipes within the borehole. Alternatively, an empirical solution for the fluid profile is available (Marcotte and Pasquier, 2008). However, this will not necessarily be appropriate for cases where significant interference occurs between the pipes, such as when they are touching (Lamarche *et al.*, 2010).

4. Thermal performance of pile heat exchangers

4.1 External thermal response

There are very few data sets available for verification of the thermal design methods for piles used as heat exchangers. Published case studies often focus on the heat pump and overall system performance and do not consider the ground thermal response. This is unfortunate as the analytical approaches used for BHE (typically less than 200 mm diameter) design all have shortcomings when applied to larger-diameter thermal piles (typically at least 300 mm in diameter). Methods that assume a line source may be valid for small-diameter holes but for piled foundations, with the heat exchange pipes fixed near to the circumference steel, there will be errors for analysis periods of less than a few days or even months. Figure 4 shows these differences non-dimensionally, with divergence between the line and cylindrical source for non-dimensional time values of less than around 10. For a 600 mm diameter pile this translates to an underestimation of the temperature change by more than 10% for times up to 5 days, and by at least 25% for up to 2 days. For a 1·2 m diameter pile these times increase to 8 days and 21 days respectively. This underestimation of temperature changes is not conservative in terms of both the thermal capacity of the system and assessing the potential for adverse thermomechanical interactions (see Section 5).

For piles with heat exchanger pipes installed in the centre of the concrete then although the heat source may more closely approximate a line, there will be two regions (concrete and ground) with different thermal properties that need to be accounted for within the thermal design.

For short piles, a steady state may develop within a few years, rather than decades as with longer boreholes. For example, for a 50 m long pile it may take 15–20 years for the error in the ILS solution to reach 10%. The corresponding figure for a 20 m long pile is only 2 or 3 years. For domestic housing piles, typically around 10 m deep, this time can be less than a year. This leads to a significant overestimation of the temperature response if an infinite source is assumed. This is conservative in terms of assessing thermomechanical interactions and thermal capacity; it does reduce the opportunities for maximising the thermal capacity of the system. Therefore it is important to use a model which considers the length of the piles when determining thermal performance.

The importance of the geometry of thermal piles is best indicated by the aspect (length to diameter) ratio (AR). Figure 7 shows aspect ratios for constructed thermal piles, which are generally in the range 10–50, in contrast to values of 500–1000 typical for BHEs. Figure 8(a) shows how the aspect ratio of a thermal pile governs its temperature response function. Figure 8(b) highlights the differences between the ILS and a finite cylindrical heat source for four different aspect ratios. This shows the small time periods for which the ILS approach gives an acceptable error range when applied to thermal piles as applied to BHEs.

Some of the differences between the models discussed above may be less important for a truly thermally balanced system, where heat extraction continues for 6 months only and is then balanced by reinjection of surplus heat from air-conditioning systems. However, it is rare for systems to be perfectly balanced and hence, depending on the actual weather conditions experienced and building usage, it is likely that there will be a net accumulation of heat (or cold) in the ground over time.

As a result of the potential for errors in predicting the ground thermal response at small and large times, considerable caution should be exercised when using any design software based on techniques developed for the assessment of BHEs. This has been highlighted by Wood *et al.* (2010a) who compared actual fluid inlet and outlet temperatures for a thermal pile test plot with values determined from commercial software using an FLS approach over a 1-year period. While the overall trend calculated was reasonable, errors of about 2°C were apparent in

Figure 7 Aspect ratios (denoted AR) of constructed thermal piles

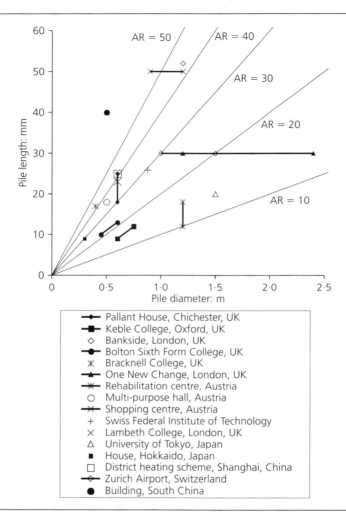

the lower ranges of temperatures, with the design software underpredicting the fluid temperature. While this might not appear significant, systems tend to operate with small temperature differences and over small temperature ranges. For example, 2°C is 40% of the total temperature variation range presented by Wood *et al.* (2010a). In this context, and given the restrictions which need to be placed on systems to avoid ground freezing, an additional 2°C margin will reduce the efficiency of the system significantly.

A design approach which has been validated for use with thermal piles is the so-called 'duct storage model' (DST) (Claesson and Hellstrom, 1981; Hellstrom, 1989). This assumes that a large number of vertical heat exchangers, or ducts, are installed close together to act as an underground thermal store. The model separates analysis of the local heat transfer around each

Figure 8 Effect of aspect ratio (AR) on ground temperature response function for thermal piles: (a) finite cylindrical source (ICS); (b) analytical methods as a percentage of ICS

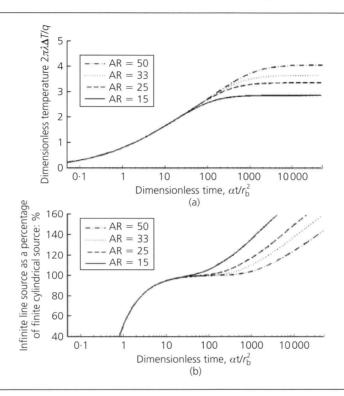

duct from global heat transfer into and out of the thermal store. For local heat transfer an ILS is applied for short-duration heat pulses. Globally and at larger times (defined as when the individual ducts are thermally interacting) a steady state is assumed within the store and subsequent heat input leads to linear changes in temperatures throughout the store. The local and global solutions are then combined to assess the overall performance of the heat store. The DST was initially validated against field data for small-diameter (<50 mm) borehole thermal stores in Sweden (Hellstrom, 1983). Subsequently, the DST approach has been implemented specifically for use with thermal piles in the software 'PILESIM' (Pahud, 2007). PILESIM has been validated against thermal pile field data from Switzerland (Pahud and Hubbach, 2007), focusing on the overall heat exchange capacity of the system. Independent analysis using time-stepping finite-element models (R. Markiewicz, personal communication, 2010) implies that for regular arrays of piles the results provided by PILESIM are appropriate. However, the DST assumes a large number of identical piles installed in a regular array within a circular plan area and it is not clear what errors result from smaller or less regular pile group arrangements that are more representative of typical foundation layouts.

The methods discussed above were all originally developed from the design of BHEs and assume a constant ground surface temperature equal to the initial average temperature in the ground. This neglects the seasonal variation of the ground surface temperature, which will affect the ground temperatures to about 10 m depth (Figure 2). For short, uncovered heat exchangers this can have a major influence on temperatures (Wood et al., 2009). For thermal piles covered by buildings, there will be no incoming solar radiation to recharge the ground temperature, but studies by Thomas and Rees (1999) show that buildings provide a small net heat flux to the ground and this may be a more appropriate long-term boundary condition. No current published methods of analysis take this into account and the topic requires further research to determine its importance.

4.2 Thermal resistance for pile heat exchangers

Theoretical values of R_b for thermal piles are given by the Swiss Society for Architects and Engineers (Table 2). These are typically smaller, by up to a factor of 2, than published values derived from either in situ thermal testing or back-analysis of system operations (Table 3). This is likely to be due to the high values of thermal conductivity for concrete assumed in the Swiss analysis ($\lambda_c = 1.8$ W/m K). In reality, for a heat exchange pile, λ_c is likely to be less than 1.5 W/m K, owing to the high cement content required for strength and the presence of admixtures which can reduce thermal conductivity (Kim et al., 2003; Neville, 1995; Tatro, 2006). However, the thermal conductivity of concrete can be improved, for example by specification of siliceous aggregates.

The total thermal resistance of a pile would be expected to be larger than for a borehole (typically in the range 0·05–0·2 m K/W, Sanner et al. (2005)) based on the geometric arrangement of the pipes. As pile reinforcement must be protected from corrosion due to groundwater there tends to be a greater concrete cover to the pipes than for BHEs. This can lead to a larger resistance, especially if the pipes are actually in the centre of the pile. On the other hand, a greater number of pipes within the cross-section would lower the resistance.

R_b is usually calculated by the separate assessment of R_c, R_{pconv} and R_{pcond} (see Figure 3). Assuming turbulent flow, R_{pconv} and R_{pcond} tend to be small, in total around 0·01 m K/W for

Table 2 Pile thermal resistance values (after SIA (2005))

Pile type	Pile diameters: m	Total thermal resistance: mK/W
Driven tube with double U-tube	0·3–0·5	0·15
Precast or cast in situ, with double U-tube attached to reinforcement	0·3–1·5	0·1–0·11
Precast or cast in situ, with triple U-tube attached to reinforcement	0·3–1·5	0·07–0·08
Precast or cast in situ, with quadruple U-tube attached to reinforcement	0·3–1·5	0·06

Table 3 Pile thermal resistance values from in situ measurement or back-analysis

Pile diameter/ type	Pipe arrangement	Total thermal resistance	Source	Comments
0·3 m CFA	Single U-tube	0·22 mK/W	Wood et al. (2010a)	Derived from combination of analytical methods and back-analysis. Laminar flow conditions
0·6 m cast in situ	Single U-tube	0·25 mK/W	Gao et al. (2008)	Bespoke thermal testing. Range of values represents different flow rates and connections between different U tubes
	Double U-tube in series	0·15–0·2 mK/W		
	Triple U-tube in series	0·125–0·15 mK/W		
0·27 m square driven	Single U-tube	0·17	Lennon et al. (2009)	Short duration (<30 h) thermal response tests
0·244 m drive steel tube	Single U-tube	0·11		

four pipes in parallel, and easy to calculate (e.g. Bernier, 2001; Marcotte and Pasquier, 2008). R_{pconv} depends on the flow conditions, captured in the heat transfer coefficient h (Equation 1). The largest component of the thermal resistance of a pile is in the concrete or grout. This is more difficult to determine than the pipe resistance and depends on the arrangement of pipes and the concrete thermal conductivity. Currently, the most practical method for determining R_c is by numerical modelling.

Minimising the total thermal resistance of the pile is important for improving thermal performance and reducing the temperature gradient across the pile. This has been the subject of targeted research for borehole design and appropriate measures include ensuring that fluid flow is turbulent, using high thermal conductivity materials (Sanner et al., 2005) and installing more pipes within the hole (Gao et al., 2008; Zeng et al., 2003). For thermal piles, maximising the number of pipes and minimising the cover to those pipes are likely to be important factors. However, as the pile diameter increases, and especially for CFA-type piles with central pipes, the contribution of the pile to heat storage and not just transfer to the ground also increases. In such cases, a steady-state resistance may no longer be valid and a two-zone transient analysis of the concrete and ground response may be required. This area has seen little attention and requires further research.

4.3 Fluid temperature profiles

Heat transfer from the fluid to the edge of the pile depends on two factors: the resistance as discussed in Section 4.2 and the temperature difference. The latter depends on the flow conditions as described in Equation 8. Profiles of fluid temperature against distance along the pipe circuit, based on Equation 8 with a pile surface temperature $T_b = 6°C$ and a fluid inlet temperature of 1°C, are given in Figure 9. The effectiveness of heat transfer will reduce substantially as the temperature difference between the fluid and the outside boundary, $T_b - T_f$, reduces around the pipe circuit. For this reason it is best to keep the circuit length to a maximum of 300–400 m depending on the flow conditions (Figure 9). Maintaining a high flow rate (and high Reynolds number) will also maximise heat transfer regardless of circuit length. However, it should be noted that, practically, the pile circumference is unlikely to remain at a uniform temperature (as assumed in Equation 8), especially for low flow velocities where there is a large temperature difference between the inlet and the outlet.

Figure 9 Exponential fluid temperature variation in pipe circuits based on Equation 8 (assumes inlet temperature of 1°C and pile surface temperature of 6°C): (a) sensitivity to flow rate; (b) sensitivity to pipe size; (c) sensitivity to pile thermal resistance

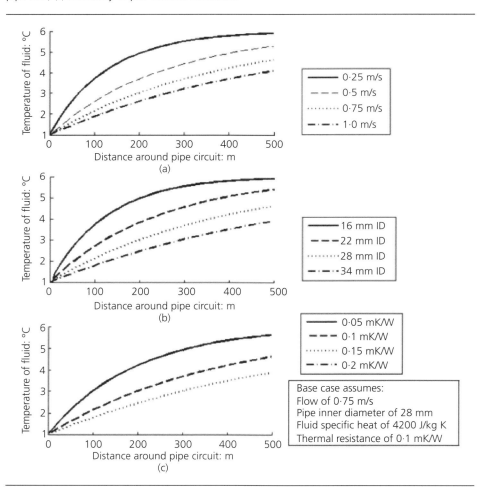

As thermal piles are much shorter than boreholes, multiple piles are sometimes connected together into a single pipe circuit. Specific arrangements will depend on the number of pipes in a given cross-section. For example, while an installation of larger-diameter 50 m deep piles may contain six pipes as one circuit, an installation of shorter 25 m deep piles of smaller diameter with only four pipes may have three piles connected in series. In the latter case, the mean temperature of the fluid in each pile may vary significantly (Figure 10, right-hand side). Hence the temperature difference relative to the ground and also the heat transfer rate may be different for each pile. This has been observed by Wood *et al.* (2010b) where in a circuit comprising four 10 m deep piles connected in series the temperature difference between each successive pile was approximately 0·5°C. For longer circuits and deeper piles these differences may be more substantial; unsurprisingly, Wood *et al.* (2010b) found that the magnitude of the temperature difference decreases at higher fluid flow rate. What is not clear is how important these effects will be for overall performance of systems and hence further research is required in this area.

Thermal interactions between individual pipes will also affect the fluid temperature profile and hence the heat transfer achieved. As the pipes in thermal piles tend to be fixed between the main steel of the pile cage, their separation is likely to be about 250–300 mm (P. Smith, personal communication, 2010) compared with less than 100 mm for typical boreholes. Consequently, less interaction between the pipes would be expected in piles than in boreholes. This is beneficial as it both maximises the heat transfer and reduces the thermal resistance. No field measurements of the fluid temperatures within the pipe circuits of thermal piles are known to have been carried out; only the inlet and outlet temperatures have been verified in situ.

Figure 10 Example mean fluid temperatures for thermal piles connected in series (calculated using Equation 8 with inlet temperature of 1°C and pile surface temperature of 6°C)

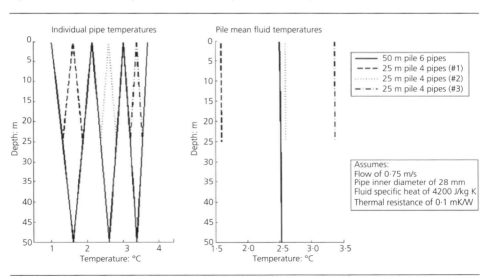

Simulation of the fluid (water) temperature profile for a 16 m long, 1·2 m diameter pile with eight pipes installed in series has been carried out by Markiewicz (2004). The profiles are replotted here (Figure 11) in terms of non-dimensional temperature in keeping with Equation 8. An average borehole wall temperature had to be estimated from the published model results (Markiewicz, 2004). Curve-fitting for the profiles was then carried out as summarised in Table 4. This assessment shows that for high flow rates (> 1 m/s) the fluid profile is sufficiently close to a straight line to allow this simplified approach to be adopted (Figure 11(a)). An exponential curve of a form matching Equation 8 is appropriate for intermediate to high velocities, between about 0·25 m/s and 1 m/s (Figure 11(b)). However, at low flow velocities (<0·25 m/s), significant interference is observed with fluid near the end of the circuit relinquishing heat energy to that at the start of the circuit (Figures 11(c) and 11(d)). In such cases an exponential type curve is not appropriate. The interference also has a detrimental effect on the thermal resistance (Table 4), significantly reducing efficiencies of the pile as a heat exchanger.

Figure 11 Normalised fluid temperature profiles from thermal pile modelling by Markiewicz (2004) (inlet temperature is 2°C, borehole wall temperature taken as 1°C higher than outlet temperature based on results of original model. For curve-fit data refer to Table 4): (a) flow of 1 m/s; (b) flow of 0·5 m/s; (c) flow of 0·25 m/s; (d) flow of 0·1 m/s

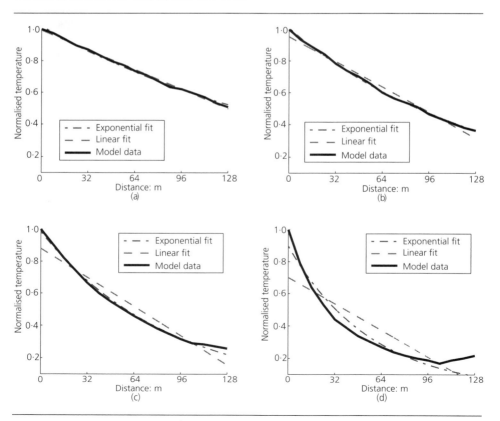

Table 4 Curve-fitting parameters for fluid profiles from Markiewicz (2004), to be read in conjunction with Figure 11

Flow: m/s	Curve type	Coefficients		Coefficient of determination	Root mean square error	R_b: mK/W[a]	Comments
		a	b				
1	$ax+b$	−0·0039	0·9938	0·9968	0·0096		Linear and exponential curves provide good and comparable fit
	$a \exp(bx)$	1·019	−0·005227	0·9965	0·0092	0·051	
0·5	$ax+b$	−0·0050	0·9551	0·9848	0·0258		Exponential curve provides better fit and temperature difference between inlet and outlet increases
	$a \exp(bx)$	1·008	−0·007956	0·9988	0·0073	0·066	
0·25	$ax+b$	−0·0056	0·8801	0·9938	0·0627		Increased errors compared to higher velocities
	$a \exp(bx)$	0·945	−0·01165	0·9957	0·0160	0·091	Some loss of fit at end of circuit due to minor interference
0·1	$ax+b$	−0·0051	0·7012	0·7490	0·1240		Significantly greater errors for linear fit
	$a \exp(bx)$	0·898	−0·01783	0·9383	0·0615	0·148	Increased errors due to interference causing poor fit

[a] Assuming $b = -1/(2R_b m S_c)$ and fluid and pipe properties as per Markiewicz (2004)

This illustrates the importance of maximising fluid flow rates while retaining pipe separation and limiting pipe circuit lengths in order to reduce interactions and hence facilitate maximum heat transfer.

4.4 Groundwater flow

Where groundwater is flowing, the temperature change in the ground adjacent to the heat exchanger will be reduced by additional advective heat transfer. While this is potentially a huge benefit in terms of the capacity of an individual ground energy system, the resulting thermal plume will travel a greater distance downstream giving the potential for interactions over a much wider area. This is evident from open-loop ground energy systems within aquifers beneath conurbations, where widespread adoption and extended use has led to significant changes in the aquifer temperatures (Ferguson and Woodbury, 2006; Gustafsson, 1993).

Design approaches for systems affected by groundwater are not well defined. Analytical solutions for the ground temperature response functions (Claesson and Hellstrom, 2000; Diao et al., 2004b; Sutton et al., 2003) are based on the principle of an infinite line heat source moving through the medium being heated and thus disregard the development of a diffusive steady state. They also do not consider characteristics of real groundwater flow, including the effects of inhomogeneity and possible fracture flow. Consequently, numerical methods are often used to assess heat transfer in the presence of moving groundwater (e.g. Gehlin and Hellstrom, 2003; SIA, 2005). While it is important to question whether a sustained and consistent groundwater flow in an urban area can be relied upon over the design life of a system, any potential for adverse effects resulting from groundwater flow must also be assessed. In particular, the capacity for inter-seasonal energy storage will be reduced by flowing groundwater, which should be accounted for in any assessment of thermal potential.

5. Thermomechanical interactions and pile behaviour

The potential for adverse thermal interactions between heat exchanger piles and the ground has led to concerns that inappropriate operation may lead to ground freezing, excessive ground deformations or additional pile stresses that cannot be safely carried by the structure. Despite these fears, no mechanical or serviceability issues with thermal piles have been reported to date, possibly as a result of conservative design and geotechnical factors of safety providing capacity within which additional concrete stresses and displacements can be accommodated. However, such factors of safety are used to account for other uncertainties (e.g. ground heterogeneity) and therefore this is not a satisfactory design approach.

Consequently, it is important that the potential for additional thermal stresses is assessed and temperature limits placed on ground energy systems to prevent structures from experiencing temperature variations which would adversely affect the geotechnical performance. The following sections discuss the theoretical framework for thermal–mechanical interactions, what can be learnt from recent case studies, and uncertainties that still remain, especially with respect to long-term cyclic loading. As temperature changes resulting from ground energy systems only occur after the building is complete and operational, the discussion will exclude early-age thermal effects in concrete. This is in keeping with recent research which argues that for piles in saturated ground, creep and shrinkage effects are insignificant compared with other loads (Bicocchi, 2011).

5.1 Behavioural framework

In principle, when a thermal pile is heated it will tend to expand and when it is cooled it will tend to contract. Free expansion or contraction will not occur because the pile is restrained, both by the surrounding soil and by any overlying structure. Consequently a proportion of the theoretical free strain will be expressed instead as a change in longitudinal stress within the pile and transferred to the ground by skin friction or end bearing. A pile that expands relative to the surrounding soil will tend to experience an increase in the axial stress (termed hereafter the 'pile axial load'), and a pile that contracts a reduction: however, the exact effect will vary, and could even be locally reversed along the length of the pile depending on the degree and nature (resilience) of the end restraints. A similar observation applies to the mobilised skin friction. Potential concerns include overstressing the cross-section, an excessive increase in base bearing pressure, or the development of negative (downward) skin friction resulting potentially in the loss of external load-carrying capacity. A useful conceptual framework for assessing this complex behaviour has been presented by Bourne-Webb *et al.* (2013) and illustrates in particular the importance of the end restraints in controlling the thermomechanical response. This framework can be used to assess potential thermal effects in terms of additional forces that should be accommodated in design. However, case studies are important for validation of the approach.

5.2 Lessons from case studies

Early observations of strain and temperature within a thermal pile were reported by Brandl (1998). While the study did not give sufficient detail to enable a full assessment of the thermomechanical behaviour, it does illustrate the consequences of excessive heat extraction. As fluid temperatures reached −5°C, ice lenses formed within the ground, causing 150 mm of heave at the surface. Relative movement between the pile and the ground would also have been expected to have altered the shaft skin friction.

This case study illustrates the importance of ensuring that the pile and ground do not freeze. The simplest way of achieving this is to specify that the fluid temperature must not fall below 0°C. Given that the fluid temperature varies around the pipe circuit and that measurement errors are possible, a safety margin is typically also allowed for. Specification of minimum fluid temperature of 2°C is therefore recommended by some bodies (NHBC, 2010; SIA, 2005). However, this is a conservative approach and will result in a failure to utilise the ground to its full thermal potential. Therefore a more sophisticated approach may be adopted whereby assessment of the pile thermal resistance and fluid temperature profiles can be made to demonstrate that lower fluid outlet temperatures will not lead to development of freezing conditions at the soil–pile interface. As well as this assessment, it is important that a suitable building control system is in place to prevent lower temperature limits from being exceeded in the case of a higher than expected heating demand.

Two systematic attempts to assess the thermomechanical response of thermal piles have recently been made. A working pile for a new building at the Swiss Federal Institute of Technology was used for thermomechanical testing, reported by Laloui *et al.* (1999, 2006) and summarised in Table 5. Before construction of the building a simple thermal test was carried out and the resulting temperature changes and strain data used to calculate the mobilised skin friction. For a 22°C temperature increase the pile expanded by 4 mm at the head, with a small

Table 5 In situ measurements of pile thermomechanical reponses

Reference	Bourne-Webb et al. (2009)		Laloui et al. (1999)	Laloui et al. (2006)
Test	Test pile – cooled	Test pile – heated	Operational pile – one storey constructed	Operational pile – seven storey constructed
Pile length	23 m		25·8 m	
Pile diameter	0·6 m		0·88 m	
Ground conditions	Made ground and river terrace deposits to 5 m overlying London Clay		Alluvium to 12 m, glacial till to 25 m, toed into sandstone	
Restraint	Non-significant		Large seven-storey building and piled toed into rock	
Temperature change	−15 to −20°C	+5 to +10°C	+22°C	+13°C
Head load	1200 kN	1200 kN	300 kN	1300 kN
Mechanical load	+1200 kN near head, zero at toe	+1200 kN near head, zero at toe	+300 kN at head	+1300 kN at head, zero at toe
Thermal load	Zero at head, −300 kN near base	+800 kN at 4 m, +200 kN at base	+1000 kN at head	+1200 kN at head, +2000 kN at toe
Additional pile head displacement	−2 mm	+2 mm	<+2 mm	<+1 m

amount of compression at the toe reflecting the high end restraint due to the embedment of the pile in hard sandstone. Near uniform heating caused between 30 kPa and 80 kPa of skin friction to develop in different soil layers. Further heating of the pile was carried out under different pile head loads (Table 5). The pile was constrained, both at the toe (by bedrock) and at its head (by the structure). Pile axial loads of up to 2 MN were induced over the full length, the largest of which were over the lower portion of the pile. This additional thermal load was greater than the mechanical pile axial load of 1·3 MN at the pile head (Table 5). However, given the pile restraint, the corresponding additional pile head displacements were small; less than 2mm of heave was recorded.

A thermomechanical load test was carried out on a sacrificial test pile at Lambeth College in London (Bourne-Webb et al., 2009). The pile was subjected to separate heating and cooling cycles while carrying an external mechanical load of 1200 kN at the head (Table 5), equivalent to the anticipated working load. The heating caused an increase in pile axial load of up to 800 kN in the upper part of the pile, while the cooling cycle led to a reduction in load of about 500 kN, mainly near the base of the pile. This smaller (up to about 70% of the original external load) and less even distribution of additional pile axial load compared with the Swiss test is due to the lower degree of restraint of the Lambeth College pile. Again the accompanying changes in pile head displacement were small at less than 2 mm (Table 5). The consequent changes in shaft friction were estimated to be up to about 50 kPa, with a maximum total of 75 kPa developing during the thermal tests, compared with a value in excess of 90 kPa developed at the ultimate limit state in subsequent destructive load testing.

Both tests indicated the thermomechanical response of the thermal pile to be largely reversible, and the pile–soil system to be acting thermo-elastically, at least over small numbers of cycles. This elastic behaviour was confirmed when a new approach for calculating the effects of thermal loading on piles using an elastic load transfer method was tested on the above case studies (Knellwolf et al., 2011). In addition to providing a good match to the experiment data, the method of Knellwolf et al. went on to assess a number of possible working scenarios. It was shown that for a pile where the head load induces skin friction close to the ultimate capacity, additional heating may cause the ultimate skin friction capacity to be reached. Conversely, cooling of the pile can cause a reversal of shear stresses and the development of negative skin friction.

The observed reversible nature of the thermal–mechanical behaviour is encouraging as the range of temperatures used in the testing is realistic compared to likely operational ranges, thereby suggesting that permanent deformation is unlikely to result from operation of ground energy systems. However, short-term testing cannot identify smaller cyclic effects that could become significant over longer timescales and larger numbers of cycles. Thus longer-term, in situ trials and/or laboratory testing will be required to confirm the soil–structure behaviour over the lifetime of a system, likely to be in excess of 25 years.

5.3   Soil thermal behaviour and cyclic loading effects

The above discussion has focused solely on the potential for volume change and induced stresses within the concrete pile. However, the temperature changes will also result in volume change in the soil and potentially in changes to the soil properties. Volume changes may occur due to both thermal expansion and temperature-induced mechanical changes to the soil structure. For normally or lightly over-consolidated clay soils, heating usually results in contraction and consolidation and large settlements are possible (Boudali et al., 1994). For highly over-consolidated clays, however, elastic expansion is typical and the overall effect on the soil is small (Cekerevac and Laloui, 2004). However, most investigations of thermally driven volume change in soils have focused on heating clays to high temperatures to simulate conditions relevant to nuclear waste disposal. The effect of smaller magnitude cycles of heating and cooling over a number of years has yet to be investigated.

Most studies of cyclic loading of piles relate to offshore structures. Poulos (1988) provides a useful discussion in this respect and highlights that two-way cyclic loading, as would probably be the case for thermal piles, is more damaging. Beyond a threshold cyclic load, typically close to the static load required to cause pile–soil slip, degradation of the shaft skin friction can occur (Poulos, 1989). Reduction in skin friction by up to 20% has been recorded (Jardine, 1991) but any individual case will depend on the soil properties, the nature of the pile, the static and cyclic loads and the loading rate. Full assessment of behaviour can be made if appropriate laboratory tests have been carried out, but caution must be exercised as thermal piles will be subject to a more uniform (with length) loading than offshore piles where the axial load is concentrated at the head.

Laloui and Cekerevac (2008) suggest that the number of mechanical load cycles required to fail a test specimen increases with temperature. Soil strength tests at elevated temperatures show varying results, but any deterioration of peak or critical state friction angle over the range of temperatures relevant to ground energy systems is likely to be small (Laloui, 2001); hence a significant reduction in the ultimate shaft capacity due to a general change in temperature is unlikely. Again, however, the effect of longer-term cyclic changes should be investigated further.

6. Practical constraints

The foregoing discussions relate to largely theoretical aspects of thermal pile behaviour. However, there are many design and construction interfaces which will affect any thermal pile scheme. While for traditional ground energy systems the layout of the heat exchangers is optimised to maximise thermal output, for thermal piles, the structural and geotechnical design will take priority. This means that the aim is to determine the thermal capacity from a given pile layout and also to check the thermomechanical effects on the geotechnical design. It is unlikely to be economic to install additional piles or increase their lengths purely to provide additional energy capacity. The ground conditions and any natural variability in their thermal properties are also a given parameter that must be accounted for in the thermal design. Currently it is usual for average thermal properties to be used in design regardless of the soil complexity. This is despite the fact, known from studies of BHEs, that stratified soil conditions can cause differences in behaviour between heating and cooling (Signorelli et al., 2007).

To some extent the layout of fluid pipes can be optimised once the pile layout has been determined. The number of pipes installed and their positions will be determined by the thermal design, as long as this is compatible with the construction process. For example with a full-depth cage the number of pipes and their locations and pipe circuit lengths can relatively easily be adjusted to maximise thermal output. However, if the pile is to be constructed by CFA techniques or has a cage over only part of its length, then it is likely that this will force installation of the pipes within the centre of the pile. It is also essential to ensure that any pipes fixed to cages during the construction process are fixed using a safe system of working and this has encouraged the placement of pipes on the outside of cages (Figure 1(a)).

If possible it will be advantageous to use concrete with a high thermal conductivity. This would mean maximising the aggregate content and using higher-conductivity aggregates such as sandstone. However, practically, the mix design is driven by the structural strength and slump requirements and it will always be more economic and more sustainable to use local sources of aggregate than to import special materials from greater distances.

Whereas construction of piles for building developments usually only interfaces with the groundworks contractor, thermal piles and the pipes which come from them have far more design and construction interfaces. It is important to protect pipes from damage at all stages of construction, from breaking out the piles, to extending the pipes beneath the building slab and ultimately to the plant room. It is essential to have redundancy in the system in case of damage during construction, but this should be coordinated by all the parties which interface with the ground energy system in order to prevent overconservatism. Pressure testing of the pipes to confirm integrity at key construction stages is essential for managing this process.

7. Conclusion

The ground is well suited to act as a thermal store and using structural piled foundations as heat exchangers is an increasingly common approach to improving the energy efficiency and reducing the carbon emissions from new buildings. The design of thermal piles has two distinct components: assessment of available heating and cooling capacity and additional checks as part of the geotechnical design to ensure that the cycles of temperature change do not have an adverse effect on the geotechnical design.

Assessment of heating and cooling capacities has often followed similar approaches to those used for the design of BHE arrays. However, care must be taken as the smaller aspect ratio of piles compared with boreholes means that thermal piles will reach a steady state more quickly. Consequently analytical methods which assume an infinite heat source will overestimate the temperature change in the ground. While conservative, in terms of assessing both the available heat output and the potential for adverse thermomechanical interactions, this approach will result in the thermal potential of the ground not being maximised. Consequently, it could potentially lead to systems being assessed as uneconomic. One of the few validated design approaches for estimating the thermal response of the ground to thermal piles is based on the duct storage model. However, this method assumes that all the piles are installed on a regular grid and it is not clear what uncertainties are introduced from more realistic pile layouts.

Thermal piles will also be significantly influenced by their internal thermal behaviour – in particular, the size of the pile, the amount of concrete cover to the pipes, and the relative positions and number of the pipes within the pile which can cause internal heat transfer. These factors are usually accounted for by the pile thermal resistance. However, there are no standard methods available for calculating the thermal resistance of piles, leading to uncertainty regarding parameter selection. The few published values of pile resistance have been derived principally from in situ tests. However, the discrepancy between these values and theoretical values suggests that more research is required in this area.

Thermal resistance is also influenced by the temperature profile of the heat exchanger fluid, which may vary non-linearly around the heat exchanger circuit. There are two typical scenarios for thermal piles, one with pipes placed around the circumference of the pile (attached to the steel cage) and one with the pipes placed centrally within the pile. The former is beneficial and will have a lower resistance as the pipes are closer to the ground. However, in the latter case there will be a large resistance, the pipes are more likely to interact adversely and questions remain as to whether a steady state approach to the pile behaviour is appropriate. These topics all warrant further research in order to assist more efficient heat exchanger design.

When multiple piles are connected in series, the change in heat transfer rate along the length of the pipe circuits can lead to each pile in the series having a different heat transfer rate to the ground. This is not accounted for in standard thermal design methods and the importance of this effect is still not known. All these uncertainties in the assessment of thermal capacity are exacerbated by the lack of high-quality monitoring data from case studies with which to validate potential new approaches.

Acknowledgements

This work has been funded by Mott MacDonald and the Engineering and Physical Sciences Research Council (EP/H049010/1). The authors would also like to thank Peter Smith of Cementation Skanska for his useful discussion on construction details.

REFERENCES

Acuna J, Mogensen P and Palm B (2009) Distributed thermal response test on a U-pipe borehole heat exchanger. *Proceedings of the 11th International Conference on Thermal Energy Storage for Efficiency and Sustainabiltiy, Effstock, Stockholm, Sweden, 14–17 June*, paper No. 18.

Banks D (2008) *An Introduction to Thermogeology: Ground Source Heating and Cooling.* Blackwell, Oxford, UK.

Bernier M (2001) Ground coupled heat pump system simulation. *ASHRAE Transactions* **107(1)**: 605–616.

Bicocchi N (2011) *Structural and Geotechnical Interpretation of Strain Gauge Data from Laterally Loaded Reinforced Concrete Piles.* PhD thesis. University of Southampton, Southampton, UK.

Bose JE, Parker JD and McQuiston FC (1985) *Design/Data Manual for Closed Loop Ground Coupled Heat Pump Systems.* American Society of Heating, Refrigeration and Air Conditioning Engineers, Atlanta, GA, USA.

Boudali M, Leroueil S and Srinivasa MBR (1994) Viscous behaviour of natural clays. *Proceedings of the 13th International Conference on Soil Mechanics and Foundation Engineering, New Delhi, India*, pp. 411–416.

Bourne-Webb PJ, Amatya B, Soga K *et al.* (2009) Energy pile test at Lambeth College, London: geotechnical and thermodynamic aspects of pile response to heat cycles. *Géotechnique* **59(3)**: 237–248.

Bourne-Webb PJ, Amatya B and Soga K (2013) A framework for understanding energy pile behaviour. *Proceedings of the Institution of Civil Engineers – Geotechnical Engineering* **166(2)**: 170–177.

Brandl H (1998) Energy piles and diaphragm walls for heat transfer from and into the ground. In *Proceedings of the 3rd International Conference on Deep Foundations on Bored and Auger Piles, Ghent* (Van Impe WF and Haegeman W (eds)). AA Balkema, Rotterdam, the Netherlands, pp. 37–60.

Brandl H (2006) Energy foundations and other thermo active ground structures. *Géotechnique* **56(2)**: 81–122.

Busby J, Lewis M, Reeves H and Lawley R (2009) Initial geological considerations before installing ground source heat pump systems. *Quarterly Journal of Engineering Geology and Hydrogeology* **42(3)**: 295–306.

Carslaw HS and Jaeger JC (1959) *Conduction of Heat in Solids*, 2nd edn. Oxford University Press, Oxford, UK.

Cekerevac C and Laloui L (2004) Experimental study of thermal effects of the mechanical behaviour of a clay. *International Journal for Numerical and Analytical Methods in Geomechanics* **28(3)**: 209–228.

Chiasson AC, Rees SJ and Spitler JD (2000) A preliminary assessment of the effects of groundwater flow on closed loop ground source heat pump systems. *ASHRAE Transactions* **106(1)**: 380–393.

CIBSE (Chartered Institute of Building Service Engineers) (2005) *Current CIBSE TRY/DSY Hourly Weather Data Set – London*. CIBSE, London, UK.

Claesson J and Hellstrom G (1981) Model studies of duct storage systems. In *Proceedings of the International Conference on New Energy Conservation Technologies and their Commercialisation* (Millhone JP and Willis EH (eds)). International Energy Agency, Berlin, Germany.

Claesson J and Hellstrom G (2000) Analytical studies on the influence of regional groundwater flow on the performance of borehole heat exchangers. In *Proceedings of the 8th International Conference on Thermal Energy Storage, Terrastock 2000, Stuttgart, Germany* (Benner M and Hahne E (eds)). University of Stuttgart, Institute of Thermodynamics and Thermal Engineering, Stuttgart, Germany, vol. 1, pp. 195–200.

Coulson JM and Richardson JF (1990) *Chemical Engineering, Volume 1, Fluid Flow, Heat Transfer and Mass Transfer*, 4th edn. Pergamon Press, Oxford, UK.

Diao NR, Zeng HY and Fang ZH (2004a) Improvements in modelling of heat transfer in vertical ground heat exchangers. *HVAC&R Research* **10(4)**: 459–470.

Diao NR, Li Q and Fang Z (2004b) Heat transfer in ground heat exchangers with groundwater advection. *International Journal of Thermal Sciences* **43(12)**: 1203–1211.

Eskilson P (1987) *Thermal Analysis of Heat Extraction Boreholes*. PhD thesis, Department of Mathematical Physics, University of Lund, Lund, Sweden.

Ferguson GF and Woodbury AD (2006) Observed thermal pollution and post development simulations of low temperature geothermal systems in Winnipeg, Canada. *Hydrogeology Journal* **14(7)**: 1206–1215.

Gao J, Zhang X, Liu J, Li K and Yang J (2008) Numerical and experimental assessment of thermal performance of vertical energy piles: an application. *Applied Energy* **85(10)**: 901–910.

Gehlin SEA and Hellstrom G (2003) Influence on thermal response test by groundwater flow in vertical fractures in hard rock. *Renewable Energy* **28(14)**: 2221–2238.

Gustafsson O (1993) Sweden. In *The Hydrogeology of the Chalk of North-West Europe* (Downing RA, Price M and Jones GP (eds)). Clarendon Press, Oxford, UK.

Hellstrom G (1983) Comparison between theoretical models and field experiments for ground heat systems. *Proceedings of the International Conference on Subsurface Heat Storage in Theory and Practice, Stockholm, Sweden, Swedish Council for Building Research, Stockholm, Sweden*, vol. 2.

Hellstrom G (1989) *Duct Ground Heat Storage Model, Manual for Computer Code*. Department of Mathematical Physics, University of Lund, Lund, Sweden.

Hellstrom G (1991) *Ground Heat Storage, Thermal Analysis of Duct Storage Systems, Theory*. Department of Mathematical Physics, University of Lund, Lund, Sweden.

Incropera FP, Dewitt DP, Bergman TL and Lavine AS (2007) *Fundamentals of Heat and Mass Transfer*, 6th edn. Wiley, Hoboken, NJ, USA.

Ingersoll LR, Zobel OJ and Ingersoll AC (1954) *Heat Conduction with Engineering and Geological Applications*, 3rd edn. McGraw-Hill, New York, NY, USA.

Jardine RJ (1991) The cyclic behaviour of large piles with special reference to offshore structures. In *Cyclic Loading of Soils from Theory to Design* (O'Reilly MP and Brown SF (eds)). Blackie, Glasgow, UK.

Kavanaugh SP and Rafferty K (1997) *Design of Geothermal Systems for Commercial and Institutional Buildings*. American Society of Heating Refrigeration and Air-Conditioning Engineers, Atlanta, GA, USA.

Kim KH, Jeon SE, Kim JK and Yang S (2003) An experimental study on thermal conductivity of concrete. *Cement and Concrete Research* **33**: 363–371.

Knellwolf C, Peron H and Laloui L (2011) Geotechnical analysis of heat exchanger piles. *Journal of Geotechnical and Geoenvironmental Engineering* **137(10)**: 890–902.

Laloui L (2001) Thermo-mechanical behaviour of soils. *Revue Française de Génie Civil* **5(6)**: 809–843.

Laloui L and Cekerevac C (2008) Non-isothermal plasticity model for cyclic behaviour of soils. *International Journal for Numerical and Analytical Methods in Geomechanics* **32(5)**: 437–460.

Laloui L, Moreni M, Fromentin A, Pahud D and Vulliet L (1999) In-situ thermo-mechanical load test on a heat exchanger pile. In *Proceedings of the 4th International Conference on Deep Foundation Practice Incorporating Piletalk, Singapore*. Deep Foundations Institute, Hawthorne, NJ, USA. pp. 273–279.

Laloui L, Nuth M and Vulliet L (2006) Experimental and numerical investigations of the behaviour of a heat exchanger pile. *International Journal for Numerical and Analytical Methods in Geomechanics* **30(8)**: 763–781.

Lamarche L, Kajl S and Beauchamp B (2010) A review of methods to evaluate borehole thermal resistance in geothermal heat pump systems. *Geothermics* **39(2)**: 187–200.

Lee CK and Lam HN (2008) Computer simulation of borehole ground heat exchangers for geothermal heat pump systems. *Renewable Energy* **33(6)**: 1286–1296.

Lennon DJ, Watt E and Suckling TP (2009) Energy piles in Scotland. Taylor and Francis, London, UK. In *Proceedings of the 5th International Conference on Deep Foundations on Bored and Auger Piles, Frankfurt* (Van Impe WF and Van Impe PO (eds)).

Marcotte D and Pasquier P (2008) On the estimation of thermal resistance in borehole thermal conductivity test. *Renewable Energy* **33(11)**: 2407–2415.

Markiewicz R (2004) *Numerical and Experimental Investigations for Utilization of Geothermal Energy Using Earth-coupled Structures and New Developments for Tunnels*. PhD thesis. Vienna University of Technology, Vienna, Austria.

Neville AM (1995) *Properties of Concrete*, 4th edn. Longman, London, UK.

NHBC (National House Building Council) (2010) *Efficient Design of Piled Foundations for Low Rise Housing, Design Guide*. NHBC Foundation, Milton Keynes, UK.

Pahud D (2007) *PILESIM2, Simulation Tool for Heating/Cooling Systems with Heat Exchanger Piles or Borehole Heat Exchangers, User Manual*. Scuola Universitaria Professionale della Svizzera Italiana, Lugano, Switzerland.

Pahud D and Hubbach M (2007) *Mesures et Optimisation de l'Installation avec Pieux Energétiques du Dock Midfield de l'Aéroport de Zurich*. Office fédéral de l'energie, Ittigen, Switzerland, final report (in French).

Philippe M, Bernier M and Marchio D (2009) Validity ranges of three analytical solutions to heat transfer in the vicinity of boreholes. *Geothermics* **38(4)**: 407–413.

Poulos HG (1988) *Marine Geotechnics*. Unmin Hyman, London, UK.

Poulos HG (1989) Pile behaviour – theory and application. *Géotechnique* **39(3)**: 365–415.

Preene M and Powrie W (2009) Ground energy systems from analysis to geotechnical design. *Géotechnique* **59(3)**: 261–271.

Rees SW, Adjali MH, Zhou Z, Davies M and Thomas HR (2000) Ground heat transfer effects on thermal performance of earth contact structures. *Renewable and Sustainable Energy Reviews* **4(3)**: 213–265.

Remund CP (1999) Borehole thermal resistance: laboratory and field studies. *ASHRAE Transactions* **105(1)**: 439–445.

Sanner B, Hellstrom G, Spitler J and Gehlin SEA (2005) Thermal response test – current status and world-wide application. In *Proceedings of the World Geothermal Congress, Antalya, Turkey*. International Geothermal Association, Bochum, Germany.

Shonder JA and Beck JV (1999) Determining effective soil formation properties from field data using a parameter estimation technique. *ASHRAE Transaction* **105(1)**: 458–466.

SIA (2005) *Utilisation de la Chaleur du Sol par des Ouvrages de Fondation et de Soutènement en Béton, Guide pour la Conception, la Réalisation et la Maintenance*. Swiss Society of Engineers and Architects, Geneva, Switzerland, Documentation D 0190 (in French).

Signorelli S, Bassetti S, Pahud D and Kohl T (2007) Numerical evaluation of thermal response tests. *Geothermics* **36(2)**: 141–166.

Sutton MG, Nutter DW and Couvillion RJ (2003) A ground resistance for vertical bore heat exchangers with groundwater flow. *Journal of Energy Resources Technology* **125(September)**: 183–189.

Tatro SB (2006) Thermal properties. In *Significance of Tests and Properties of Concrete and Concrete Making Materials* (Lamond J and Pielert J (eds)). Portland Cement Association, Skokie, IL, USA.

Thomas HR and Rees SW (1999) The thermal performance of ground floor slabs – a full scale in situ experiment. *Building and Environment* **34(2)**: 139–164.

VDI (2009) *Thermal Use of the Underground – Ground Source Heat Pump Systems*. The Association of German Engineers (VDI), Dusseldorf, Germany, VDI 4640, Part 2.

Wood CJ, Liu H and Riffat SB (2009) Use of energy piles in a residential building, and effects on ground temperature and heat pump efficiency. *Géotechnique* **59(3)**: 287–290.

Wood CJ, Liu H and Riffat SB (2010a) Comparison of a modeled and field tested piled ground heat exchanger system for a residential building and the simulated effect of assisted ground heat recharge. *International Journal of Low Carbon Technologies* **5(12)**: 137–143.

Wood CJ, Liu H and Riffat SB (2010b) An investigation of the heat pump performance and ground temperature of a pile foundation heat exchanger system for a residential building. *Energy* **35(12)**: 3932–3940.

Xu X and Spitler JD (2006) Modelling of vertical ground loop heat exchangers with variable convective resistance and thermal mass of fluid. *Proceedings of the 10th International Conference on Thermal Energy Storage – EcoStock 2006, Pomona, NJ, USA*, paper 4A-3.

Zeng H, Diao N and Fang Z (2003) Heat transfer analysis of boreholes in vertical ground heat exchangers. *International Journal of Heat and Mass Transfer* **46(23)**: 4467–4481.

Index

ABAQUS software, 186
Abramovich, H., 66
absorber pipes, 100, 103, 132, 133, 135–7, 138, 140, 143, 151
Adam, D., 206
air panels PV-thermal model, 119, 120–3, 125–7, 128
Allis, R.G., 50
Amis, T., 224, 234
AR (pile aspect ratio), 249, 261–3, 275
Aratiatia Dam, 43, 45
ArcGIS software, 208, 215, 218
Argentina, 3, 5, 6, 7–19
ASC (asphalt solar collectors), 63, 65, 72
Ashrae (American Society of Heating, Refrigerating, and Air-Conditioning Engineers), 79, 81–2, 85, 89, 90, 94, 96, 236, 238
Australia, 85, 87–96, 97
Austria, 110, 133, 134, 182, 206, 224

Baggs, S., 12
Bakker, M., 115, 116
Banks, D., 230
Barla, M., 103–4, 105, 213
Beck, J., 229, 230
Bernier, M., 81
BHEs (borehole heat exchangers)
    and CFA piles, 185, 197, 200–201
    chalk aquifer TRT study, 158, 160, 165, 168, 173, 174, 175
    external response, 256–8
    fluid temperature profiles, 258–60

    internal response, 258
    and pile heat exchangers, 252, 256–60, 264, 275
    Republic of Ireland TRT study, 224, 225, 227, 230, 231, 236, 243
Bidarmaghz, A., 9
Biotto, C., 145
Boltzmann function, 51, 52, 53
borehole thermal resistance, 81, 159, 161, 165–7, 174, 230–1, 256, 260
Bourne-Webb, P., 213, 234, 271, 272
Bowen, L., 66
'bowls', subsidence, 42–5, 46–56
Brandl, H., 206, 210, 271
Bray to Cork Gas Pipeline, 227
Brettmann, T., 224, 234
brine circulation systems, 25, 27
Brockbank, K., 46
Bromley, C.J., 50
Brown, D.A., 183
Brusaw, Julie, 63–4
Brusaw, Scott, 63–4

Cantor, J., 126
carbon emissions, 80, 182, 275
Carbon Free Group, 128
Carboniferous rocks, 25, 27, 29, 30, 33
Carraigtwohill site (Republic of Ireland), 227, 228, 233–4, 239–43
Cauchy boundary condition, 104
Cecinato, F., 186, 193
Cekerevac, C., 274
Cengel, Y., 234

281

CFA (continuous flight auger) piles, 181, 182, 202, 250–1, 265
　cases considered, 187–90
　central steel bar impact, 190, 195, 201, 202
　concrete resistance, 197, 199–200, 201–2, 203
　construction techniques, 183–4, 201
　convection-diffusion equation, 185–6
　discussion, 201–2
　energy performance, 190–7
　finite-element model, 186–7, 190–1
　fluid velocity, 185, 188, 189, 190–1, 193
　numerical model, 181, 185–7, 188, 197–201, 202
　pile heat exchangers, 274–5
　pipe arrangements/positioning, 190, 191, 195–7, 198, 201
　pipe-to-pipe interaction, 181, 185, 195, 197–201, 202
　results, 190–201
　and rotary bored piles, 182, 183, 184–5, 188–90, 193–5, 201–2
　sensitivity analysis, 187–8, 189, 190, 191–3
　simulation model, 184, 186–91, 192, 193, 196, 197–201, 202
　thermal conductivity, 184–5, 186, 189, 190, 203
　thermal performance assessment, 184–90
　thermal resistance, 197, 200, 201–2
　thermal response testing, 186, 224, 227
　U-loops, 183, 185, 186–8, 190, 197–8, 200, 201
cGPS (continuous global positioning system), 44–5
chalk aquifer TRT study *see under* TRT
Channel Tunnel, 132
CIBSE (Chartered Institution of Building Services Engineers), 138
city-scale analysis (Warsaw thermoactive structures), 205, 206–7, 213, 220–1
　analysed data set, 207–12
　existing and planned infrastructure map, 210, 212
　general layout of Warsaw City, 207–8
　GIS analysis, 206, 207, 208, 213, 214, 215–19, 220
　GSHPs, 206, 210, 220, 221
　high-rise buildings, 208, 209, 210, 211, 213–14, 216, 218, 219–20, 221
　increasing demand for structures, 205, 206–7, 220
　metro lines, 206, 207–8, 210, 212, 213, 215, 218, 219, 220
　results, 219–20
　simplified thermoactive foundation model, 213–15
　underground infrastructure, 208, 210, 211, 215, 216, 218, 219, 220, 221
climate change, 42, 133, 146
closed-loop systems, 4, 133, 152, 154, 158, 224, 236, 243, 250
compressibility testing, 45–6, 48–50, 52–3, 54–5, 56
computational models, 24
Comsol Multiphysics, 9
concrete resistance, 197, 199–200, 201–2, 203
constant parameter settings, 189
Contact Energy Ltd, 44, 45
contact resistance, 256, 258
control system PV-thermal model, 119, 120–3, 125–7
convection, 9–10, 83, 109, 113
　convection-diffusion equation, 185–6
　parsimonious numerical modelling, 27, 32, 37–8
　pile heat exchangers, 252–4
conversion efficiency, 61, 64, 65, 66, 67, 68, 69, 70, 71–2, 73
COP (coefficient of performance), 115, 126, 129, 154
Córdoba (Argentina), 3, 5, 7–9, 11, 12, 16, 17–19
Cork Docklands site (Republic of Ireland), 225–7, 228, 234–9, 242–4
Crossrail tunnel energy segment system *see* TES system
Crown bowl site (New Zealand), 42–3, 44, 45, 47, 48, 49, 50–1, 52–3

CXB (hydrothermal eruption breccia), 47, 48, 49, 51
cyclic loading effects, 273–4

Darcy's law, 10, 24, 26–7, 102, 106, 107, 230, 254–5
DB (dry bulb) temperature measurement, 145
deep geothermal reservoirs, numerical modelling of *see* parsimonious numerical modelling
design decision parameters, 85, 88, 89–90
Di Donna, A., 213
diaphragm walls, 206, 213, 214–15, 218, 224–5
DInSAR (differential interferometric synthetic aperture radar), 45
drawdown, 26, 29, 30–1, 37, 38, 42, 47, 231–2
DST (duct storage model), 262–3, 275
Duarte, F., 69
Dupray, F., 213

Eastgate (UK), 23, 24–31, 34–9
Eastgate Geothermal Field, 23, 39
EED (Energy Earth Designer database), 226–7
electromagnetic technology, 63, 66–70, 71, 72, 73
electromechanical harvesting systems, 63, 67–9, 71, 72, 73
Elizabeth Blackburn School of Sciences (Melbourne), 85
energy generation parameter, 61, 66, 67, 69, 70, 71–2, 73
energy harvesting, road pavement *see* road pavement energy harvesting
Energy Intelligence, 69, 71
energy piles
  continuous flight augers piles *see* CFA piles
  geothermal energy in loess, 3, 5, 9, 11, 14, 17, 19
  pile heat exchangers *see* pile heat exchangers
  Republic of Ireland TRT study, 223, 224–5, 243–4
    analysis, 227, 229–34

borehole heat exchangers, 224, 225, 227, 230, 231, 236, 243
Carraigtwohill site, 227, 228, 233–4, 239–43
Cork Docklands site, 225–7, 228, 234–9, 242–4
CPTU cone resistance and friction ratio, 235–6
discussion, 243–4
energy pile installations, 225–7
GPM model, 229–30, 238, 240, 242, 243, 244
groundwater flow, 230–1, 236, 238, 242, 243–4
heat addition and loss, 233–4
laboratory thermal measurement of soil, 238
line source method, 225, 227, 229, 230, 233, 234, 237–8, 239–41, 242–4
pumping recovery test, 231–2
pumping test, 231–2, 233
results, 234–42
thermal conductivity, 224–5, 226–7, 229–30, 231, 234, 237–8, 240, 242–3
thermal diffusivity, 229, 233, 237–8
thermal recovery test, 225, 230–3, 234, 238, 239, 240–1, 242, 243
Warsaw thermoactive structures study *see* city-scale analysis
energy tunnels, 99, 100–101
  Crossrail tunnel energy segment system *see* TES system
  finite-element models, 99, 102, 103–5, 107, 113
  ground conditions, 99, 102, 106–13
  ground temperature influence, 107, 108, 109, 110, 111–12, 113
  groundwater flow velocity, 102, 106, 107, 108–9, 110, 111–12, 113
  installation costs, 101
  Metro Torino line tunnel, 104–5, 106, 108, 112–13
  monitoring data comparisons, 109–10
  numerical model, 102–5, 111, 113

parametric numerical analyses, 106–11
preliminary design charts, 111–13
summer mode, 105, 106–10, 111–13
technological aspects, 101–2
thermal conductivity, 103, 105–6, 107–11, 112–13
thermos-hydro mathematical formulation, 102–3
winter mode, 105, 106–10, 111–13
Engineering-Geological Database of Warsaw, 207, 214, 215
Eskilson, P., 257–8
Eulerian coordinate system, 102
European Commission, 206
European Union, 182, 203, 206, 207
Everloc permanent couplings, 137
EVS (effective vertical stress), 48–9, 50, 52, 53, 57

Feflow software, 102, 103
FILM subroutine, 186
finite-element models
    CFA piles, 186–7, 190–1
    geothermal energy in loess, 3, 9–11
    geothermal subsidence study, 51, 54
    ground conditions for energy tunnels, 99, 102, 103–5, 107, 113
    pile heat exchangers, 263
    TES system, 140, 143–4, 147–50
fire safety, 135, 137, 150–1
Fisher Street case study, 152–4
FLAC code, 53
FLS (finite line source) model, 257, 261, 262
fluid flow, 9–10, 90, 103, 185, 236, 252
    geothermal subsidence study, 51, 54
    parsimonious numerical modelling, 24, 27, 28–31, 34, 38
    pile heat exchangers, 258–9, 265–7, 267–70
    see also groundwater flow
fluid temperature, 81, 84, 102
    CFA piles, 186, 190, 193–4
    chalk aquifer TRT study, 159, 162, 164–6, 168–70, 171, 173

geothermal energy in loess, 13, 14, 16–17, 18
pile heat exchangers, 251, 258–60, 262, 265–70, 271
Republic of Ireland TRT study, 224, 225, 232, 236, 239–40
TES system, 140, 144, 146
fluid velocity, 10, 105–6, 110–11, 185, 188, 189, 190–1, 193
fluvioglacial sands, 215
Fossa, M., 96
foundation slab, 213, 214, 215, 218
Fourier's law 230, 254, 255
fractures/fracture systems, 46, 172–4, 270
    parsimonious numerical modelling, 25, 26, 27–8, 29–31, 34, 38
Franzius, J.N., 110, 132, 224
'free cooling', 252
Frodl, S., 224
'full-extent' models, 28, 30, 34

Gaia system, 65
Geertsma, J., 52
Gehlin, S., 225
Gelder, A., 225
Genziko, 66, 71
Geological Survey of Ireland Quaternary, 227
geothermal subsidence study (Wairakei–Tauhara), 41–2, 56–7
    background deformation, 44–5
    benchmark subsidence levels, 43, 44, 45
    compressibility testing, 45–6, 48–50, 52–3, 54–5, 56
    core property correlations, 48
    Crown bowl site, 42–3, 44, 45, 47, 48, 49, 50–1, 52–3
    deformation surveys, 44
    geology and subsidence, 46–7
    geotechnical investigation, 45–8
    Rakaunui bowl site, 42–3, 44, 45, 47, 50, 52, 53
    reservoir pressure changes, 41, 42, 47, 48, 50, 51–7
    simulation modelling, 41, 50, 51–5, 56

Spa bowl site, 42–3, 44, 45, 47, 48, 50, 51, 52, 53
subsidence mechanisms, 48–55, 57
subsidence mitigation, 55–6, 57
subsidence modelling, 51–5
Wairakei bowl site, 42–3, 44, 45, 47, 48, 50, 51, 52, 53–5
Wairakei–Tauhara history, 42–4
well measurements, 47
Germany, 110, 132, 133, 182, 224
GHEs (ground heat exchangers), 101, 116, 132, 182, 200, 206, 252, 260
   design uncertainties, 79, 80–1, 96–7
      borehole characteristics, 93, 96, 97
      case study, 85, 87–96, 97
      design decision parameters, 85, 88, 89–90
      design length uncertainty, 81–4, 93–6
      design parameter quantification, 85, 86
      design parameter uncertainties, 88–91
      ground thermal load, 90, 92, 96
      grout thermal properties, 82, 85, 86, 88, 89, 92–3, 95, 96
      methodology, 81–4
      probability density function, 82, 85–6, 88–91, 93, 94
      sensitivity analysis, 82, 92–3, 94–5
      site condition parameters, 85, 88, 90
      thermal demand predictions, 81, 85, 87, 90, 96, 97
      thermal resistance, 81, 82, 83–4
      thermal response test, 81, 86, 88, 94–6
   geothermal energy in loess, 3, 4–5, 9–11, 13–19
   PV-thermal cost comparison, 127–8
GIS (geographic information system), 139, 141
   Warsaw thermoactive structures study, 206, 207, 208, 213, 214, 215–19, 220
glycol flows, 115, 116, 119, 120, 124, 126–7, 128

Gnielinski correlation, 175
GPM (geothermal properties measurement) model, 229–30, 238, 240, 242, 243, 244
greenhouse gas emissions, 80, 182, 275
ground conditions (for energy tunnels), 99, 102, 106–13
ground thermal load, 90, 92, 96
groundwater flow, 9–10, 24, 31
   chalk aquifer TRT study, 157, 159, 161, 164, 165, 167, 169, 171–3, 174
   ground conditions for energy tunnels, 102, 106, 107, 108–9, 110, 111–12, 113
   pile heat exchangers, 253, 254, 255–6, 270
   Republic of Ireland TRT study, 230–1, 236, 238, 242, 243–4
grout thermal properties, 161, 175, 231, 258, 265
   GHE design uncertainties, 82, 85, 86, 88, 89, 92–3, 95, 96
GSHPs (ground source heat pumps)
   geothermal energy in loess, 3, 4–6
   GHE design uncertainties *see under* GHEs
   PV-thermal modules simulation *see* PV-thermal modules simulation
   and TES system, 131, 133, 152, 154
   Warsaw thermoactive structures study, 206, 210, 220, 221

H&V (heat and ventilation) design, 135
Hagen-Poiseuille law, 103
Harb, A., 62, 69
Harper, G., 126
Hasebe, M., 64
HDPE (high-density polyethylene) pipes, 4, 6, 9–10, 11, 86, 116, 119
He, M., 200–201
header pipes, 132, 137, 138–9, 140, 145, 151, 152, 154
heat extraction rate
   energy tunnels, 111–13
   TES system, 138, 140, 144–6, 147, 149–50, 152, 154

Warsaw thermoactive structures study, 208, 213, 218–20
heat injection phase/rate, 225, 233, 242
    chalk aquifer TRT study, 159, 162, 163–4, 166–71, 172, 173, 174
    energy tunnels, 112–13, 116
heat load *see* thermal load
heat transport modelling, 23, 24–5, 37, 31–3, 34–6, 37–8, 39
Hecht-Mendez, J., 32, 33
Hellstrom, G., 197
Helpin, V., 127
Hemmingway, P., 224, 225, 234, 236, 237, 238, 243
Hendrickson, B., 67
HFF (Hula Falls formation), 46, 47, 48
Highway Energy Services, 68, 71
Hill, B., 66
hoop stress, 149–50
Horianopoulos, D., 67
Horianopoulos, S., 67
horizontal dispersivity, 32, 33, 37
hydraulic diffusivity, 255, 256
'hydraulic skin', 230–1
hydraulic/pneumatic road pavement energy harvesting, 63, 67, 71, 72, 73

ICAX, 65
ICS (infinite cylindrical heat source), 257
IEE Sepemo project, 124
IGSHPA (International Ground Source Heat Pump Association), 6, 81, 89, 224
ILS (infinitely long line heat source), 256–7, 261, 263, 270
IM (installation method), 70, 71–2
induction heating, 63
Innowattech, 66, 71
installed power, 63–4, 70, 71, 72, 206
International Tunnelling Awards, 155
IPEG (Innowattech Piezo Electric Generator), 66
IVT Industrier, 116

Japan, 64, 65
Jenbach tunnel, 110, 134, 224

Kapuściński, J., 206
Katzenberg tunnel, 110, 224
Kavanaugh, S.P., 88–9, 96
Kelk, 64
KinerBump, 67
KinergyPower International Corporation, 67, 71
Kjellsson, E., 115–16
Knellwolf, C., 273
Korea Institute, 63
Külah, H., 69

Lachal, B., 115
Laloui, L., 213, 271–3, 274
Lamarche, L., 197
Lambeth College, 273
Lambeth Group, 147, 148, 160–1, 172
Laminated Beds, 172–3
Lee, C., 110
'limited-extent' models, 28, 29, 30
line source approach
    CFA piles, 197
    chalk aquifer TRT study, 159–60, 165, 168, 170–1, 174
    pile heat exchangers, 255, 257, 261, 263
    Republic of Ireland TRT study, 225, 227, 229, 230, 233, 234, 237–8, 239–41, 242–4
linear isotherm sorption, 32
linear motion conversion, 68–9
loess, 3, 17, 19
    case study
        conclusion, 17, 19
        finite-element model, 3, 9–11
        geometry, initial and boundary conditions, 11–12
        numerical modelling, 9–10
        results and discussions, 12–17, 18
        simulation modelling, 3, 9–12, 14
        site description and thermal characterisation, 7–9
    characteristics of, 6–7
    ground heat exchangers, 3, 4–5, 9–11, 13–19

ground-source heat pump systems, 3, 4–6
thermal conductivity, 3, 6–7, 9–10, 12, 15–17, 19
London Basin, 160, 173–4
London Clay, 132, 147, 148–9, 160, 169, 172, 272
London heat map, 139, 141
London Underground tunnel energy segment system *see* TES system
Long, M., 224, 225, 234, 236, 237, 238
longitudinal dispersivity, 32, 33, 37, 103, 105, 106
Loveridge, F., 182, 186, 193, 197
low enthalpy geothermal systems, 100, 131, 206
'Low-temperature geothermal energy in Poland and worldwide', 206
LS-Dyna code, 140, 143, 145, 147–8
Lund, J.W., 65
LYBRA, 69

macro-energy harvesting, 62
Malenković, I., 124
Marcotte, D., 165
Markiewicz, R., 206, 268, 269
MEMS (microelectromechanical systems), 63, 69–70, 72, 73
Mersenne twister, 82
'methods of moments' analysis, 82
Metro Torino line tunnel, 104–5, 106, 108, 112–13
micro-energy harvesting, 62
micropiles, 3, 5, 17
Mitcheson, P.D., 69
MODFLOW code, 24, 28–31, 37, 38
Mohr–Coulomb model, 147
Monte Carlo simulation methods, 82, 90, 92, 93
motion conversion systems, 68–9
MotionPower, 67
MT3DMS code, 23, 24, 31–6, 37, 38

National Renewable Energy Centre, 127
National Research Institute (Poland), 207

Near, C., 66
Netherlands, 64, 65
New Energy Technologies, 67, 71
New Zealand Wairakei–Tauhara study *see* geothermal subsidence study
Newton's law of cooling, 253
Nusselt number, 175–6

Ochota, 207, 220
Office of Architecture and Spatial Planning of Warsaw Municipality, 207
Ooms International Holding, 65
open-loop systems, 270
Optical Energy Technologies Inc., 128

Pahud, D., 115
parsimonious numerical modelling (of deep geothermal reservoirs), 23–5
  background of study, 25–7
  conceptual modelling, 23, 24, 27–34
  drawdown, 26, 29, 30–1, 37, 38
  Eastgate drilling history, 25–7
  fluid flow, 24, 27, 28–31, 34, 38
  fracture systems, 25, 26, 27–8, 29–31, 34, 38
  geological setting of study, 25
  heat transport modelling, 23, 24–5, 37, 31–3, 34–6, 37–8, 39
  MODFLOW code, 24, 28–31, 37, 38
  MT3DMS code, 23, 24, 31–6, 37, 38
  recommendations for future work, 38–9
  sensitivity analysis, 36–7, 38
  simulation modelling, 23, 24–5, 29–30, 34–6, 37, 38
  study discussion and conclusions, 37–8
  well doublet system, 24, 29, 34–6
Pasquier, P., 165
patents, 64, 66, 67–70
pdf (probability density function), 82, 85–6, 88–91, 93, 94
Peclet number, 38
Pender, M., 46
Perino, A., 213
PE-Xa polymer pipe, 134, 135–6, 146, 150–1
Philippe, M., 81–2, 96

287

photovoltaic systems
    PV-thermal modules simulation *see*
        PV-thermal modules simulation
    road pavement energy harvesting, 63–4,
        71, 72, 73
piezoelectric technology, 63, 65–6, 70, 71,
    72, 73
pile axial load, 271, 273
pile heat exchangers, 249, 250–2, 275–6
    behavioural framework, 271
    and borehole heat exchangers, 252,
        256–60, 264, 275
    case studies, 271–3
    convection and conduction, 252–4
    duct storage model, 262–3, 275
    external response, 260–4
    fluid flow rate, 258–9, 265–7, 267–70
    fluid temperature profiles, 265–70, 276
    groundwater flow, 253, 254, 255–6, 270
    heat transfer concepts, 252–6
    increased adoption of, 249, 250, 275
    pile aspect ratio, 249, 261–3, 275
    practical constraints, 274–5
    seasonal temperature variation, 251, 264
    soil thermal behaviour and cyclic loading
        effects, 273–4
    thermal conductivity, 254–6, 264, 265,
        275
    thermal diffusivity, 255–6
    thermal performance, 249, 252, 260–70
    thermal resistance, 254, 256, 258, 260,
        264–5, 266–8, 271, 275–6
    thermomechanical interactions and pile
        behaviour, 270–4
pile-raft foundation, 213
PILESIM software, 263
pipe-to-pipe interaction, 181, 185, 195,
    197–201, 202, 258
Pirisi, A., 68–9
Plaxis software, 51
p-linear temperature, 164, 165, 166–7, 168,
    170
Pliocene clays, 215, 216, 217, 218
PLT (point-load test), 48

Poland
    Warsaw thermoactive structures study *see*
        city-scale analysis
Polish Geological Survey, 207
PORT PC (Polish Organization for
    Development of Heat Pumps
    Technology), 219
Poulos, H.G., 274
Powrie, W., 182, 197
Pralle, N., 110, 132, 224
Prandtl number, 176
pressure testing, 6, 137, 275
primary loess, 4, 6, 8
probability theory, 82
pumping test, 26, 30, 165–6, 231–2, 233
PV-thermal array models, 119, 120–7, 128–9
PV-thermal modules simulation, 115–16, 128–9
    accuracy of simulations, 124, 126–7
    air panels model, 119, 120–3, 125–7, 128
    ambient temperatures, 116, 118
    control system model, 119, 120–3, 125–7
    discussion, 124–8
    electricity savings, 124
    energy inputs and outputs, 120, 122
    GHE cost comparison, 127–8
    glycol flows, 115, 116, 119, 120, 124,
        126–7, 128
    heat load, 119, 121
    model, 116–19
    results, 119–24
    seasonal performance factor, 124, 125, 126
    sources of heat, 120, 123, 124
    test house, 115, 116, 117–18, 124, 126–7
    thermal array models, 119, 120–7, 128–9

QW (Quantum Well) materials, 64

Rafferty, K.D., 88–9, 96
Rakaunui bowl site (New Zealand), 42–3, 44,
    45, 47, 50, 52, 53
Ramsay, G., 46
Raymond, J., 225
Rees, S.J., 200–201
Rees, S.W., 264

Republic of Ireland TRT study *see under* TRT
reservoir pressure changes (geothermal subsidence study), 41, 42, 47, 48, 50, 51–7
retardation factor, 31–2, 33
Reuss, M., 225
'reverse return' arrangement, 137, 138–9, 153
Reynolds number, 176, 191, 254, 266
Rey-Ronco, M., 229–30
@Risk software, 82
Road Energy Systems, 65
road pavement energy harvesting, 61–2, 72–3
　ASC technology, 63, 65, 72
　comparison of technologies, 71–2
　electromagnetic technology, 63, 66–70, 71, 72, 73
　electromechanical systems, 63, 67–9, 71, 72, 73
　energy harvesting technologies, 62–70
　hydraulic or pneumatic systems, 63, 67, 71, 72, 73
　macro-energy harvesting, 62
　MEMS harvesting systems, 63, 69–70, 72, 73
　micro-energy harvesting, 62
　patents, 64, 66, 67–70
　photovoltaic technology, 63–4, 71, 72, 73
　piezoelectric technology, 63, 65–6, 70, 71, 72, 73
　solar radiation, 62–5, 71–3
　technical analysis, 70–2
　thermoelectric technology, 63, 64, 71, 72, 73
　vehicle mechanical energy, 62–3, 65–70, 71–3
Rodzoch, A., 206
Rosenberg, M.D., 46
rotary bored piles, 182, 183, 184–5, 188–90, 193–5, 201–2, 250
rotational motion conversion, 68

RQD (rock-quality designation), 46
Rubik, M., 219
rural-urban migration, 61

Sanner, B., 236
Save Your Energy Stafford Area, 128
Schneider, M., 224
Schreier, M., 64
secondary loess, 4, 6
Seebeck effect, 64
sensitivity analysis
　CFA piles, 187–8, 189, 190, 191–3
　geothermal energy in loess, 11
　GHE design uncertainties, 82, 92–3, 94–5
　parsimonious numerical modelling, 36–7, 38
　TES system, 140, 144
Seocheon energy tunnel, 110
SERSO (Solar Energy Recuperation from the Road Pavement) system, 65
Sharp Energy Solutions, 119
Shonder, J., 229, 230
Signorelli, S., 168
simulation modelling
　CFA piles, 184, 186–91, 192, 193, 196, 197–201, 202
　geothermal energy in loess, 3, 9–12, 14
　geothermal subsidence study, 41, 50, 51–5, 56
　GHE design uncertainties, 82, 90, 92, 93
　ground conditions for energy tunnels, 103–7
　parsimonious numerical modelling, 23, 24–5, 29–30, 34–6, 37, 38
　pile heat exchangers, 268, 269
　PV-thermal modules simulation *see* PV-thermal modules simulation
　TES system, 140, 143–4, 145, 149
site condition parameters, 85, 88, 90
Slitt Vein, 24, 25–7, 28–31, 34–6, 37
'smart' cooling, 154
solar radiation, 62–5, 71–3, 116, 118, 264
Solar Roadway, 63–4, 71
SolaRoad, 64
South Korea, 110

Spa bowl site (New Zealand), 42–3, 44, 45, 47, 48, 50, 51, 52, 53
SPF (seasonal performance factor), 124, 125, 126
Spitler, J., 225
SPT (Standard Penetration Test), 8
Śródmieście, 207, 220
Stuttgart Metro U6, 132, 133
Sullivan, C., 65
Sweden, 263
Swiss Federal Institute of Technology, 271
Swiss Society for Architects and Engineers, 264
Switzerland, 65, 263, 264, 271

TBM (tunnel boring machine), 137
tectonic deformation, 44–5, 48, 215, 216
TEGs (thermoelectric generators), 63, 64, 71, 72, 73
tensile stress, 149–50
TES (tunnel energy segment) system, 131–2, 154–5
   absorber pipes, 132, 133, 135–7, 138, 140, 143, 151
   assessment of heat inside the tunnel, 132–3
   carbon savings, 151–2, 154
   case study, 152–4
   cost savings, 151, 154
   Crossrail experience, 135
   design of, 133–9
   durability and operational considerations, 146–51
   finite-element modelling, 140, 143–4, 147–50
   fire safety, 135, 137, 150–1
   header pipes, 132, 137, 138–9, 140, 145, 151, 152, 154
   heat extraction rate, 138, 140, 144–6, 147, 149–50, 152, 154
   heat revenue, 152
   modelling of heat transfer, 140, 143–4
   operational and commercial benefits, 151–2
   pipe durability, 146
   pipework hydraulics and plant room, 138–9
   potential market for tunnel heat, 139–40, 141–2
   previous systems, 133, 134
   'reverse return' arrangement, 137, 138–9, 153
   risk assessment, 135, 150–1, 154
   schematic diagrams, 132, 137–8
   sensitivity analyses, 140, 144
   surface connection points, 140, 142
   thermal effects on tunnel structure, 147–50
   thermal energy segments, 135–8
   train frequency, 133, 134
   trapped air removal, 138–9
   tunnel cooling study, 144–6
   ventilation system, 131, 135, 140, 144–5, 146, 151, 152, 154
Thanet Sands (UK), 171, 172
thermal conductivity
   CFA piles, 184–5,186, 189, 190, 203
   geothermal energy in loess, 3, 6–7, 9–10, 12, 15–17, 19
   GHE design uncertainties, 81, 85, 86–9, 92–3, 94–5, 96
   ground conditions for energy tunnels, 103, 105–6, 107–11, 112–13
   pile heat exchangers, 254–6, 264, 265, 275
   Republic of Ireland TRT study, 224–5, 226–7, 229–30, 231, 234, 237–8, 240, 242–3
   TES system, 143, 146
   thermal response testing *see* TRT
thermal demands, 13, 15, 81, 85, 87, 90, 96, 97
thermal diffusivity, 12, 37, 159, 165, 255–6
   GHE design uncertainties, 88–9, 92–3, 95
   Republic of Ireland TRT study, 229, 233, 237–8
thermal load
   geothermal energy in loess, 12–14, 16–17
   GHE design uncertainties, 82, 85, 90, 92, 96
   pile heat exchangers, 272–3

PV-thermal modules simulation, 119, 121
TES system, 139–40, 144
thermal needle probe method, 9
thermal recovery testing, 225, 230–3, 234, 238, 239, 240–1, 242, 243
thermal resistance
  borehole thermal resistance, 81, 159, 161, 165–7, 174, 230–1, 256, 260
  CFA piles, 197, 200, 201–2
  chalk aquifer TRT study, 159, 161, 162, 165–8, 174, 175–6
  GHE design uncertainties, 81, 82, 83–4
  pile heat exchangers, 254, 256, 258, 260, 264–5, 266–8, 271, 275–6
  Republic of Ireland TRT study, 225, 230–1
thermistors, 160–1, 162–4, 170–3
thermoactive structures, Warsaw *see* city–scale analysis (Warsaw thermoactive structures)
Thomas, H.R., 264
TNO (Toegepast Natuurwetenschappelijk Onderzoek), 64, 65, 71, 72
tornado diagrams, 36–7, 92, 95
Tottenham Court Road station, 152–4
TOUGH2 code, 51, 52, 54
TPO (Taupo ignimbrite), 47, 48, 49, 50, 55
transient conduction, 186, 254, 255
TRLs (technology readiness levels), 70, 71–2
TRNSYS software, 116, 118, 124–7, 128
TRT (thermal response test), 110, 157, 158
  CFA piles, 186, 224, 227
  chalk aquifer study, 157, 160, 174
    average fluid temperature, 165, 166, 168–9
    borehole instrumentation, 170–3
    discussion, 173–4
    dynamic interpretation, 167–9
    groundwater flow, 157, 159, 161, 164, 165, 167, 169, 171–3, 174
    heat injection phase, 159, 162, 163–4, 166–71, 172, 173, 174
    line source approach, 159–60, 165, 168, 170–1, 174
    p-linear temperature, 164, 165, 166–7, 168, 170

  procedure, 162–3
  recovery phase, 165–6, 167, 168–9, 172, 174
  results, 163–73
  single-value interpretation, 160, 165–7, 173
  site description, 160–1
  temperature profiles, 162, 163–4, 169
  test rig, 161–2
  thermal resistance, 159, 161, 162, 165–8, 174, 175–6
  thermistors, 160–1, 162–4, 170–3
GHE design uncertainties, 81, 86, 88, 94–6
limitations of, 157, 159–60, 174
Republic of Ireland energy piles study, 223, 224–5, 243–4
  analysis, 227, 229–34
  borehole heat exchangers, 224, 225, 227, 230, 231, 236, 243
  Carraigtwohill site, 227, 228, 233–4, 239–43
  Cork Docklands site, 225–7, 228, 234–9, 242–4
  CPTU cone resistance and friction ratio, 235–6
  discussion, 243–4
  energy pile installations, 225–7
  GPM model, 229–30, 238, 240, 242, 243, 244
  groundwater flow, 230–1, 236, 238, 242, 243–4
  heat addition and loss, 233–4
  laboratory thermal measurement of soil, 238
  line source method, 225, 227, 229, 230, 233, 234, 237–8, 239–41, 242–4
  pumping recovery test, 231–2
  pumping test, 231–2, 233
  results, 234–42
  thermal conductivity, 224–5, 226–7, 229–30, 231, 234, 237–8, 240, 242–3

291

thermal diffusivity, 229, 233, 237–8
thermal recovery test, 225, 230–3, 234, 238, 239, 240–1, 242, 243
Tunnels & Tunnelling International Awards, 155

U loops
  CFA piles, 183, 185, 186–8, 190, 197–8, 200, 201
  chalk aquifer TRT study, 160–1, 165
  geothermal energy in loess, 6, 11, 15
  GHE design uncertainties, 83, 86, 87, 88, 89, 93, 96
UK (United Kingdom)
  chalk aquifer TRT study *see under* TRT
  CFA piles, 182
  Crossrail tunnel energy segment system *see* TES system
  Eastgate geothermal study, 23, 24–31, 34–9
  PV-thermal modules simulation, 115, 116, 117–18, 124, 126–7
  road pavement energy harvesting, 65, 68
  thermal response testing, 224–5
Underground Power, 69, 71
University College Dublin (UCD), 223, 224, 225, 233, 236, 242, 243
UPE (under platform exhaust), 151, 154
URDFIL subroutine, 186
USA (United States of America), 66, 69–70

Vahdati, M., 115
variable parameter settings, 188–9
Varney, K., 115
vehicle mechanical energy harvesting, 62–3, 65–70, 71–3

vertical dispersivity, 32, 33, 37
Vienna LT22 testing plant, 133
Vistula River, 207, 208

Wairakei bowl site (New Zealand), 42–3, 44, 45, 47, 48, 50, 51, 52, 53–5
Wairakei–Tauhara study *see* geothermal subsidence study
Wanninayake, Ajitha, 53
'Warsaw City action plan goals for sustainable energy use', 207, 220
Warsaw City Office of Architecture and Spatial Planning, 208
Warsaw thermoactive structures study *see* city–scale analysis (Warsaw thermoactive structures)
water table, 8, 9–10, 105, 168, 216
Waydip, 69, 71
Waynergy Vehicles, 69
Weardale granite, 24, 25–7, 30, 33
well doublet system, 24, 29, 34–6
Wischke, M., 66
Witte, H., 225
Wola, 207, 220
Wood, C.J., 261, 262, 267
Wu, G., 64

Xiong, H., 66
X-ray diffraction, 46, 51

Yu, X., 64

Zervantonakis, I., 225
Zhang, Y., 206, 220
Zhao, H., 65, 66
Zorlu, Ö., 69